Farm Animal Well-Being

STRESS PHYSIOLOGY, ANIMAL BEHAVIOR, AND ENVIRONMENTAL DESIGN

SOLON A. EWING
Iowa State University

DONALD C. LAY JR.
Iowa State University

EBERHARD VON BORELL
University of Halle

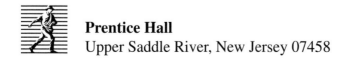

Prentice Hall
Upper Saddle River, New Jersey 07458

Library of Congress Cataloging-in-Publication Data

Ewing, Solon A.
    Farm Animal well-being : stress physiology, animal behavior,
  environmental design / Solon A. Ewing, Donald C. Lay Jr., Eberhard
  von Borell.
        p.   cm.
    Includes bibliographical references (p. 333) and index.
    ISBN 0–13–660200–2
    1. Livestock--Behavior. 2. Livestock--Housing. 3. Livestock-
-Effect of stress on. 4. Veterinary physiology. 5. Animal welfare.
I. Lay, Donald C. II. Von Borell, Eberhard. III. Title.
SF456.7.E94   1999
636'.089698--dc21   99-9445
                                                                    CIP

Acquisitions Editor: Charles Stewart
Assistant Editor: Kate Linsner
Production Editor: Lori Harvey, Carlisle Publishers Services
Production Liaison: Eileen M. O'Sullivan
Director of Manufacturing & Production: Bruce Johnson
Managing Editor: Mary Carnis
Production Manager: Marc Bove
Interior Design/Compositor: Carlisle Communications, Ltd.
Printer/Binder: R.R. Donnelley & Sons Co.
Cover Art Director: Jayne Conte
Cover Designer: Miguel Ortiz
Cover Artist: Alan Leiner

 © 1999 by Prentice-Hall, Inc.
Simon & Schuster/A Viacom Company
Upper Saddle River, New Jersey 07458

Printed in the United States of America

10   9   8   7   6   5   4   3   2   1

0-13-660200-2

Prentice-Hall International (U.K.) Limited, *London*
Prentice-Hall of Australia Pty. Limited, *Sydney*
Prentice-Hall Canada, Inc., *Toronto*
Prentice-Hall Hispanoamericana, S.A., *Mexico*
Prentice-Hall of India Private Limited, *New Delhi*
Prentice-Hall of Japan, Inc., *Tokyo*
Simon & Schuster Asia Pte., Ltd., *Singapore*
Editora Prentice-Hall do Brasil, Ltda., *Rio de Janeiro*

Contents

Preface

Intensification of animal agriculture and increasing concern for animal care and well-being have heightened interest on the part of livestock producers and the public in basic and applied sciences related to animal well-being. Pet owners show an increasing interest as well in matters pertaining to providing an appropriate environment for their charges.

These interests are served by a better understanding of the basic principles of environmental stress and its impacts on both physical and mental states of animals. Such understanding gives a better appreciation for the likelihood of negative impacts of stress occurring and when these are stressors readily accommodated by the animal. A better understanding also provides ideas and concepts for improved animal care and broadens the search for such improvements.

A goal for consideration of stress, behavior, animal well-being, and related management and care is to enhance sensitivity to animals' needs, with the assumption that if such basic needs are met, animals might produce more efficiently, be more comfortable, and enjoy a state of well-being during their lifetime. This sensitivity should also enhance the long-standing relationship between humans and animals and the effectiveness of that interdependence.

Animals are clearly important to society as companions and providers of food, power, and enjoyment in a great variety of sports. Humans are important to animals because almost all animals, even in the wild, are managed by limits, allowances, and protection.

Opinions differ as to how animals should be managed, maintained, and utilized. A better understanding of the relationship of animals to their environment will hopefully result in a more constructive approach on the part of all parties having divergent views on these issues.

To a great extent, the well-being of humans depends on the well-being of animals. The typically well-known benefits of pleasure, companionship, food, and power must be expanded to include the economic importance of the infrastructure serving the livestock industry, pets and their owners, and the sports and entertainment sectors. Finally, an appreciation for the beneficial relationship between animals and people must recognize that in many cases measures for conservation of natural resources required in a truly sustainable agriculture depend on this relationship.

A greater understanding of issues and basic principles involved in the relationship between animals and their environment should provide a better appreciation for these interactions and related impacts. It can enhance important dimensions in the relationship between animals and humans. The result will be a more effective quest for a better world for all of its inhabitants.

This book was developed in designing an undergraduate course with the broad objective of enhancing the understanding of animals' basic needs and proper approaches to effective animal care. To accomplish this objective, some specific goals were established:

- To understand the animal's basic needs and its ability to cope with varying levels of stress in maintaining a steady state of functioning and, in turn, a general state of well-being.
- To develop an increased awareness of the animal's behavioral characteristics, how these relate to elements of its environment including physical, dietary, and social dimensions, and the importance of behavioral responses in overall well-being.
- To consider the basic biological and behavioral characteristics of the animal as these relate to the animal's needs and how these influence basic concepts in designing and providing an appropriate environment.
- To gain an appreciation for the extensive amount of objective information that exists to assist in designing environments, care procedures, and production practices that are consistent with our obligation to ensure both animal well-being and efficiency in related enterprises.

To relate to the mentioned broad goals, topics are organized to consider the following:

- The issue of animal well-being as a concern of society and individuals directly involved in animal care as livestock producers or those maintaining farm animals as companions or for sports activities.
- Environmental characteristics that are commonly associated with stress in animals, how they are equipped to cope in stressful situations, and the limits of the biological systems involved in responding to demanding environments.
- Basic characteristics of farm animal behavior and the relationship of the physical, dietary, and social environments with normal and aberrant behaviors.
- Environmental design characteristics reflecting the physical, dietary, and social dimensions to meet the animal's needs.

Acknowledgments

We are deeply indebted to a long-standing collaborative relationship with many leaders in the livestock industry, which has provided a clear definition of priority industry needs. One of these needs is for a better understanding among producers, consumers, service industries, and regulators of objective information treating the area of animal well-being. This book was developed, in part, as a means of improving this information exchange.

Appreciation is extended to Iowa State University for allowing a period of time for the concentrated effort required to develop this text as a basis for enhanced learning by both students and those in the animal industry. Appreciation is extended also to the Animal Science Department, University of California–Davis, for hosting me during preparation of the early manuscript and to members of that faculty for their support. In terms of individual faculty at UC–Davis, appreciation is extended to Gary Moberg and Ed Price for their counsel and to Eric Bradford for extending the invitation and support of this exceptional department. Appreciation is also extended to the following colleagues in the Animal Science Department, Iowa State University: Lloyd Anderson, Palmer Holden, Douglas Kenealy, Lee Kilmer, Daniel Loy, Daniel Morrical, Emmett Stevermer, and Renee Knosby for early review and suggestions on specific topics. We thank the Committees on Animal Nutrition and species-related subcommittees of the National Research Council, which assembled and published dietary requirements of farm animals and those that developed the *Guide for the Care and Use of Laboratory Animals*; committees of Midwest Plan Service for developing requirements and facility designs for the various species and publishing these as major reference documents; committees appointed by the National Pork Producers Association to summarize and publish information on environmental requirements of swine in the *Swine Care Handbook;* and numerous committees of extension professionals who developed the *Pork Industry Handbook* series of technical reference documents. The mentioned documents have not only provided sound technical information but have also contributed enormously to standardization of the great range of recommendations relating to animal care and well-being.

The valuable work and publication of the Consortium for Developing a Guide for the Care and Use of Agricultural Animals for Agricultural Research and Teaching is acknowledged. Appreciation is also extended to the Commission of the European Communities, Scientific Veterinary Committee (Animal Welfare Section)Directorate-General for Agriculture for their work in examining welfare issues in farming systems and in publishing relevant information.

Photographs, in addition to those taken by us, were generously supplied by E. O. Price, Carolyn Stull, and Lisa Holmes, University of California–Davis; the American Angus Association, St. Joseph, Missouri; J. D. Hudgins Ranch Inc., Hungerford, Texas; George Brant, Liesl Hohenshell, Paul Brackelsberg, Jay Harmon, and Hongwei Xin, Iowa State University; Brian Nielson, Michigan State University; and Jean LeDividich and photographer Jacky Chevalier, Institut National de la Recherche, Agronomique, Saint-Gilles, France. Richard Willham, Iowa State University, sketched the horse and rider to illustrate a behavioral characteristic.

Eberhard von Borell and Donald C. Lay contributed generously as coauthors in areas of the text relating to animal behavior and through review and comment on the balance of the publication. I am also indebted to each of them for joining me in developing a new undergraduate course in the area of stress physiology and behavior. Team teaching with them has given me an expanded appreciation for the field of animal behavior and related sciences.

Special appreciation is extended to Dorothy Ewing for her assistance and counsel as the manuscript was developed. I am deeply indebted also for her many hours of creative effort in processing the original manuscript and in formatting the final copy for editing. Appreciation is extended also to Donna Watson and Julie Roberts of the Animal Science Department, Iowa State University, for incorporating several revisions into the manuscript during the process of development. Photographic and image processing was enhanced by the efforts of Bill Pyle and Mark Hawley.

The authors acknowledge the constructive suggestions provided by the following reviewers: Murn Nippo, University of Rhode Island; Stephen L. Pottorfs, Barton County Community College; and Scott Whisnant, Texas Technological University. The professional skills and efforts of the following who contributed to the preparation of an improved final manuscript and its efficient production on a timely basis are gratefully acknowledged: Lori Harvey, project editor; Kathy Davis, associate project editor; Susan Brehm, copyeditor; Mark Wyngarden, paging specialist; and Charlotte Grass, illustrator. Appreciation is also extended to Kathleen Linsner, assistant editor Career and Technology, who guided the authors through the numerous steps in preparation and submission of the manuscript.

S. A. Ewing

PART I

ANIMAL WELL-BEING: THE ULTIMATE GOAL

Except in the opinion of a few isolated thinkers, the anticruelty ethic and laws expressing it were by and large considered adequate to social concerns about animal treatment until the mid-twentieth century. At this time a series of significant social changes brought inchoate but gradually more precise new social thought about animals.

BERNARD E. ROLLIN (1995)[1]

In his book, *Farm Animal Welfare—Social, Bioethical, and Research Issues,* Bernard Rollin suggests that the social concern about the well-being of animals and the apparent persistence of society's interest in this regard represents a migration of what historically was largely a personal ethic to one described as a social ethic. He makes a compelling case for this view. If animal care is becoming more a social issue, those of us responsible for animal care can expect continuing interest in and surveillance of the practices we use in maintaining animals for food, fiber, sport, entertainment, and companionship.

Those associated with animals generally respect them and empathize with them as we do with humans when they experience negative influences. In some cases, however, we may not fully appreciate when the animal's environment compromises its well-being. Objectivity in examining what we do with animals and how we care for them and utilize them is often lacking because we are so familiar with and accustomed to the processes that have been used for centuries. Although many view animal welfare activism as an interference, it is quite evident that such pressures have caused widespread reevaluation of many aspects of the animal industry, regardless of the sector. Such activism can now be judged as a component in the transformation of matters relating to animal care and use from largely a personal concern to a much broader social concern in many areas of the world. Thus, we are appropriately advised to hear carefully and to consider what society is thinking and seek to achieve a higher level of objectivity in evaluating our total relationship with animals and how we pursue an appropriate symbiosis.

[1]Rollins, B.E., *Farm Animal Welfare—Social, Bioethical, and Research Issues* (Ames, Ia.: Iowa State University Press, 1995), p.8, by permission. ©1995 by Iowa State University Press

CHAPTER 1

Animal Well-Being: The Issue

The goal to be achieved by an enhanced understanding of the basic biology of stress, animal behavioral responses, and related environmental influences is to ensure appropriate strategies for animal well-being. Such understanding creates a sensitivity to the issue of adequate animal care and, hopefully, increases the conscious effort of those responsible for animals to provide a satisfactory environment.

Pet owners are sensitive to the fact that animals need care and take great pride in the fact that pets are, in many cases, considered to be members of the family. Actions in caring for such animals, however, often reflect a lack of understanding of the basic environmental requirements of the animal to avoid damaging stress effects.

Livestock producers are sensitive to animal care because they are normally perceptive of animal needs and because the environment provided influences animal performance that, in turn, may affect the producers' livelihood through negative economic impacts. A thorough understanding of stress effects, behavior, and related impacts on performance may offer ideas for improving animal performance and well-being simultaneously through enlightened approaches to facilities and practices that make up the animal's environment.

ANIMAL WELL-BEING

Terms such as *animal well-being* and *animal welfare* are often used interchangeably. The term *animal welfare*, for some, has a broader meaning than well-being. Use of the terms is intended to describe a state in which an animal is existing within a range of acceptable environmental specifications. Biological scientists dealing with the physiological, biochemical, and physical effects of an animal's surroundings are more inclined to use the term *well-being*. Social-behavioral scientists appear to favor the term *welfare*. Some have observed that *welfare* is the term of choice in Europe and *well-being* the term of choice in North America. In this text, except where direct quotes are cited, the term *well-being* is used.

The goal of scientists and producers should be to provide an environment that minimizes the negative aspects of stress. In this regard, decisions must be made by pet owners and livestock producers relating to the level to which certain needs can be met. An important dimension in the decision process is recognizing that the presence of acute stress does not mean that an animal's well-being is necessarily compromised, because biological mechanisms are in place to cope with the effects of stressors. The issue of well-being comes into play when such stress is prolonged and is likely to create a pathological condition or aberrant behavior. Most would agree that persistent or chronic stress resulting in one or more pathological conditions violates an animal's well-being. There is a philosophical difference of opinion, however, as to whether an animal's well-being is compromised when aberrant behavior appears. The different views hinge on whether certain performance measures such as rate of gain, rate of egg lay, or amount of milk produced are adequate to reflect well-being. This argument is likely to persist because some feel that normal performance criteria are simply not the only important considerations. A thorough understanding of stress physiology, behavior, and related economic factors by all parties to this argument will likely bring viewpoints on these issues closer together.

The typical dictionary definition of *welfare* suggests the presence of happiness, healthfulness, and prosperity. While this may be a good general definition, it is not considered sufficient in the context of adequate environments for animals. Except for healthfulness, the previously mentioned terms describing welfare or well-being can bring one to an anthropomorphic position of attempting to judge happiness and prosperity on the part of animals as humans perceive these in our own lives. Hurnik et al. (1995) provide a more useful definition of well-being, believing that it is a condition in which physical and psychological harmony exists between the organism and its surroundings. These authors suggest that the most reliable indicators of well-being are good health and manifestation of a normal behavioral repertoire.

A compelling argument states that if you ascertain and measure the most stress-sensitive biological phenomenon, it provides an index of well-being. Moberg (1985b) proposed that the reproductive process is sufficiently sensitive to stress to provide a reflection of well-being (i.e., if animals are performing optimally in terms of reproduction, then an acceptable state of well-being exists). Various measures of reproductive adequacy are possible at different stages of the reproductive cycle for females; factors such as libido and semen quality are measures of reproductive status of males. Steers in the feedlot that are gaining weight at rates normally expected, based on various predictive models, reflect a state of well-being. We, and more importantly animals, are fortunate in that an enormous body of research data exists on which to base predictive models that provide a benchmark of what may be described as normal levels of performance. In spite of such examples, some argue that even satisfactory, normal, or expected levels of performance may occur while animals are exhibiting some aberrant behaviors; they may thus conclude that the state of well-being is subminimal. This approach moves any discussion of well-being back into the philosophical mode and the associated impasse. It leads, also, to concepts related to animal rights.

A much broader view of evaluating environments favoring animal well-being is presented by Hurnik (1992). This perspective is based on the philosophy that some traditional practices utilized in animal enterprise management, while acceptable from an economic standpoint, may warrant further evaluation by giving greater attention to the animal's psychological and social needs. The thesis reflected is that, to be acceptable, conditions in which an animal is maintained should assure harmony between the genetic character of the individual and the environment in which it exists. The design of an appropriate environment, then, requires a good understanding of behavioral characteristics, cognitive abilities, learning capabilities, and physical characteristics, along with a variety of other factors that establish requirements. This thesis suggests that an optimum environment is one that provides the most appropriate combination of factors that meet a variety of needs supporting normal biological function. It does not propose, however, that maximum performance is necessarily associated with an acceptable level of well-being.

In addition to considering the animal's physiological needs, Hurnik emphasizes the importance of meeting needs related to psychological well-being in achieving environment-animal harmony. Important in this concept is that animals have expectations of their surroundings and that the extent to which these expectations are satisfied will influence the level of harmony that exists. If the environment is limiting, relative to satisfying such expectations, it is very likely that the animal's well-being is compromised. Stated another way, if the animal's environment precludes its ability to satisfy normal behavioral needs, the animal is likely to respond by expressing behaviors that reflect environmental inadequacy. Hurnik suggests that an animal's total environment consists of a variety of factors that can be viewed as stimuli and that animals will normally try to acquire or to avoid some of these. Such stimuli were classified and are summarized in four basic categories to provide a systematic means of clearly recognizing fundamental animal needs in considering environmental adequacy. The proposed classes are attraction, aversion, deprivation, and placation.

The preceding classes of stimuli are clearly involved in commonly recognized behavioral repertoire of animals. All farm animals are ambitious in exploring their environment for those things to which they are attracted and which they will consume, utilize, or associate with. Examples are dietary elements, mates, and physical characteristics and limits of the area in which they reside. Skilled caretakers recognize these normal behavioral characteristics and when such demands are satisfied or, stated in another way, when the expectations for these elements are met. Such observations are important in judging environmental adequacy. Caretakers observe animals as they express behaviors related to avoiding aversive elements of the environment and in their relationships with other animals. Such behaviors offer the opportunity to assess how the animals feel about their environment, individually or collectively, and whether different management approaches may be considered to enhance well-being. Similarly, those responsible for animal care must be able to recognize behavioral characteristics associated with environments that deprive the animal of one or more elements that are important to allow the animal to perform certain typical behavioral repertoire. These are also reflective of environmental inadequacies in terms of meeting innate behavioral needs, which if not met, may result in the development of aberrant behaviors. Thus, both Hurnik (1992) and Duncan (1996) emphasize the importance of behavioral influences in any definition and assessment of animal well-being. Correspondingly, those that strongly suggest the importance of physiological parameters in the definition and assessment of well-being (Moberg, 1985b, 1996 and McGlone, 1993) recognize the importance of behavioral evaluation as well. Thus, it is clear that the scientific community is in agreement that animal well-being is likely compromised when animals demonstrate either physiological or behavioral aberrations. It is also recognized that the environment that results in psychological stress results in physiological parameters that are typically associated with stress regardless of cause.

Introducing the concept of animal expectations and feelings in evaluating environmental adequacy for animals is not readily accepted by some. However, it is clear that those involved in the animal industry are rapidly developing an appreciation for the significant impacts that animals' behavioral characteristics and needs can have on enterprise economics as well as animal well-being. This is both simultaneous and consistent with the transformation of concern about animal care from a personal to social ethic described by Rollin (1995).

Recognition of the classes of environmental stimuli described by Hurnik (1992) and an understanding of associated behavioral responses provide a basis for considering the needs and desires of animals. The classification does not, however, provide a means for separating the two effects involved in behavioral responses (i.e., the effects of needs and desires are confounded). Consequently, this system is useful in partially evaluating the degree to which animals are in harmony with their environment and, in turn, judging the adequacy of the surroundings. Even with the shortcomings cited, the process suggested provides a useful approach in considering the issue of well-being. Some specific relationships between environmental deficiencies and aberrant behaviors are considered in part 3 and serve to establish more specific associations.

Duncan and Poole (1990) described the importance of physical health and freedom from injury in animal well-being and added to this, as emphasized by Duncan (1996), the importance of how an animal feels about its condition, how it perceives its environment, and how aware it is of factors related to its well-being. This approach introduces cognitive processes that become important considerations in the overall issue of animal well-being. Such factors are undoubtedly involved in the psychological health of animals; experiences related to frustration and the absence of important stimuli, for example, result in typical behavior patterns that are often judged to be reflective of poor well-being. Duncan and Poole suggest that much of the difficulty associated with the issue of animal well-being arises from the fact that "it lies at the intersection of science, ethics and aesthetics." Thus, while we may be able to measure certain parameters reflecting well-being, society or individuals may conclude that some practices are ethically unacceptable or are simply in poor taste.

HISTORICAL PERSPECTIVE: A REFLECTION OF CONCERN BY SOCIETY AS WELL AS ANIMAL CARETAKERS

Society, in many parts of the world, has typically provided some basis for legal rights of animals that are aimed at prevention of neglect and abuse. This concern for animal well-being is reflected in many laws and regulations that impose certain legal requirements on owners and caretakers of animals. In general these civil requirements are directed at gross neglect to prevent undue suffering. There are many examples of owners of animals being prosecuted by civil authorities, but generally only after such neglect or abuse has occurred. Most civil requirements have not included specifications as to animal facilities, frequency of observation, and other procedures involved in the day-to-day care of animals. More recently, however, laws that specify certain requirements of animal care have been passed and enforced. In the use of animals for research, scientists must adhere to very strict regulations dealing with housing and the related environmental parameters, routine maintenance procedures, experimental treatment, and the method by which animals are euthanized when sacrificed. These new laws enacted over the past decade or so reflect a growing concern among the general public about animal care. Political activity and organized interest-group activity have combined to create an increasing level of awareness and concern for the numerous issues surrounding animal well-being. This activity is reflected in frequent demonstrations by concerned individuals and organizations and efforts on the part of animal owners and their organizations to examine practices utilized in animal-enterprise management. It is likely that this increased level of activity has heightened the sensitivity of pet owners as to the well-being of companion animals.

Rowan (1993) provides a concise review of the history of public issues surrounding animal well-being. The following is a brief summary of these issues.

Laws protecting farm animals emerged in the United States early in the 1870s and the American Humane Association was founded in 1877.

During the 1960s increasing public concern relative to intensive farm animal systems was more evident in Britain than in the United States. In 1964 *Animal Machines* by Ruth Harrison focused on systems of animal production that involved confinement production systems.

In Britain the Brambell Commission issued a report on animal farming systems in 1965. The Brambell report formed the basis for regulations relative to practices in animal agriculture. European countries also enacted laws dealing with animal facilities, care, and professional evaluation of such procedures.

The 1970s were characterized by frequent activity of individuals and groups whose purpose related to increased awareness and sensitivity to animal well-being issues in the United States and Canada. In 1975 Peter Singer completed the book *Animal Liberation,* which is credited by Rowan (1993) as an important factor in providing a more useful logic related to the movement.

In 1980 Jim Mason and Peter Singer collaborated on the book entitled *Animal Factories*. This publication had an impact on many sectors of society that had shown little interest in the subject previously. Thus, during the 1980s the issue of animal well-being became increasingly important to the general public. Several books published in the 1980s reflect activity related to the animal welfare movement. Among these titles are those by Dawkins (1980), Frey (1980), Rollin (1981, 1989), Regan (1983), and Fox (1986).

In the late 1980s and the 1990s various organizations involving livestock producers and the general public concerned about animal care have had increasing interaction, and both are involved in examining existing and new technologies and procedures involved in animal well-being. During these decades scientific organizations, government agencies, and animal industry groups in North America and in The European Community became increasingly active in examining the body of research relating to animal well-being, developing care guidelines, and in encouraging more research relative to these issues. Publications such as *Guide for the Care and Use of Laboratory Animals* (NRC, 1996), *Guide for Care and Use of Agricultural Animals in Agricultural Research and Teaching* (CDDGCAA, 1998), *Report of the Scientific Veterinary Committee on the Welfare of Calves* (SVCAWS, 1995), *Report of the Scientific Veterinary Committee on the Welfare on Laying Hens* (SVCAWS, 1996) and *Swine Care Handbook* (NPPC, 1992) are examples.

Specific groups formed over the years to focus on issues related to animal well-being are numerous (Rowan, 1993)[2] :

1951	Animal Welfare Institute
1954	Humane Society of the U.S.
1976	Animal Rights International
1978	Animal Legal Defense Fund
1979	Animal Rights Network Magazine
1979	People for the Ethical Treatment of Animals
1980	Action for Life
1980	Association of Veterinarians for Animal Rights
1980	Farm Animal Reform Movement
1982	Farm Animal Concerns Trust
1984	Humane Farming Association
1986	Farm Sanctuary

Any mention of organizational influences in the animal welfare movement must emphasize that many agricultural organizations are actively engaged in the issue from the standpoint of supporting research in design and evaluation of practices used in animal agriculture. The National Pork Producers Council established the Animal Welfare Council and publishes animal care guidelines. The American Farm Bureau Federation is active in educational activities related to these issues. The National Cattlemen's Beef Association has emphasized animal care and well-being in numerous national meetings. One could

---

[2]Rowan, A.N., *Food Animal Well-Being* (Lafayette, Ind.: Purdue University, 1993), p.26, by permission. ©1993 by Purdue Research Foundation

cite many such activities by other organizations at national, state, and local levels as well.

ANIMAL RIGHTS: LEGAL AND MORAL

Concepts related to rights are typically based on one or both of the primary thrusts of most efforts to define the term.

On the one hand is the concept of legal rights. Animals may have few if any legal rights in terms of human rights. Certain laws do exist, however, as protection for animals, such as statutes aimed at prevention of neglect and inhumane treatment. Legal restrictions on animals also exist to protect the animal and rights of property owners and other individuals from damage that may be inflicted by animals. Such laws dictate where animals may or may not be located. Leash laws and prohibition of animals in municipal areas, parks, highways, and so forth are examples.

Therefore, to a degree, animals are treated in the legal system to some protection (i.e., the right to a reasonable level of care and freedom from neglect). Animals are treated as property in a legal sense in that legal liabilities related to damage, trespass, and so forth transfer to the animal's owner. The liability of society for damages inflicted by so-called "wild" animals is much less clear, even though many of these animals and herds are essentially managed by the rules and regulations imposed by society. They may well be wild in terms of temperament and ability to move about to some extent, but many other dimensions of their existence are carefully controlled.

The greater disagreements are in terms of animal rights related to the definition suggesting that one has a moral obligation to see to an animal's needs and, more significantly, to interpret what such needs are. Most would agree and accept that we have a moral obligation for proper care of animals, even though disagreement exists as to what constitutes "proper" care, what an animal's needs may be, and the level to which these needs must be met. Animal rights philosophy may also include the concept that animals have the right to live, and thus decisions to terminate the life of an animal violate this right.

Disagreement about the issue of an animal's freedom from restraint also occurs, often taking two forms. One is related to the idea that animals should be able to move and explore, unrestricted by physical barriers (e.g., pens and fences). As a practical matter this is basically a nonissue in that laws provide for the assessment of damages for trespass effects including such things as property damage, personal injury, unwanted pregnancies, and the like. The other dimension of the freedom issue relates to the level of restricted movement practiced in various production systems. Examples are gestation crates, tethering of sows, calf crates, and small pens. A related dimension includes such factors as access to bedding and pasture. Some have the perception that such restrictions and environments reflect the failure of humans to meet moral obligations to provide what the animal needs. The other side takes the position that even within such environments, if performance and productivity of the animal are unaffected, the animal's needs for well-being are met and therefore rights have not been violated. Thus, the argument about both well-being and rights goes full circle. In general, however, most would agree that an animal's needs should be met as they relate to a quality of life reflected in good health and a reasonable level of comfort to avoid the occurrence of damaging stress. Elimination of occurrences of acute stress is likely not a reasonable goal for farm or household animals. Fortunately most animals have well-developed biological mechanisms for dealing with short-term stresses. The role of the manager of such animals is to provide an adequate environment to ensure that these defensive mechanisms are not exhausted to the detriment of longer-term well-being.

The frustration from the circular nature of such debates prompts the scientific community to continue to refine measures of well-being and to develop models for animal well-being and potential stress control. Assuming, however, that a quantitative evaluation of well-being or welfare has been accomplished, the question remains, in a range of well-being from poor to good, what level is acceptable. Consequently, scientifically devised environmental specifications are of critical importance—first to the animal and second to those engaged in the continuing debate.

FREEDOMS FOR ANIMALS: A CONCEPT RELATED TO WELL-BEING

We have long conditioned ourselves to speak of freedoms as those choices to which we are entitled in order that life is fulfilling and that our needs are met. Needs can also become confused with wants, which introduces some danger in transferring concepts of human freedoms and human needs to those of animals.

Webster et al. (1986) described five freedoms for animals, including freedom from hunger and malnutrition, freedom from thermal or physical distress, freedom from fear, freedom from disease and injury, and freedom to express most normal behaviors. In general, only the latter seems to stir any real controversy. Society typically has elected to restrict certain animal behaviors in the interest of management, population control, and liability.

The listed freedoms, however, provide conceptual guides for design of animal environments. Assessing whether an essential level of a freedom is being met requires both a determination of the so-called essential level and an objective measure to reflect the extent to which that level is achieved.

Regardless of how matters of freedoms or needs are expressed, the manager of animal enterprises and those responsible for animal care must approach the issue of animal well-being in terms of environmental design. And design of animal environments giving consideration to physical, dietary, and social requirements is a complex undertaking. Further, one must expect to encounter a variety of views and a highly variable level of understanding relative to environmental needs of animals in all of the mentioned dimensions of environmental specification.

Society imposes certain expectations on those caring for animals, expectations that become part of our moral fabric. It is quite likely that major dimensions of how society feels about the acceptability of animal care practices relates to how we view the level of cognition in animals and the degree to which the animal may experience or feel pain and suffering, including what humans refer to as mental anguish or psychological stress.

COGNITION IN ANIMALS: MOTIVATION TOWARD ENVIRONMENTAL ADEQUACY

Reviewing a body of literature developed over the years dealing with the ability of animals to reason, make decisions, plan and execute actions in advance of challenges, and seek to mitigate stressful situations leads one to the conclusion that animals utilize cognitive processes. The ability to make choices demonstrates the presence of feelings, and the pursuit of goals is rather compelling evidence that animals, in this regard, are more like people than we may care to recognize. Although it is not the intent to review this literature here, one need only to reflect on the great body of data to conclude that animals are goal oriented and make great efforts to achieve such goals. Repeated failure to achieve goals may lead to a variety of psychological problems associated with abnormal behavior.

All of the information suggests a significant level of cognition in animals that prompts the conclusion that humans and animals have a great deal more in common than just anatomical and metabolic similarities. A high level of appreciation for the level of cognition that exists in animals is almost certain to increase our empathy for animals and in turn the motivation to ensure a state of well-being.

Evidence of Cognition in Animals

Kendrick (1992) suggests that the term *cognition* refers to the mental facility of knowing and that the study of cognitive behavior covers the individual's acquisition and use of knowledge about self and the environment that results in achieving the most from its interaction with the environment. This concept suggests that animals have some capability in the choice of how available resources might be used and how relationships with other animals might be maintained and conducted. It also introduces some aspects of economy in how an animal may view environmental relationships, including those of a social nature. Understanding more about the cognitive behavior of animals is important in developing environments for animals because interactions with such surroundings relate to animals' physical and mental well-being.

The view that animals are essentially passive relative to environmental influences is inadequate in terms of conceptualizing and designing the environment to be provided. This view does not recognize even the routine experiences one has in observing the behavior of animals as they interact with their surroundings, humans, and conspecifics. Such observations suggest that animals are persistent in achieving objectives; they often avoid perceived hazardous encounters, which shows some level of predicting future outcomes based on experience; they show collaborative efforts; they will seek alternatives for achieving goals; they recognize benefits of rewards; they demonstrate mapping ability and memory related to exploration of resources important to survival; and they also use exploratory experience in a variety of relationships with other animals. This array of commonly observed characteristics suggests that animals employ at least some important cognitive processes.

Duncan and Poole (1990) concluded that animals are aware of internal events through feelings and external events by perception. Acceptance of this conclusion forces one to conclude that animals utilize some important cognitive processes as they interact with environmental influences.

A sizable body of research has centered around the use of various techniques to measure how an animal feels about its surroundings by allowing the animal to choose among alternatives. Such evaluations are referred to as preference studies. For example, sows might be given a choice between a resting area bedded with straw and one with no bedding. Hens might be given a choice between cage floors made of wire or plastic. In such studies efforts are made to ensure that variables other than those under study are equal and that the animal has complete freedom of choice as to the environment preferred. Duncan and Poole (1990) note some problems with this approach. If animals are given but two choices they may be selecting the better of two inadequate environments. Thus, providing a greater number of choices may provide a more useful evaluation in some cases. Another shortcoming suggested by these authors is that animals may not make choices in their best interest in terms of well-being and still another is the fact that a shorter time spent in one environment does not necessarily mean that the period there is not essential. Animals may in some cases make choices that are indeed the best for the short term but not for the long term. With such limitations, however, properly designed preference studies may assist in useful criteria to consider in animal facility design. Even a limited review of research in this area suggests that animals make considered choices relative to their surroundings. They may not make what one would consider the best choice in the interest of well-being, but the fact that such choices are made, and consistently in many cases, provides added confidence that animals have significant cognitive capabilities.

Some evidence suggests that animals relate to expectations. Goldman et al. (1973) found greater circulating concentrations of corticosteroids when less-than-expected rewards were given and lower than base levels of corticosteroids when greater-than-expected rewards were given. The higher levels of corticosteroids suggest that the animals are stressed. The stress is assumed to be associated with failure to experience the expected result.

Studies of neural cell activity reported by Kendrick (1992) were summarized by stating that olfactory and visual systems are probably the most important factors in an animal's recognition of other animals and humans. These stimuli result in nerve cell coding that essentially creates a memory in cells of the brain that provides reference for future recognition of animals and humans. Olfactory-related coding of such cells is commonly recognized by the example of dam-offspring identification. Less clear has been the ability of animals other than primates to make visual recognition of individuals. Work with sheep, however, led Kendrick to conclude that this species, at least, can learn and remember and later identify categories of animals and humans by their faces and body shapes. The ability to identify specific individuals by visual contact only is not clear. Kendrick concludes that the existence of this ability in animals underscores the importance of interactions with other animals and humans in considerations related to animal behavior. Since sensory recognition involves learning and memory, and since these processes are based in areas of the brain normally associated with emotional responses, it is logical to conclude that cognitive processes are involved.

Observations relative to how animals learn and apply memory in routine functions suggest cognitive characteristics. The fact that learning is an important factor in how an animal views and reacts to its environment is clearly established by many activities. An example is the priority placed on early exploration of surroundings that establishes in the animal's memory the location of a great many environmental elements such as fence boundaries, shade, water, and feeding facilities, to name a few.

Other examples, such as a cow searching for a missing offspring, a bird dog searching for a downed quail, or a steed constantly showing a desire to turn and return to its home base, could be cited. These examples cause one to ponder just how much animals think and plan and how much they are aware of their circumstances in relation to surroundings. Such common examples, along with the mentioned conclusions by scientists studying the response of animals to stimuli influencing cognitive areas of the brain, are sufficient to suggest that animals possess a substantial level of cognition. As a result, they are subject to negative influences of stressors having potential psychological impacts. This recognition on the part of animal managers and pet owners increases their sensitivity to the needs of animals and encourages efforts to ensure that stress is minimized.

CONSIDERATIONS RELATIVE TO STRESS IN ANIMAL MANAGEMENT

Those responsible for animal well-being deal constantly with a range of factors influencing the status of their charges. Thus, management strategies must recognize potentially stressful environments and the ability of animals to cope with stress effectively within a normal biological range. An animal's ability to survive and perform effectively (cope successfully with both acute and chronic influences) may be referred to as fitness. This is broader than the classical definition of fitness, which reflects the concept that animals are fit or unfit based on the ability to reproduce their kind. Those not adapted or unable to adapt to the environment in which they exist will eventually become extinct because reproductive failure will prevent survival. Although this definition of fitness is appropriate for livestock enterprises based on reproduction and is certainly correct from an evolutionary perspective, it is not adequate from a practical standpoint to cover the range of important production environments involved in the animal industry. The concept of fitness is equally important in enterprises based primarily on growth of meat animals such as stocker and finishing operations in the beef industry, the growing-finishing sector in the swine industry, and similar operations in the broiler industry. In such enterprises, given lines of animals simply may not be able to adapt to the environmental conditions imposed. Such animals are unfit for a given environment; consequently, the manager must seek animals that are better suited to that production system or change the system to be compatible with the animals involved.

Fitness reflects the two dimensions from which it arises. These dimensions are, first, the genetic capability to perform effectively in a given environment and, second, environmental conditioning. Plomin and Bergman (1991) and Pinel (1993), writing on genetics of human behavior, use the term *natural* for genetic influences and the term *nurture* for environmental influences. These terms, while reflecting the same concepts, may be more meaningful to some. An understanding of such descriptive terms and their interactions, along with economic considerations, can encourage effective management reflecting both animal well-being and fiscal soundness of animal enterprises. Examples of the importance of conditioning are the tragic experiences encountered when cattle are relocated from a warm climate to a Great Plains or Midwestern feedlot and encounter a blizzard before they have time to become acclimated. Although these animals may have the genetic capacity to perform very well in a cold climate, their current status contributes to the failure to adapt. An animal's status at any given time is the net effect of both genetic and environmental influences. The ability to sustain a given status has an economic dimension in that management of a livestock enterprise involves many instances of weighing costs versus benefits.

Genetic Influences: Factors Contributing to the Animal's Inherited Capacity

An animal's genotype is the major component in its ability to adapt to, survive, and perform effectively in a given environment. Genetic capacity may be referred to by several commonly used terms. Examples are *innate* or *inherent capability.* The term *constitution* is also used to refer to the genetically controlled capability to cope with environmental influences. Other terms used to reflect genetic and environmental influences are, respectively, *nature* (inherited) and *nurture* (acquired).

If the animal cannot adapt, then it cannot perform the function of procreation, and the species, subspecies or genetic line will eventually disappear. In practical livestock production, animals that are not adapted to a given system of production will fail to perform and producers will avoid selecting similar animals in the future. In enterprises involving reproduction, producers will avoid lines that characteristically show reproductive failure. Many enterprises do not involve reproduction as a performance criterion, however. Steers in the feedlot or growing-finishing pigs being fed for slaughter show varying levels of genetic merit for their ability to gain, to utilize feed efficiently, and to produce desirable products. Thus, effective managers are aware of the inherent ability of certain genetic lines to provide superior animals that are best adapted to the enterprise. An enormous effort has been under way for many years to identify lines of animals having genetic merit to accomplish objectives of specific systems. Some genetic factors related to animals' ability to adapt and perform desired functions are discussed briefly in the following sections.

Heat and Cold Tolerance
Breed and species differences in the ability of animals to adapt to changes in climate are well established. Tropical breeds have evolved to have superior heat resistance or tolerance. In general, animals maintained for meat and milk production are better equipped to withstand extremes in cold than in heat.

Insect Resistance

An example of the importance of insect resistance is seen in the advantage cattle of Zebu breeding have in hot, humid regions where insect infestation can be devastating.

Disease Resistance

Research is now establishing that certain genetic lines of animals are more resistant to some diseases and stresses. Thus, immunogenetics is an extremely important field of science in terms of animal well-being.

Spatial Requirements

Most farm animals are described as contact species. These allow close physical contact with other animals except in specific situations. Species differ in spatial-related behavior, and individual variations exist in terms of aggressiveness when space parameters are violated.

Sexual Behavior

Significant differences exist among species in terms of sexual behavior, and variations may also exist among lines within a species. Such observations have led animal breeders to select for certain characteristics related to successful performance of sexual activity.

Learning Ability

Differences in learning ability have been demonstrated among species of animals. Individual animals also vary in their ability to learn. Such differences may be important in the animal's ability to adapt to demands placed on it by its environment. In cases where an animal simply cannot learn to exist satisfactorily in a given set of circumstances, frustration may result and behavioral problems may follow.

Environmental Influences: Factors Contributing to the Animal's Acquired Characteristics

Experience

Animals learn to avoid certain stressors. Avoidance does not mean that they have adjusted to the stressors psychologically or physiologically, however. If the animal does not adapt, acute or chronic stress develops and leads to abnormal behavior that may impact the efficiency of the animal's performance. An increasing importance of psychological stressors is being recognized in the design of animal environments.

Nutrition

Nutrition and associated feeding behavior are very important factors in the management of an animal enterprise to meet demands related to weather, housing, stage of growth, product composition, and production systems, to mention a few. A detailed review of factors important in designing the nutritional environment for animals is presented in part 4.

Macroenvironment

Some animals have the ability to adjust, within limits, to variable climatic conditions. An example is presented in detail in part 4 to illustrate changes in the dietary energy requirement per unit of metabolic size for maintenance as beef cattle condition to colder or warmer temperatures.

Space Requirements

An animal's condition may be affected through both physical and mental pathways if spatial requirements are such that stress results. Access to feed and water must not be restricted by crowding. In some cases such restriction may lead to increased aggression within a group of animals and result in loss of condition through injury.

Temperature and Humidity

Adhering to appropriate specifications is a critical consideration in environmental design to ensure animal health through minimizing environmental stress.

Light

Light can have a profound influence on performance, feeding behavior, sexual behavior and maturity, reproductive cycles, and aggression. Light management in poultry facilities has developed into a highly sophisticated technology for managing performance and in some cases general health and well-being of birds.

Animal Handling Facilities

Enlightened design criteria have increased the ease with which animals are handled. Similarly, safety for both animals and handlers can be enhanced by an understanding of animal behavior in confinement and movement of animals. Improved safety enhances well-being.

Materials Handling

Facility design, methods and equipment for handling waste, feed, and water are critical in providing an adequate environment for the animals involved.

Sanitation

Effectiveness and efficiency of cleaning are important considerations in facility design and must be included as an integral part of the enterprise management and animal health plan.

Safety

Adequate attention to safety issues is important relative to maintenance of animal condition in case of emergency or during day-to-day operations. Facility management plans should include those for evacuation and protection of animals.

Security

Adequate security of premises where animals are maintained is essential in some cases to prevent unlawful entry and potential animal abuse. In addition, biological security is a critical consideration in preventing transmission of disease within the unit as well as from other premises by visitors and employees. Rodent, bird, and insect control are also important considerations in maintaining biological security.

Transportation Equipment

Animals may be subjected to some of the most stressful conditions possible during shipment from one production facility to another or to market. These conditions can greatly damage an animal's condition in a very short time. Requirements for minimizing stress during such activities are rather well established and promoted by various agencies, including the Livestock Conservation Institute, which supplies the industry with current guidelines for environmental management during such periods. These guidelines involve space allowances, safety considerations, length of travel, rest periods, and related matters (See Grandin, 1992b).

Economic Considerations in Animal Care and Enterprise Management

System Requirements

Generally, cost considerations are a major factor in the level of environmental conditions that can be provided within an economically viable system. Fortunately a great body of data exists to form the basis for facility design specifications (see part 4). Several argue, appropriately in instances, that some commonly used animal facility designs, while providing for good performance, may not provide an adequate social and psychological environment. Those responsible for developing design specifications, as well as livestock producers, are increasingly aware of the importance of psychological well-being to the condition of animals and are now giving much greater consideration to such issues.

Flexibility

Because of cost and often space and design limitations, producers seek to have facilities that are effective yet flexible in terms of production systems that may be accommodated by a given set of facilities. These limitations may present some degree of risk in terms of animal condition in that such facilities may not be totally adequate for all production occurring there. Still, the concept of flexibility of use is an important economic consideration and makes facility design a more challenging issue. It is indeed common practice currently in the livestock industry to design highly sophisticated facilities for a very specialized animal system and related marketing strategies. In such cases flexibility to accommodate other systems is sacrificed to gain the efficiency of the specialized unit. This concept may also allow for enhanced environmental conditions for the animals simply because of the specialized design.

METHODS OF ASSESSING WELL-BEING

Behavioral and Physiological Evaluation of Well-Being

Duncan (1981) emphasized the importance of utilizing a combination of criteria including health, physiological, and biochemical parameters; productivity; and behavior. This author described the scenarios typically used to assess animal well-being, including (1) observing animals for unusual or inappropriate behavioral changes and demonstrating independently that these are indicative of reduced welfare, (2) allowing the animal to choose among variable environmental factors, and (3) subjecting the animal experimentally to stressful environments such as deprivation, frustration, or fright and comparing the resulting behavior with that occurring under commercial conditions.

While it would be desirable to have easily administered physiological tests and behavioral evaluation standards for level of animal well-being under practical farm conditions, no current tests are sufficiently consistent or objective to be totally useful as single parameters. The complexity of the phenomena being measured and the difficulty in interpretation suggest the challenge in developing such tests. Rushen (1991) describes difficulties associated with the use of physiological parameters in assessing animal well-being that emphasize combinations of behavioral, physiological, and biochemical measures. Craig and Craig (1985) also reported examples of evaluations where plasma corticosteroid levels did not accurately reflect low welfare observed as lower survivability and productivity in laying hens. Moberg (1985a, 1996), McGlone (1993), and Curtis (1985b) emphasize the importance of multiple-component assessments of well-being. Ladewig and von Borell (1988) concluded that behavioral evaluations should be supplemented with physiological measures to more fully understand cause and effect relationships involved in animal well-being. The use of such complex evaluations beyond observation for aberrant behaviors and some easily measured phenomena such as body temperature and various diagnostic procedures for diseases is not often practical under farm conditions.

Behavioral Evaluation

Observational approaches to the assessment of well-being include preference tests and testing environmental characteristics relative to level of aversion the animal may reflect and the animal's motivation or level of demand to accomplish a behavioral scenario that may be restricted or precluded by some environmental factor. Behavioral evaluations often emphasize whether animals are exhibiting what are described generally as normal or abnormal behaviors. Among those reflecting the possibility of an inadequate environment are fear responses, replacement and incomplete behaviors, stereotypies, and various forms of undesirable socially oriented behaviors that may inflict damage to self or other animals. Observation for aberrant behavioral scenarios for clues related to environmental problems is a practical management tool.

Physiological Evaluation

A variety of physiological evaluations are made in efforts to evaluate well-being. Such tests are based on a variety of body processes to reflect function of physiological systems involved in the animal's responses to environmental influences.

Evaluation of neural responses may be based on measures of neural activity including both neural impulse activity and biochemical phenomena involving neurotransmitters. Endocrine evaluation usually necessitates measures of various hormones, neurohormones, and neurotransmitters involved in stress responses. Levels of one or more corticosteroid hormones in blood, saliva, or urine along with response duration and variability are often determined to assess stress status of the animal. Another endocrine evaluation technique is challenge by injection of adrenocorticotropic hormone (ACTH), in which adrenal cortical response to the corticotropic hormone is measured. Other measures of physiological phenomena include determinations for various neurotransmitters or neurohormones in body tissues or fluids. Examples are tests for levels of epinephrine, norepinephrine, dopamine, or serotonin.

Stressful environments may be accompanied by an increased production of morphinelike substances within central nervous system tissues (possibly in other tissues as well) that have analgesic or pain-reducing effects. Because of their structure and effect, they are termed *opiates*. Examples of these neuroactive peptides are metenkephalin, leuenkephalin, dynorphin, and beta-endorphin. These opiates as a group may also be referred to as endorphins. Measurement of this class of compounds in body tissues and fluids may assist in the assessment of the animal's stress-related condition.

Certain physical phenomena can reflect physiological problems and are relatively easy to measure. Examples are heart rate, respiration rate, and blood pressure.

Emphasis in assessing well-being is also placed on the use of measures reflecting status of the immune system. There is evidence that hormonal influences engaged as a result of stress may have both inhibitory and enhancing effects on the immune system through cell formation and function. Likewise, recent evidence has exposed an inhibitory effect of immune cell activation on the hypothalamic-pituitary-adrenal (HPA) stress-response function (Rivier, 1993). Mal et al. (1991) demonstrated that horses isolated from others for several hours showed an increased inflammatory reaction. Furthermore, Minton et al. (1992) showed that repeated restraint and isolation in sheep caused a reduction in lymphocyte proliferation and altered some populations of leukocytes. Some examples of immune function characteristics that have been used to evaluate the deleterious effects of stress are ratio of relative populations of lymphocytes, neutrophils, and eosinophils; the ability of both T and B cells to proliferate in response to mitogens; the ability of macrophages and neutrophils to phagocytose bacteria; and the production of specific cytokines by immune cells.

Since all of the mentioned evaluations require specialized skills and equipment, of greatest utility in planning for well-being is good definition of science-based environmental specifications that can be used to design satisfactory environments for animals. Well-established design specifications provide a high level of confidence in the resulting facilities, feeding programs, and handling methods. Adequate design specifications, along with useful performance measures and astute observation for aberrant behaviors, are likely to continue to be the most practical approach to ensure that the animal's needs are met.

Performance Evaluation

Although there are some problems in performance assessments related to definition of the terms *normal* and *optimal,* adequate data are available in most cases to make reasonable assessments of performance level as one measure of well-being. Examples of such measures are lactation level, reproductive performance, rate of gain, and efficiency of gain. Baseline data and predictive models, along with appropriate computer applications, are highly refined with respect to several important performance criteria. Dairy and poultry enterprise managers can avail themselves of performance data of contemporary and competitive enterprises for comparison. Various enterprise records services for several farm animal species provide comparative performance data. In some enterprises well-designed records systems provide continuing performance information relative to individual performance.

In using performance information to assess well-being, one should consider the fact that when animals are performing poorly the level of well-being is probably low. On the other hand, animals performing at a maximum level may not necessarily be in a high state of well-being, at least in the short term. A satisfactory state of well-being can exist if the animal's basic needs are met for health and comfort, even though performance may not be at the animal's maximum level in a given production system. Thus, a good manager must develop the skill to judge well-being in the context of the expectations the animal is to meet. Finally, well-being must be evaluated on an individual animal basis if the system is to be completely acceptable in terms of animal care. Evaluation of well-being on a group basis alone is inadequate.

PRODUCTION PRACTICES AND ANIMAL WELL-BEING

An important aspect of assessing the effect of production practices on animal well-being is consideration of alternative practices commonly in use or available to the livestock industry. A major challenge in this regard to the ethologist and physiologist is to determine whether practices, such as beak trimming and ear notching, cause undue stress to livestock. Because life without stress is not possible, it is not simply a question of determining whether production practices cause stress. The question must be, do production practices cause unacceptable stress? Because humans have differing values relative to animal care, it is obvious that individuals may come to opposing conclusions on the acceptability of a potentially stressful procedure. Even so, the charge of the ethologist and physiologist is to provide objective information on which sound decisions can be made. Critical in this decision process is weighing both positive and negative aspects of such practices. Many of these management practices involve acute pain for the animal. If scientists devise less stressful alternatives to presently used procedures, they must be considered as potentially preferred methods.

The extent to which an animal experiences pain is indeed an issue in making judgments about animal management practices. Such practices are usually weighed in terms of longer-term benefits to the animal, its conspecifics, and animal caretakers. Swanson (1995) summarized numerous research efforts relative to management practices commonly used in animal production. Typical practices reviewed include castration (Chase et al., 1995; Molony and Kent, 1997; Morrow-Tesch and Jones, 1997;

Robertson et al., 1994), tail docking (French et al., 1994; Mellor and Murray, 1989; Shutt et al., 1988), hot-iron and freeze branding (Lay et al., 1992a,b,c; Schwartzkopf-Genswein, 1996), and beak trimming (Breward, 1984; Duncan et al., 1989; Lee and Craig, 1991). Generally, it is concluded from these studies that the procedures cause pain; if they need to be performed, they should be done at an early age because this seems to impose the least amount of stress. Of the above, the only procedure known to produce chronic pain is beak trimming (Breward, 1984; Duncan et al., 1989; Lee and Craig, 1991), a procedure performed to decrease the damage that poultry inflict on one another in intensive production systems. Beak trimming illustrates the point that the alternatives to painful procedures may decrease well-being as well.

To logically address issues of livestock production practices, such as castration, confinement, and branding, one must have the basic knowledge of (1) how the procedures are accomplished, (2) results from objective studies identifying the cost and benefits, and (3) the alternatives currently available. Only by objectively evaluating the issues from these perspectives can an intelligent decision be made.

ENVIRONMENTAL DESIGN AND ANIMAL WELL-BEING

Approaches to environmental design to assure well-being must consider the following issues.

Design of the Physical Environment

Shelter is recognized as a basic need and is a critical issue in terms of both adequate care and economics. Physical environment includes factors such as housing, space allowances, equipment, access to food and water, and protection or exposure related to terrain, trees, or other characteristics of the natural environment.

Design of the Dietary Environment

Food (and, especially in farm animals, type of food) is recognized as a basic need. Because the species differ so distinctly in type of digestive tract and because goals of production systems are so divergent, designing appropriate diets to ensure good performance and well-being is a highly technical endeavor. Swine are monogastrics and omnivores; they cannot accommodate high-fiber, roughage feeds as can ruminants and horses. Cattle and sheep are ruminants and herbivores. Horses have a large cecum for digestion of high-fiber feeds. Both cattle and horses must have a bulky diet to prevent stress related to type of diet. Thus, the design of the nutritional environment is critical in terms of meeting both the animal's requirements and the economic needs of the enterprise.

Design of the Social Environment

The social and psychological needs of animals are of increasing concern as such requirements become better understood. Designing the social environment requires an understanding of relationships between animals, their environment, and behavioral responses.

While the previous features mentioned (physical, dietary, and social) reflect major components of designed environments for livestock production systems, numerous other operational plans are critical. Examples of those that impact the environment in the management sense are plans for health, facility maintenance and sanitation, marketing, and accommodation of varied production systems that might occur in the physical environment provided.

FUTURE PERSPECTIVES: DEBATE, RESEARCH, DEFINITION, AND CRITERIA TO ENSURE ANIMAL WELL-BEING

Debate concerning animal well-being will undoubtedly continue; however, all sides should be aware that the best solutions to problems are those developed collectively over time and adjusted as scientific knowledge is expanded. Those in the business of producing animal products must recognize that the proponents of what are perceived as improved methods are sincere in their empathy for animals. Producers and processors must be keenly aware of methods to enhance both the environment and the animal's inherited and conditioned resistance to periodic stress. Pet owners must realize that many pets are subjected to stressful environments and practices. Society, in general, must become more informed about problems, both physical and economic, that producers face in terms of meeting the real environmental needs of animals. We must all be careful not to assume that human needs, often confounded with wants, are the same as those for animals.

In 1993 a major effort was made by representatives of a broad spectrum of interests (scientists, livestock producers, public interest groups, and government agencies) to define high-priority research areas related to animal well-being. These interests provide some insight into areas where additional information is critical to appropriate decision making in the future as these regard animal care and production practices. Filling voids in the bank of objective information relative to animal care is essential for better decisions by producers, regulators, and the general public. The priority areas are described as reported by Mench (1993).

Bioethics and Conflict Resolution

Our current level of understanding of public attitudes toward food animal well-being is inadequate, and research should be directed toward assessing these attitudes. In addition, methods need to be developed to improve communication and resolve conflicts among the various interested parties. One important component of improved communication is more comprehensive education about animal well-being, ethics, and animal agriculture in general at all educational levels.

Responses of Individual Animals to the Production Environment

Arriving at meaningful and interpretable measures of well-being is critical. Research is needed to determine what constitutes normal behavior and to analyze the interrelationships among normal and abnormal behaviors and immunological, neuroendocrine, production, and health indicators of well-being. The influences of genetic background and previous experience on the animal's responses to the production environment also need to be assessed.

Stress

The term *stress* is used here in the broadest possible sense to include the evaluation of any practices that might result in physical or psychological discomfort or pain to the animal. Examples are the so-called "special" categories of standard agricultural practices (e.g., beak trimming and dehorning), genetic selection or bioengineering to improve performance, transportation, slaughter, handling, feed or water restriction, limitation of movement, and social isolation.

Social Behavior and Space Requirements

Although the research described previously will yield information about factors affecting the well-being of individual animals, most food animals are managed in groups. Studies are needed to investigate the effects of group size, group composition, and the expression of social behaviors like aggression and play on well-being and to determine how these factors affect the use of space by agricultural animals.

Cognition

Objectively designed studies to determine subjective feelings of animals are necessary. One method of learning about these subjective feelings is to allow animals to tell us how they perceive the environment and what their needs are by using preference testing.

Alternative Production Practices and Systems

The information derived from basic studies can be used to evaluate current management practices and production systems and to develop alternative or modified systems where necessary. However, a large variety of factors need to be weighed when evaluating production systems and determining their costs and benefits. A partial listing of some of these factors appears in table 1.1. They include not only animal well-being but human health implications, economic and environmental impacts, policy issues, and consumer acceptability as well. Complex modeling approaches will be required to accomplish this analysis.

Based on the mentioned criteria for evaluating practices, it is clear that the issue of animal well-being, in the context of broad societal interests, is an extremely complex issue and many answers will be generated slowly in terms of the broad range of interests.

Recognizing that the issue of animal well-being is of sufficient concern to the public in general, the scientific community must press ever harder to develop more objective measures of animal condition in relation to the environment. Without adequate information, there is a serious threat to the livestock industry and the

TABLE 1.1 Considerations in Evaluating Practices in the Animal Industry[a]

Issues	Ethical and Technical Considerations
Consumer acceptability	Ethical acceptabilty
	Product safety
	Product quality including appearance and nutritional value
	Product cost
Animal well-being	Behavior
	Health
	Growth and reproduction
	Pain and distress
	Physical comfort
Worker health and safety	Indoor air quality
	Zoonotic diseases
	Hygiene
	Noise
	Hazards
	Psychological acceptability including human-animal interactions and ease of management
Environmental impact	Air pollution
	Water pollution
	Land availability
	Impacts on wildlife
	Sustainability
Economics	Domestic markets
	International markets
Policy considerations	Policy making including regulations and guidelines
	Economic impacts including costs of implementation and subsidies
	Impact on farm structure
	Impact on rural communities

[a] *Source:* Mench, J.A., *Food Animal Well-Being* (Lafayette, Ind.: Purdue University, 1993), p.18, by permission. © 1993 by Purdue Research Foundation

various relationships by which humans and animals are interdependent. Development of such information requires an adequate investment in research and education. The answers provided by such research will aid in determining well-being or a level of well-being on some scale from satisfactory to unsatisfactory. Questions about the morality of how animals are maintained and utilized will remain as a matter of individual decision based on interpretation of level of well-being and other matters in an array of ethical considerations. It is most important to ensure that animals are cared for properly in an adequate environment, however long and intense the ethical debate.

PART II

STRESS: A CHALLENGE TO WELL-BEING

The coordinated physiological processes that maintain most of the steady states in the organism are so complex and so peculiar to living beings—involving, as they may, the brain, nerves, heart, lungs, kidneys and spleen, all working coopera-tively—that I have suggested a specific designation for these states, homeostasis.

W. B. CANNON (1932)[1]

Stress has such pervasive effects impacting on an animal's comfort, health, and performance that it may potentially be the most costly of negative influences considered by those who manage enterprises or care for animals. Many attempts to improve animal care deal with symptoms of the basic problem rather than seeking an improved understanding of the basic causes of stress or the actions that might be taken to minimize them. Thus, our goal is to encourage the search for the underlying reasons of the many things that can influence animals adversely and to view control of stressors as a major approach to improved animal care and management. First, one must understand characteristics of the environment that may present a challenge sufficient to engage the animal's biological systems to resist. Next, it is important to understand the systems that are available for resisting stressors, as well as how these systems are controlled and how they operate to maintain normal function of the organism and ultimately life. Finally, to ensure well-being, one must be a judge of limits—limits to which the animal can resist negative influences, limits to which the animal can adapt successfully, and limits to which the environment can be feasibly altered to accommodate the animal's needs.

[1]Cannon, W. B., *The Wisdom of the Body* (New York: W. W. Norton, 1932), p.24, by permission

CHAPTER 2

Biology of the Stress Response: Stressors and Control of Response Systems

Stress is defined as a condition in an animal that results from the action of one or more stressors that may be of either external or internal origin. Examples of external stressors are severe climatic conditions in which the animal encounters a challenge to maintain body temperature within a normal range. An internal stressor, for example, could be related to an abnormal biological function that results in an above-normal level of a blood component such as glucose that requires the reaction of regulatory systems to reduce the level through metabolism or excretion. Effects of such influences challenge the body to maintain a normal range of a variety of biological parameters involved in sustaining life.

Stressors, either physical or psychological, cause the body's stress response systems to mount an effort to maintain a stable internal environment, referred to as homeostasis. This process must be functional for the normal existence of all living organisms. Homeostasis is, thus, essential for the well-being of animals.

The following conclusions by early, distinguished physiologists reflect the concept of homeostasis.

Claude Bernard (1878)[1], the French physiologist, in writing about various biological aspects of experimental medicine, commented on the importance of recognizing that animals and humans live in two environments that he described simply as external and internal. He emphasized that the external environment in which an animal exists varies greatly depending on efforts to modify natural conditions. And, that the internal environment is regulated by biological mechanisms to limit variation in tissue parameters such as temperature, pH, and the numerous organic and inorganic components. He emphasized further that this metabolic ability to maintain uniformity is the basic characteristic that allows such organisms

---

[1]An English translation of Bernard (1878) is provided by Hoff et al. (1974). See references section for the source of the translation

a high level of independence in moving about and living in widely variable environments. These writings are likely the first that clearly outline the concepts involved in homeostasis, although this specific term was not used.

W. B. Cannon (1932)[2] used the term *homeostasis* in the following statement: "The coordinated physiological processes that maintain most of the steady states in the organism are so complex and so peculiar to living beings—involving, as they may, the brain, nerves, heart, lungs, kidneys and spleen, all working cooperatively—that I have suggested a specific designation for these states, homeostasis." Cannon is also credited with an early description of the fight-flight response to stressful conditions, which will be discussed later.

The biology of the stress response relates to the mechanisms involved in the maintenance of stability and balance in numerous characteristics of the internal environment. Examples are acid-base balance of blood and lymph and level of components of these body fluids such as glucose and calcium.

The three phases of the stress response are generally described as (1) alarm, followed by (2) resistance, which if unsuccessful results in (3) exhaustion of the response system and eventually death. Major negative impacts on well-being and performance of animals is obviously related to the exhaustion stage. Less obvious but of major importance are economic losses associated with the resistance phase of the response common to stressful conditions that may exist at times in some animal enterprises.

The body's alert system consists of cognitive sensing of potential stressors and numerous other tissue sensors located throughout the body that do not have cognitive influences. These are designed to monitor biological characteristics. Sensors monitoring physical characteristics are those sensitive to body temperature reflected in thermal characteristics of the blood, osmotic pressure, and blood pressure. Other sensors monitor chemical components of body fluids, such as blood glucose and blood calcium levels. Once these sensors detect an abnormal condition, a neural alert signal normally results in both neural and humoral responses to correct the problem by restoration of the level to a normal range or the homeostatic state. If such systems cannot endure the continuing influence of a stressful situation, the ability of the response is exhausted. Thus the final phase of the response can result in death.

STRESSORS: SOME EXAMPLES OF TYPICAL CHALLENGES TO WELL-BEING

Physical Stressors

The combination of characteristics making up the animal's environment will challenge its ability to cope and maintain homeostasis. If the environment is chronically stressful, the normal progress of stress impacts will occur. These responses include an immediate alarm reaction to maintain the normal ranges of biological parameters such as body temperature, sustained resistance over an extended time to maintain these parameters, and exhaustion of the metabolic ability to sustain the response and finally death.

[2]Cannon, W. B., *The Wisdom of the Body* (New York: W. W. Norton, 1932), p.24, by permission

Restricted Movement

In some cases crowding or inadequate facility design may lead to restricted access to feed or water. While such limitations may influence the performance potential of the animals, they may also result in stress related to frustration and associated abnormal behaviors. Thus, stressors of a physical nature may result in psychological influences as well.

Injury or Pain

Animals experiencing pain and injury are subject to secondary infections as a direct result of lesions and stress related to such conditions. Both may reduce the general resistance of the animal over time through influences on the immune system.

Thermal-Related Demands

Among the most common stressful situations encountered in animal production and, in many cases, pet care are demands related to heat and cold. As a result, design of facilities, management scenarios, and dietary characteristics become major considerations in ensuring that animals have the proper resources to meet environmental demands. Acclimatization is an important phenomenon in considering management of animals moving into an environment of increased thermal demand. This adaptation to more severe cold results from an increase in smooth muscle tone and vasoconstriction in peripheral blood vessels, which aids in regulating blood flow to the outer tissues such as skin and muscle; this regulation enhances the effectiveness of these tissues in terms of insulation. The process is gradual and may require two to three weeks to reach maximum development. Thus, the level of protection from severe cold, as an example, for animals being relocated to more demanding environments is an important management consideration.

Dietary Limitations

Food and water are in fact physical resources like other utilities available to animals. Inadequate dietary design and quantity available are common problems in management of both farm and companion animals. Environmental demands are often related to climatic conditions. Effects of drought on the feed supply of grazing animals is a common example. Dietary deficiencies are also encountered frequently because of lack of knowledge relative to dietary requirements, mistakes and oversights in allowances and availability, and, in some cases, simply lack of resources on the part of the provider to ensure adequate care.

Psychological Stressors

Spatial Intrusion

Individual space desires of animals may vary with species and individual animals. Violation of space requirements normally increases tendency toward aggressive behavior. The common species of farm animals are considered to be social in nature and thus are quite flexible in terms of spatial requirements and contact with other animals. They are, however, subject to aberrant behaviors associated with crowding and social pressures over extended periods. Such responses represent coping mechanisms such as those discussed later in part 3.

Fear

Fear generates the tendency to the so-called fight or flight response. This tendency can result in chronic stress effects that adversely influence performance or behavior.

Restricted Movement

Restricted movement may result in behaviors normally associated with boredom. Animals may respond with activities to substitute for normal movement, such as excessive chewing, tail biting, hair pulling, abnormal standing or lying, and horses perform a variety of stable vices.

Frustration

Animals show goal orientation. Inability to achieve such goals leads to frustration-related stress. The concept of goal orientation in animals introduces the issue of animals' level of cognition. While there is some debate on this subject, it is increasingly evident that cognitive characteristics are important issues in design of physical and social environments for animals.

Learning Limitations

Species and individual animals differ in learning ability. Different systems for maintaining animals pose different learning demands. Limitations in the ability of animals to adapt to the physical environment provided may result in chronic stress and possibly undesirable behavior.

Interactions between Physical and Psychological Stressors

Clearly physical and psychological factors operate jointly in terms of impacts on the whole animal. Environmental planning must consider the integrated impacts of both on the animals involved. An already large body of data dealing with the physical requirements of animals is being expanded rapidly by major research and development efforts around the world. Less clear and therefore more debatable are specifications to meet the social needs of animals. Those designing animal production systems must consider both aspects in relation to animal well-being and economics.

STRESS EFFECTS

Stress may result in one or more of the general effects listed in table 2.1. Presence of the effects indicate that an animal's environment is inadequate. Indicators of stress may also be associated with major economic effects on the animal enterprise as a result of broad effects on biological efficiency of the animals, behavioral impacts, and increased death loss.

TABLE 2.1 Effects and Impacts of Stress

Typical Effects of Stress	Possible Biological Impact	Potential Economic Impact
Adaptation/habituation problems	Increased variability in animal performance; animal discomfort	Increased cost of production
Altered nutrient requirements	Reduced efficiency in nutrient use	Increased feed cost and related cost of production
Reduced nutrient utilization	Reduced efficiency in nutrient use	Increased feed cost and related cost of production
Reduced reproduction rate	Lower conception rate or neonatal survival rate; reduced semen quality	Increased cost of production
Altered body or product composition related to altered performance	Altered ratio of tissue components or changes in other tissue characteristics	Reduced product value
Increased possibility of injury to the animal	Tissue damage; reduced performance	Reduced product value and increased cost of production
Self-isolation	Reduced performance	Increased cost of production
Listlessness, abnormal gait, lameness, restlessness	Reduced performance	Difficult management Increased production costs
Increased potential for injury of handlers and damage to equipment and facilities		Increased operational and employee liability costs
Reduced immune response and the related increased susceptibility to disease and ultimately death	Reduced biological efficiency	Reduced product value and increased cost of production

THE NERVOUS SYSTEM IN THE STRESS RESPONSE

The nervous system interprets signals that reflect impending stress. Cognitive brain centers such as the cerebral cortex perceive external threats and act to initiate response mechanisms. Noncognitive centers of the brain and, in some cases, other tissues throughout the body detect adverse changes within the body that may pose problems related to normal functioning of various biological systems. These sensors respond by initiating neural and hormonal (humoral) actions important in mechanisms for coping with stressors. Responses result from both voluntary and involuntary actions. Voluntary responses include resisting offenders, seeking shelter, returning to a herd, and similar activities to relieve stressful situations. Involuntary responses represent a variety of physical and metabolic responses that support the animal's stress resistance. In all cases, however, the nervous system is critically involved in the process of detecting, signaling, and maintaining response mechanisms.

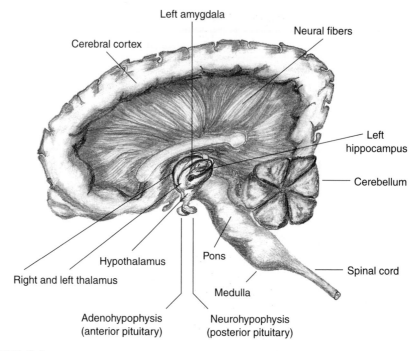

FIGURE 2.1 Components of the brain primarily involved in an animal's responses to stress

In the broadest classification, the nervous system consists of the central nervous system, including the brain and spinal cord, and the peripheral nervous system, which includes cranial nerves, spinal nerves, autonomic nerves, and various ganglia that provide connections (synapses) between nerve fibers (neurons). Figure 2.1 is a graphic representation of some important anatomical regions of the brain that are important in the stress response.

ORGANIZATION OF THE NERVOUS SYSTEM

Components of the nervous system are reflected in the schematic presented in the section on organization of the autonomic nervous system. While the entire nervous system may be involved in both recognition and resistance to stressors, the autonomic nervous system and its component sympathetic and parasympathetic divisions are of particular interest in understanding the animal's response to environmental demands. The relative functions and balancing effects of these divisions represent remarkable biological phenomena related to sustaining well-being during periods of stressful external and internal challenges to the individual organism.

The central nervous system is made up of the brain and spinal cord and provides key detection of threats and coordination of responses to stressors. The somatic division of the peripheral nervous system is largely directed to detection of sensory stimuli and activity of skeletal muscle. The autonomic division of the peripheral nervous system is largely involved in control and coordination of functions of secretory glands important to endocrine systems and smooth muscle activity of glands and internal organs associated with glandular secretion, digestive and circulatory functions, and metabolic regulation to maintain homeostasis. Innervation of functional components involved in these processes will be discussed later.

ANATOMICAL COMPONENTS OF THE CENTRAL NERVOUS SYSTEM AND ENDOCRINE TISSUES CLOSELY ASSOCIATED IN THE CRANIAL REGION

Along with an outline of components of the central nervous system (brain and spinal cord) a general description of function of these tissues and cranial nerves is included below. However, important functions of several components listed here are discussed in more detail later since these functions are involved in the animal's specific responses to stressful environmental influences.

I. *Prosencephalon (forebrain).*
   A. *Telencephalon.* The telencephalon includes components of the central nervous system that are involved in cognitive processes. It also includes numerous sensory mechanisms and controls a wide variety of central and peripheral nervous system functions.
      1. *Cerebrum.* The cerebrum makes up the largest component of the brain and consists of paired divisions termed right and left hemispheres. Components are the cerebral cortex, medullary substance, and basal ganglia. Olfactory bulbs are closely associated anatomically.
         a. *Cortex.* The cortex is the outer layer of the cerebrum and is made up largely of gray matter. Subdivisions are called lobes or regions. The general functions of each area or lobe are as follows:
            (1) *Prefrontal lobe.* Functions of the prefrontal lobe relate to cognition, emotions, and some cognitive influences on the autonomic nervous system.
            (2) *Frontal lobe.* This area provides motor related functions.
            (3) *Parietal lobe.* These tissues are involved in numerous sensory functions related to receiving information reflecting physical stimuli and pain from the body's periphery including relative intensity of such stimuli.
            (4) *Temporal lobe.* A function of the temporal lobe is related to hearing.
            (5) *Occipital lobe.* A function of the occipital lobe is related to vision.
         b. *Medullary substance.* This region is made up of masses of neural fibers connecting the two cortical hemispheres and the cortex with other parts of the brain and the spinal cord.
         c. *Olfactory bulb.* Functions of these structures are related to smell.
         d. *Basal ganglia.* These components are located below the anterior area of the cortex and consist of large collections of nuclei and neurons that initiate and transmit neural impulses to other parts of the central and peripheral nervous systems. Neuron connections between the cortex and other central nervous system components are clearly critical in providing cognitive input into both neural and endocrine

responses to environmental influences including those related to stressful environments.

e. *Amygdala.* Some functions of the amygdala are related to behavioral characteristics of animals and may involve aggressive reactions to environmental influences such as interactions with other animals and humans. This subcortical region is also thought to be related more generally to feelings and emotional responses to psychological stressors. The amygdala and hippocampus work in concert to enable the animal to learn from emotional experiences.

f. *Hippocampus.* Hippocampal functions are thought to be related to behavioral characteristics such as responses related to space allowances or stress related to population density. This subcortical component is also involved in memory formation, which may be a factor associated with spatial behavior of animals including mental mapping of the animal's environmental resources.

g. *Origin of cranial nerve I* (olfactory nerve).

B. *Diencephalon.* The diencephalon includes several bodies that are critical in both neural and endocrine (humoral) responses to stressors. Major centers for sensing internal signals that reflect threats to homeostatic conditions exist in this area, along with centers that transmit signals for initiation and maintenance of the stress response. Components represent major elements making up the limbic system. Anatomical components of the diencephalon are shown in the list that follows. Included also is the pituitary gland and its components, which are attached to the hypothalamus by the pituitary stalk. While the pituitary tissues are not considered a part of the central nervous system, they are critically involved in both neural and endocrine functions related to stress responses through stimulation and coordination by the hypothalamus. The pituitary gland, also called the hypophysis, is made up of the adenohypophysis (anterior pituitary), medial hypophysis (intermediate pituitary), and neurohypophysis (posterior pituitary).

1. *Epithalamus.* This component anatomically is closely associated with the pineal gland and contains nuclei involved in auditory function.

2. *Thalamus.* The thalamus provides important neural connections between the cortex and the hypothalamus and is critically involved in coordinating the animal's response to stressors.

3. *Hypothalamus.* The hypothalamus provides critical functions in coordinating a variety of neural and endocrine functions throughout the body. The hypothalamus controls the pituitary gland (hypophysis) and, as a result, is involved in both neural and endocrine functions related to homeostasis, reproduction, and growth. Hormones synthesized in the hypothalamus cause production and(or) release of hormones by the pituitary gland.

4. *Hypophysis (pituitary gland).* Components of the pituitary gland provide major control of the endocrine system and, thus, are essential for normal functioning of homeostatic functions and numerous other noncognitive systems. Because of its widespread endocrine function, the pituitary gland is often referred to as the "master gland."

a. *Adenohypophysis (anterior pituitary).* The anterior pituitary is involved in a variety of physiological functions. A critical relationship to an animal's response to stressful conditions is the release of the hormone ACTH, which stimulates the release of corticosteroids by the adrenal cortex. This group of cortical hormones is essential in successful resistance to stressors. The gland also produces β endorphin which can assist the animal in coping with stressors as a result of its opiate effect.

b. *Medial (intermediate) hypophysis.* This structure is a likely source of endorphins that have opiate effects.

c. *Neurohypophysis (posterior pituitary).* The posterior pituitary is critical in the initial and rapid physical and metabolic response to stressors through neural stimulation of the adrenal medulla to release epinephrine (adrenaline). This structure is the source of oxytocin, the hormone involved in milk release, and also releases vasopressin, a hormone involved in water and mineral homeostasis.

5. *Pineal gland.* Pineal tissue is the source of melatonin, the hormone involved in some behaviors related to photoperiod (day length). These include effects of light on sleep characteristics and seasonal influences on reproductive behavior in some species. The gland normally contains serotonin, a neurotransmitter involved in "feelings" or how organisms feel about the environment in which they exist. Serotonin is an intermediate in the synthesis of melatonin.

6. *Origin of cranial nerve II* (optic nerve).

II. *Mesencephalon (midbrain)*

A. *Inferior colliculi.* Functions of the inferior colliculi are related to auditory stimuli.

B. Superior colliculi. Functions of the superior colliculi are related to visual stimuli.

C. *Origin of cranial nerves III (oculomotor nerve) and IV (trochlear nerve).* These are associated, respectively, with ciliary and circular muscles of the eye and with the dorsal oblique muscle of the eye.

D. *Periaqueductal gray matter.* This area likely mediates the analgesic effect of opiates produced by the body in response to stress and is, therefore, thought to influence the animal's ability to cope with stressful environments.

E. *Substantia nigra.* This structure is a source of dopamine, an important neurotransmitter. Like the red nucleus, this tissue is involved in the sensorimotor neural system controlling some muscles.

F. *Red nucleus.* This area is an important synaptic site for neurons involved in motor function of facial and limb muscles.

III. *Rhombencephalon (hindbrain)*

A. *Metencephalon.* Tissues of the metencephalon are extensively involved in receiving and synthesizing information relative to balance and thus are critical to independent and appropriate body movements. The region also serves as a conduit for a large number of neural fibers as well as the origin of cranial nerves.

1. *Cerebellum.* The cerebellum is located to the rear of the cerebrum and consists of two hemispheres. It serves basically as a collector of information reflecting physical conditions throughout the body and as a coordinator of muscular responses to make corrections. Information is received from sensors in joints and muscles, inner ear, and visual centers relative to equilibrium of the body which in turn send impulses to provide motor responses by muscles to make skeletal adjustments that restore balance.
2. *Pons.* The pons is located ventral to the cerebellum. It connects the hemispheres of the cerebellum and serves as the route for numerous descending and ascending neural tracts.
3. *Origin of cranial nerve V (trigeminal nerve).* This nerve is sensory to the eye and face and motor to muscles of mastication.

B. *Myelencephalon.* The myelencephalon serves as a conduit for nerve fibers and gives rise to a number of cranial nerves.

1. *Medulla oblongata.* This component is also referred to as the medulla. The structure connects the pons and the spinal cord and therefore accommodates numerous descending and ascending neurons.
2. *Origin of cranial nerves VI* (abducent nerve—eye muscles), *VII* (facial nerve—ear and tongue), *VIII* (acustic or vestibulocochlear nerve—hearing), *IX* (glossopharyngeal nerve—pharynx and salivary glands), *X* (vagus nerve—pharynx, larynx, and visceral structures of the thoracic and abdominal cavities), *XI* (accessory nerve—shoulder and neck muscles), and *XII* (hypoglossal nerve—tongue muscles).

IV. *Spinal cord.*

A. *Spinal cord white matter.* The outer component of the spinal cord accommodates the numerous ascending (afferent) and descending (efferent) nerve tracts.
B. *Spinal cord gray matter.* The gray matter column surrounded by white matter is the origin of neurons and therefore contains numerous nerve cell bodies and synaptic junctions connecting neural fibers.

COMPONENTS OF THE LIMBIC SYSTEM

The limbic system is a group of central nervous system components that assists animals in meeting stressful challenges. Thus, the organization and function of this system have special significance in understanding an animal's ability to respond to stressors.

Several components of the brain make up the limbic system. (See Figures 2.2, 2.3, and 2.4.) These include areas of the cerebral cortex and subcortical regions including the thalamus, hypothalamus, hippocampus, and amygdala. The system is recognized as being critically involved in both behavioral and homeostatic responses to stressful environments. For example, the amygdala is involved in behavioral characteristics related to aggression, emotions, and feelings of well-being. It is also thought to be involved in memory formation related to such behavioral characteristics. It is likely involved in how animals perceive adequacy of space and spacing among animals or what might be referred to as the spatial environment. The hip-

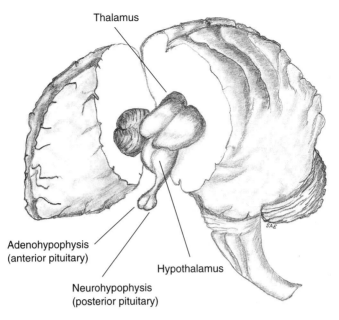

Thalamus

Adenohypophysis
(anterior pituitary)

Hypothalamus

Neurohypophysis
(posterior pituitary)

FIGURE 2.2 A sectioned view of the thalamus, hypothalamus, and pituitary

pocampus is involved in spatial-related behaviors and in memory formation associated with space characteristics, as well as mapping of areas occupied by animals and location of available resources. Both the amygdala and hippocampus are probably involved in memory formation related to experiences and social encounters. The hypothalamus is involved in initiating stress responses. The pituitary gland is critically involved in both neural and endocrine aspects of the stress response. For example, the posterior pituitary is involved in initiating neural stimulation of the adrenal medulla, which releases epinephrine (adrenaline) as an important component of the animal's initial response to a stressful condition. The anterior pituitary releases adrenocorticotropic hormone (ACTH) in response to a stressor. This hormone, in turn, stimulates the release of corticosteroids by the adrenal cortex, which enhances and extends metabolic responses to withstand stressful conditions.

The collection of structures usually referred to as the brainstem include those making up the diencephalon, mesencephalon, metencephalon, and myelencephalon.

Spinal Cord

The spinal cord is the conduit for numerous spinal tracts that transmit impulses to and from the brain. These tracts connect with spinal nerves by synapses to conduct impulses to and from the periphery of the body. The spinal tracts and connecting spinal nerves are responsible for both voluntary and involuntary innervation, resulting in action by various tissues that are responsible for the stress response. A variety of involuntary actions are innervated by the sympathetic and parasympathetic

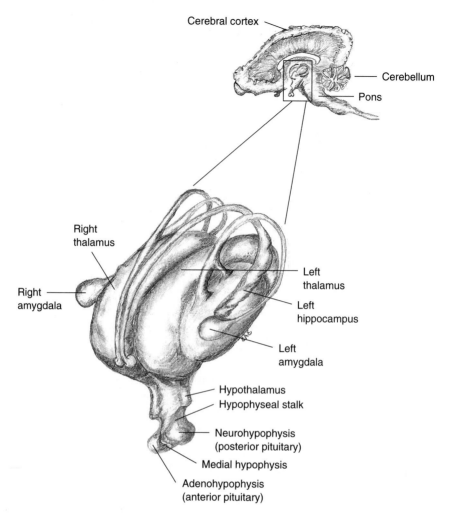

FIGURE 2.3 Major components of the limbic system in relation to cortical structures

components of the autonomic division of the nervous system. Tracts and neurons of sympathetic origin arise from the thoracic and lumbar sections of the spinal column, and those of parasympathetic origin arise from the cranial region near the base of the skull and the sacral area of the spinal column.

Anatomically the spinal cord is made up of a central, gray matter column with many nerve cell bodies. This area is surrounded by a white matter column in which the numerous spinal neural tracts exist. Figure 2.5 depicts components of the spinal cord. Spinal nerves innervating the periphery arise from cell bodies in the ventral horn of the gray matter column. These nerve cells conduct impulses to tissues remote to the spinal column and are termed efferent neurons. Impulses arising

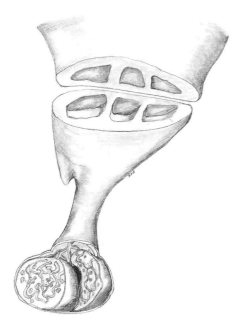

FIGURE 2.4 The hypothalamus and hypophysis (pituitary) reflecting major hypothalamic nuclei and the profuse vascular system of the hypophysis

from the periphery are conducted by afferent neurons that synapse in the dorsal horn of the spinal column gray matter with afferent neurons existing in spinal tracts that will conduct the impulse to the brain.

Nerve Cell (Neuron) Types

Nerve cells extending through the spinal column are components of both the somatic and autonomic nervous systems. Figure 2.6 illustrates the components of a nerve cell.

Somatic Neurons   Neurons of the cranial and spinal nerves innervate striated muscles of the skeleton and some organs and are thus important in voluntary functions of the body such as locomotion and muscular reflexes. (Reflex action of muscles may occur without involvement of the brain; such actions are referred to as spinal reflexes.)

Autonomic Neurons   Neurons of the autonomic nervous system exit the spinal column to become components of either the sympathetic or parasympathetic system. They are involved in many functions of an involuntary nature such as digestion, blood flow, heart rate, respiration, and other functions related to the stress response.

Cerebrospinal Fluid

The fluid component of the central nervous system is responsible for the transport of hormones, neurotransmitters, and nutrients to, and waste products from, the various tissues within the system. It serves the spinal cord and the brain by

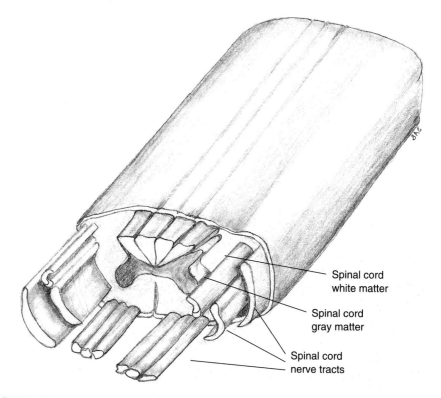

Spinal cord
white matter

Spinal cord
gray matter

Spinal cord
nerve tracts

FIGURE 2.5 Spinal gray matter and example spinal tracts of the white matter component of the spinal cord

occupying the space in the ventricles of the brain and the subarachnoid space surrounding the brain and spinal cord. The presence of cerebrospinal fluid around the brain provides effective cushioning for components within the cranium. The composition of cerebrospinal fluid is higher in water than blood plasma, generally lower in organic components, and variable in relative mineral content as compared with blood plasma.

ORGANIZATION OF THE AUTONOMIC NERVOUS SYSTEM

Sympathetic neurons arising from the thoracic and lumbar vertebral regions serve ganglia that innervate postganglionic neurons connecting with the various effector tissues, the exception being the adrenal medulla, which is innervated by preganglionic fibers (fig. 2.7).

Parasympathetic neurons arising from the midbrain, pons, medulla regions of the brain (cranial region), and sacral vertebral region connect with ganglia located in the various effector organs.

Dual innervation of effector organs served by the autonomic nervous system is important in maintaining functional balance in an organ's performance. In general,

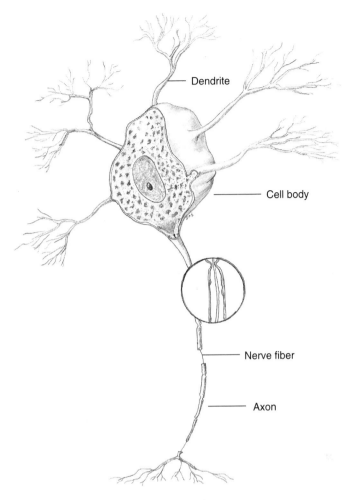

Dendrite

Cell body

Nerve fiber

Axon

FIGURE 2.6 Components of the nerve cell (neuron)

the sympathetic and parasympathetic influences have opposite effects. Prime examples are the tonic effect on smooth muscle in blood vessels, coordinated muscle contraction and relaxation in gut peristalsis, and constriction or relaxation of bronchi in respiratory tract function. Although sympathetic and parasympathetic systems generally have opposing actions in a given target effector tissue, the response to each source of stimulation may be different in another effector tissue. Different effects by the same neurotransmitter are dependent on the nature of receptors existing in a given target tissue (Ahlquist, 1948). Examples are α-adreno-ceptors and β-adrenoceptors existing in target smooth muscle. These receptors generally prompt contraction and relaxation of the muscle, respectively, when stimulated by epinephrine; however, in a given tissue, the effect may be reversed.

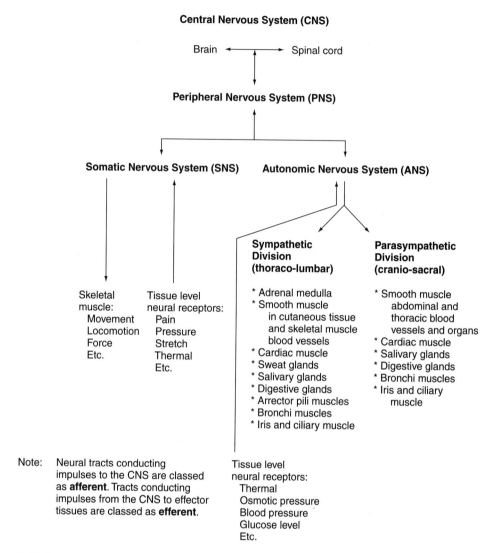

FIGURE 2.7 Organization of the nervous system

This phenomenon is referred to as epinephrine reversal. The net effect depends on a complex system whereby one type of receptor dominates the other.

Divisions of the Autonomic Nervous System

Sympathetic Nervous System

Preganglionic neurons located in the spinal column exit the column in the thoracic and lumbar regions to connect with ganglia that are generally located at some distance from the various target organs (autonomic effector organs). Postganglionic neurons arise from these ganglia and continue to the effector organ (figs. 2.8, 2.9,

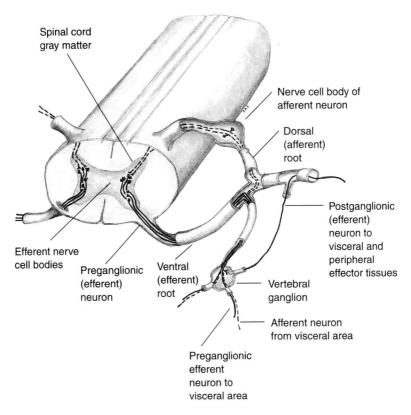

Spinal cord
gray matter

Nerve cell body of
afferent neuron

Dorsal
(afferent)
root

Postganglionic
(efferent)
neuron to
visceral and
peripheral
effector tissues

Efferent nerve
cell bodies

Preganglionic
(efferent)
neuron

Ventral
(efferent)
root

Vertebral
ganglion

Afferent neuron
from visceral area

Preganglionic
efferent
neuron to
visceral area

FIGURE 2.8 Spinal cord and example neural connections involved in innervation of the sympathetic nervous system in the thoracolumbar region

and 2.10). The nerve impulse from the preganglionic fiber is transmitted to the postganglionic cell in the sympathetic system by release of acetylcholine at the synapse. The neurotransmitter between the postganglionic neuron and the effector tissue in the sympathetic system is norepinephrine. Thus, the effector organ is stimulated by norepinephrine, with one exception: the effector secretory cells of the adrenal medulla are innervated without postganglionic fiber involvement. Thus the neurotransmitter in this case is acetylcholine. The medullary cell responds in the same way a postganglionic fiber responds (i.e., by the release of the adrenergic neurotransmitters epinephrine and norepinephrine). Embryologically the adrenal medulla develops from the same class of tissue giving rise to postganglionic sympathetic nerve fibers.

The relationship between the sympathetic division of the autonomic nervous system and the medulla of the adrenal gland, called the sympathoadrenal (SA) system, is a major consideration in the stress response of animals. The neurotransmission results from stimulation of receptors in the central nervous system. The hypothalamus then originates nerve impulses that are transmitted by way of the neurohypophysis directly to secretory cells in the adrenal medulla with release of acetylcholine at the site. The adrenal medulla responds by release of

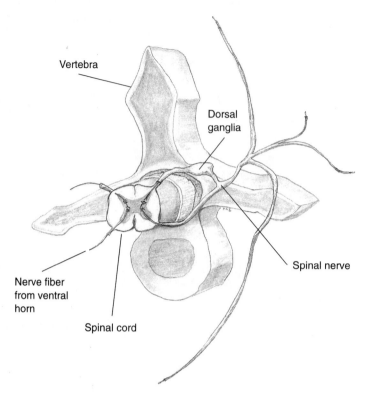

Vertebra

Dorsal
ganglia

Spinal nerve

Nerve fiber
from ventral
horn

Spinal cord

FIGURE 2.9 Section of the spinal cord reflecting the relationship to vertebral column components

epinephrine and norepinephrine into the bloodstream. The adrenal medulla (the central core of the adrenal gland) consists largely of A cells (epinephrine storing) and N cells (norepinephrine storing) that release their products upon neural stimulation. Synthesis of epinephrine from norepinephrine is dependent on the presence of glucocorticoids, exemplifying a symbiotic relationship between the adrenal medulla and cortex. The reaction requires the enzyme phenylethanolamine N-methyltransferase (PNMT), which is active in the presence of glucocorticoids from the cortex. This relationship establishes a role of the anterior pituitary in the adrenal medullary response. The adrenal cortex surrounds the medulla and is the source of glucocorticoids and mineralocorticoids. The adrenal cortex and its hormones are a part of the hypothalamic-pituitary-adrenal (HPA) stress-response system. The system responds to stimuli, causing the hypothalamus to release corticotropin-releasing hormone (CRH) into a portal vascular system that stimulates the anterior pituitary to release ACTH, which enters the bloodstream and ultimately stimulates the adrenal cortex to release its hormones into the blood. This system is humoral in nature (i.e., transmitted by the blood) after initial stimulation by the hypothalamus as compared with neural transmission outlined previously for the SA system.

1. Thalamus
2. Hypothalamus
3. Spinal cord - cervical vertebra C1-C8
4. Spinal cord - thoracic vertebra T1-T13
5. Spinal cord - lumbar vertebra L1-L7
6. Spinal cord - sacral vertebra S1-S3
7. Vertebral ganglia (sympathetic trunk)
8. Preganglionic neurons - sympathetic
9. Preganglionic neurons - parasympathetic
10. Eye and lacrimal gland
11. Salivary gland area
12. Skin
13. Postganglionic neurons - sympathetic
14. Vagus nerve
15. Lungs
16. Heart
17. Stomach
18. Intestine
19. Adrenal gland
20. Kidney
21. Bladder
22. Celiac ganglion
23. Mesenteric ganglion

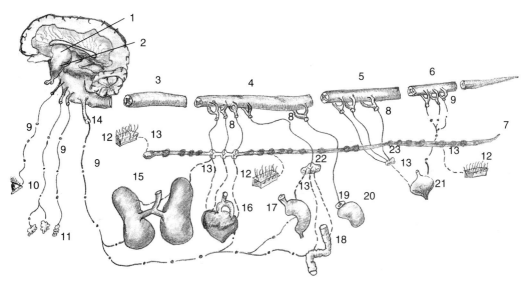

FIGURE 2.10 Innervation of tissues and organs by the autonomic nervous system

Parasympathetic System

Cranial preganglionic neurons arise from nuclei in the brain and exit the cranial area rather than following tracts of the spinal cord. A second component of parasympathetic innervation involves fibers that follow spinal tracts and exit the cord by synapsing with preganglionic fibers in the sacral vertebral region (fig. 2.10). These preganglionic fibers end at ganglia generally located in the effector organs served. Short postganglionic neurons originating at the ganglia connect with cells of the effector tissues served. The synaptic neurotransmitter at the ganglion and at the site of action on tissues of the effector organ is acetylcholine.

The cranial nerves having parasympathetic effects are of special note due to their involvement in a variety of functions providing routine maintenance of the animal and in some cases the stress response. The vagus nerve (cranial nerve X) serves all major organs in the thoracic and abdominal regions and is involved in respiration, digestion, and elimination of waste products. Cranial nerve III also has parasympathetic influences through innervation of the ciliary and circular muscles of the eye. Cranial nerves VII and IX innervate salivary glands.

Other cranial nerves have motor and/or sensory functions related to vision, taste, smell, eye movement, tongue movement, speech, and some muscular movement in the head and neck region. These cranial nerves are not considered to be part of the parasympathetic nervous system but may be involved in certain aspects of responses to stressful conditions.

Actions of the Autonomic Nervous System

Nerve cells (neurons) of the autonomic nervous system supply functional innervation to skin and all visceral organs (e.g., heart, blood vessels, gastrointestinal tract, pancreas, adrenals, and kidneys). Effects of such innervation include the following.

Vasomotor (Smooth Muscle) Effects

Innervation of smooth muscle of the blood vessels causes constriction or dilation by respective contraction or relaxation of the smooth muscle layer in the wall of the vessel. This action aids in regulation of blood pressure and body temperature. The specific response depends on the dominant receptor type in a given tissue.

Secretomotor Effects

Innervation causes secretion of products of various glands. Release of saliva and pancreatic juice are examples of secretomotor effects.

Organ (Smooth Muscle) Motor Effects

Major organs of the thoracic and abdominal regions are innervated by the autonomic nervous system. This innervation results in contraction and relaxation of smooth muscle in the wall of the stomach and both large and small intestine to control the mixing of ingesta and its movement through the tract. Innervation of smooth muscle in the wall of the bladder is involved in control of urine excretion.

Cardiac Muscle Effects

Regulation of blood flow is critical in response to stressors. Heart rate, force of contraction, and volume of ventricular filling are influenced by autonomic innervation. This combination of effects clearly relates to an animal's ability to respond to emergencies.

Example responses to the two divisions of the autonomic nervous system are shown in table 2.2.

Function of the Divisions of the Autonomic Nervous System

Sympathetic Nervous System

The sympathetic system regulates processes related to emergencies by adjusting mechanisms to solve internal problems. This system is critically involved in an organism's response to stress and preparation for the fight or flight syndrome. It is

TABLE 2.2   Examples of Tissue Effects of Sympathetic and Parasympathetic Nervous Innervation

Tissue or Function	Sympathetic Effects	Parasympathetic Effects
Heart rate	Accelerated	Decreased
Heart volume	Increased	Decreased
Blood vessels		
Cutaneous	Constriction	—
Muscle	Dilation	—
Coronary	Dilation	Constriction
Salivary glands	Constriction	Dilation
Cerebral	Dilation	Constriction
Abdominal and pelvic viscera	Constriction	—
Iris	Dilation	Constriction
Smooth muscle		
Bronchi	Dilation	Constriction
Small intestine	Inhibition	Increased motility
Large intestine	Inhibition	Increased motility
Bladder	Inhibition	Contraction
Urinary sphincter	Contraction	Relaxation

also involved in longer responses to chronic stressors, which constitutes a continuing resistance mechanism referred to as the general adaptation syndrome (GAS). Prime examples of such effects are increased heart rate and increased blood flow to muscle, brain, and heart. Simultaneously, blood flow is decreased to the skin, digestive tract, and kidneys.

Parasympathetic Nervous System

The parasympathetic system is primarily concerned with body functions related to maintenance of routine processes such as digestion, movement of food through the digestive tract, and transport, use, and storage of nutrients in liver, adipose tissues, and muscle. In many functions the parasympathetic effect is opposite that of the sympathetic system and results in a balancing mechanism.

Synergism between Sympathetic and Parasympathetic Systems   Actions of acetylcholine and norepinephrine, while they tend to be opposite in effect, work together (figs. 2.11 and 2.12) to maintain a balance in tissue response. An example of these influences is peristaltic movement of the digestive tract, which pushes the contents along during the process of digestion and elimination. Acetylcholine (parasympathetic) causes the smooth muscle of the gut to contract; norepinephrine (sympathetic) likely causes the same muscle to relax ahead of the contraction. Thus, a wave of contraction coordinated with a wave of relaxation along the gut is the basis for peristalsis. Segmented contractions in the small intestine mix the contents to enhance digestion. (It should be noted that other regulators such as cholecystokinin and secretin are also involved in gastrointestinal activity.) Another example is the constant adjustment in tone of smooth muscle in blood vessels (e.g.,

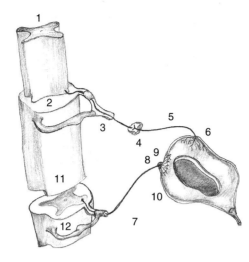

1. Spinal gray matter column
2. Spinal white matter column
3. Preganglionic sympathetic neuron
4. Hypogastric ganglion-synapse (neurotransmitter is acetylcholine)
5. Postganglionic sympathetic neuron
6. Sympathetic synapse with effector organ smooth muscle
7. Preganglionic parasympathetic neuron
8. Pelvic ganglion-synapse (neurotransmitter is acetylcholine)
9. Postganglionic parasympathetic neuron
10. Parasympathetic synapse with effector organ smooth muscle (neurotransmitter is acetylcholine; effect is contraction)
11. Spinal cord-lumbar region
12. Spinal cord-sacral region

FIGURE 2.11 Autonomic innervation of bladder smooth muscle as an example of effector tissue

the balance between contraction and relaxation is the mechanism for maintaining blood pressure). Example responses to the two divisions of the autonomic nervous system are shown in table 2.2.

Balancing Impacts of Autonomic Innervation in Homeostasis Homeostasis was defined by Cannon (1932) as the tendency to uniformity or stability in the normal body states (internal environment or fluid matrix) of the organism. It is an important biological phenomenon that forms the basis for animals to adapt to changes in their environment. Stress challenges the homeostatic state. Thus the stress response includes a complex of responses to maintain a steady state.

The autonomic nervous system, consisting of sympathetic and parasympathetic components (fig. 2.10), is central in homeostasis because it coordinates and conducts the biological balancing acts involved in altering the animal's ability to survive and perform under variable conditions. These regulatory mechanisms are primarily involuntary and include body temperature control, excretion, water balance, heart rate regulation, and related factors such as blood pressure and oxygen-carbon dioxide exchange (figs. 3.16 through 3.20).

The concept of environmental stress effects requiring response of the central nervous system through the HPA pathway was a major component of explanations offered by Hans Selye, who is considered by many to be the founder of stress research. Selye also hypothesized that animals could meet unusual demands to maintain homeostasis for a limited period of time, beyond which a variety of serious stress effects would occur and eventually lead to death.

Selye concentrated on the role of the stress-response system involving the pituitary and the adrenal cortex. This system became known as the general adaptation syndrome (GAS). Cannon demonstrated the importance of the sympathetic ner-

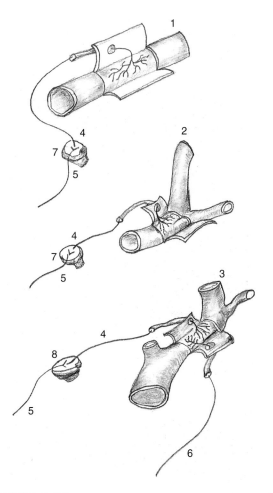

1. Cutaneous blood vessel
2. Skeletal muscle blood vessel
3. Cardiac blood vessel
4. Postganglionic-sympathetic neuron (effect is adrenergic, e.g., E or NE)
5. Preganglionic neuron (sympathetic)
6. Preganglionic-parasympathetic neuron (synapses in the effector tissue, effect is cholinergic, e.g., acetylcholine)
7. One of several vertebral ganglia
8. Cervical ganglia

FIGURE 2.12 Autonomic innervation of smooth muscle in blood vessels

vous system innervating the adrenal medulla in emergency situations, which is now referred to as the fight-flight syndrome.

J. W. Mason is credited with pioneering work to emphasize the importance of psychological or nonphysical stressors. Such neurobiological and cognitive aspects of stress are being given increasing priority in contemporary research efforts related to animal well-being.

Irrespective of the classification of mechanisms involved in an animal's response to stress, the origin of such states, or the period of time resistance mechanisms must endure, the body's ability to restore and maintain stable conditions through homeostasis is quite remarkable. Interactions between physical and psychological stressors are the norm in terms of ultimate effect on the whole animal and its well-being. An example of this relationship is the animal's perception of a stressful physical environment and the resultant psychological impact expressed as aberrant behaviors.

CHAPTER 3 ▐████████████

Biology of the Stress Response: Physiology and Function

The response to stressors requires a progression of events beginning with sensing and signaling the animal's various biological mechanisms that a threat exists. These events are followed by activation of neurophysiological mechanisms to mount a biological effort to resist and prevent major damage. The various sensory detectors not only receive the information but transform that information into neural signals that are transmitted to either or both cognitive and noncognitive centers of the nervous system to generate a coordinated response to the challenge. The coordinated response represents a magnificent example of the sophistication and automation involved in biological phenomena. The mechanisms respond so rapidly, so effectively, and so efficiently that in many if not most cases the individual is totally unaware of the activation and operation of the response. In other cases cognitive centers are involved in recognizing and providing some coordination of the response. The response systems are typically described in three phases: (1) alert or alarm, followed by a period of (2) resistance, and finally (3) exhaustion of the system. Severe negative influences of stress may be apparent in any of the stages, and death may be associated with the exhaustion phase. Demands in the resistance phase can have negative effects both physically and psychologically on the animal, as well as on the economics of animal enterprises. Severe demands in responding to stressful conditions, and particularly when such demands deplete response systems to the point of exhaustion, are issues of major significance in animal well-being.

NEURAL RECEPTORS INITIATING THE ANIMAL'S STRESS-RESPONSE SYSTEMS

Indicators of stressful conditions may arise externally to the animal and thus be recognized by one or more of the senses such as sight, hearing, or smell. Other indicators of impending stress may be changes in the internal environment such as glucose or oxygen level of the blood, body temperature, and blood pressure. Thus, systems for detecting such challenges to the individual involve nerve cells or neurons that are designed to accomplish the receptor function and to transmit information to the central nervous system. The presence of a challenge, whether it be an external or internal influence, is recognized by specialized dendritic nerve cell endings. Neural transmission then occurs through the neuron's axon and ultimately synapses with other neurons in the particular system involved. Components of the brain, in turn, deal with the issue through cognitive processes and noncognitive scenarios that respond to the challenge. Higher levels of the central nervous system such as the cortex, thalamus, and hypothalamus may coordinate the response, or, in the case of spinal reflexes, the response mechanism may involve only the spinal cord in receiving information that a challenge exists and in transmitting neural signals to reduce the effect of the challenge. Responses may involve single or any combination of physiological functions influencing metabolic, physical, and behavioral characteristics.

Frequently discussions of stress responses deal only with external and internal challenges to animal well-being that are of a physical nature. Examples are responses to the thermal environment or responses to some internal environmental change such as the pH of body fluids or oxygen tension of the blood. As will be discussed later, an important dimension to stress-related aspects of animal well-being involves psychological stressors. These stressors are influences or challenges that cause the animal to perceive some threat such as fear. Another example would be the social pressure of crowding or other restrictions that prompt the animal's stress-response systems to be activated and sustained as a means of coping with something perceived as undesirable relative to the animal's condition.

Aberrant behaviors may, in some cases, result from factors causing psychological stress and represent a coping response to the stressor. In such cases, cognitive processes involving the cerebral cortex and noncognitive processes involving the limbic system are involved in the response mechanism. Thus, the receptor system involved must deal with issues of how the animal perceives its condition physically and psychologically. The animal is made aware of its condition by sensory mechanisms involved in visual, auditory, and olfactory detection or perhaps through systems involved in feelings of frustration and depression. Cognition aids the animal in perceiving the condition and may prompt some response such as movement or other action to minimize the threat. Noncognitive components of the brain making up the limbic system are involved in responses to the perceived condition. As a result, physiological stress-response mechanisms are activated. Associated with this activation may be one or more aberrant behaviors or other effects that influence the effectiveness of the animal for its intended function, whether it be performance or even social interaction with humans or other animals.

Receptor Types

Types of tissue receptors involved in detecting and transmitting neural information to support stress responses are broadly described as those that detect external stimuli and those that detect internal stimuli. Although such distinction simplifies categorization, one should be mindful that in many cases there may be significant interaction among the different types. It should be noted also that information as to location of some receptor cells is not complete.

Neural receptor type	*Primary location*
External stimuli receptors	
Visual	Retina
Olfactory	Mucosa of upper nose
Auditory	Organ of Corti
Gustatory	Tongue and interior of mouth
Thermal	Skin, and mouth
Tactile	Skin
Nociceptors (pain)	Body surfaces and other peripheral
Mechanical	tissues, as well as deep tissues, such
Thermal	as joints and tendons
Pressure	
Visceral	
Internal stimuli receptors	
Thermal	Hypothalamus
Pressure	Blood vessels and heart
Osmotic pressure (osmoreceptors)	Hypothalamus, kidney, aorta, carotid artery, and respiratory centers of the medulla
Chemical (chemoreceptors)	Hypothalamus, and medulla
Oxygen	
Carbon dioxide	
Hydrogen ions (pH)	
Glucose	
Nociceptors (pain)	Visceral and peripheral tissues

CENTRAL AND AUTONOMIC NERVOUS SYSTEMS: INITIATION OF STRESS RESPONSES

Sympathoadrenal System

The hypothalamic–adrenal medullary stress-response system (SA) involves the hypothalamus, neurohypophysis, the sympathetic neural pathways to the adrenal medulla, and the release of epinephrine by this component of the adrenal gland. This stress-response system (fig. 3.1) was originally proposed by W. B. Cannon and is referred to as the fight-flight syndrome. It is a rapid response system to perceived threats to the animal's well-being and is characterized by a chain of neural and humoral events to prepare the animal for an impending emergency. The events are initiated by higher brain centers including the cerebral cortex, which synthesizes information relative to the stressor and sends neural signals to the neural hypothalamus. Corticotropin-releasing hormone (CRH), which is highly involved in the longer-term stress response (to be discussed later as the general adaptation syndrome [GAS]), is produced in the hypothalamus and may also be involved in stimulation of hypothalamic nuclei to transmit nerve impulses through the neurohypophysis and spinal nerve tracts to preganglionic sympa-

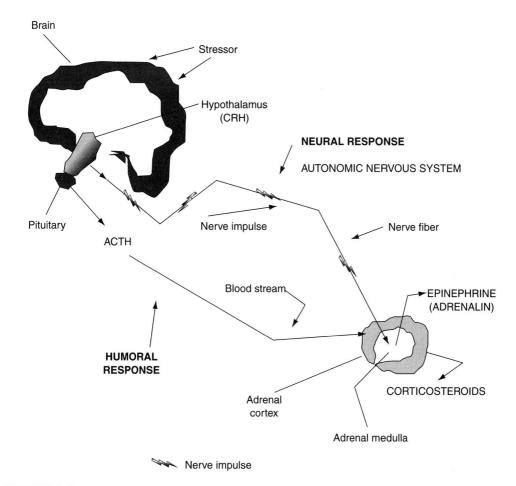

Brain

Stressor

Hypothalamus
(CRH)

NEURAL RESPONSE

AUTONOMIC NERVOUS SYSTEM

Pituitary

Nerve impulse

Nerve fiber

ACTH

EPINEPHRINE
(ADRENALIN)

Blood stream

**HUMORAL
RESPONSE**

CORTICOSTEROIDS

Adrenal
cortex

Adrenal medulla

Nerve impulse

FIGURE 3.1 Pathways of humoral and neural responses to stressors

thetic neurons with cell bodies located in the spinal cord. These neurons conduct impulses directly to the adrenal medulla and activate effector cells in this part of the adrenal gland. Direct innervation of the adrenal medullary cells by preganglionic nerve fibers is unique in terms of sympathetic innervation of glands of the abdominal region. In general, abdominal and thoracic glands and organs are innervated through a series of preganglionic and postganglionic neurons.

The cells of the adrenal medulla act much the same as postganglionic fibers in the production of adrenergic hormones, which in this case are epinephrine (fig. 3.2 illustrates the broad effects of epinephrine) and norepinephrine. An embryonic basis for this phenomenon exists in terms of common embryonic tissue giving rise to the adrenal medulla and other neural tissue that secrete similar neurohormones. This innervation stimulates the adrenal medulla to release primarily two catecholamines mentioned earlier—epinephrine (sometimes called adrenaline) and norepinephrine (sometimes called noradrenaline). Epinephrine has dramatic influences on systems

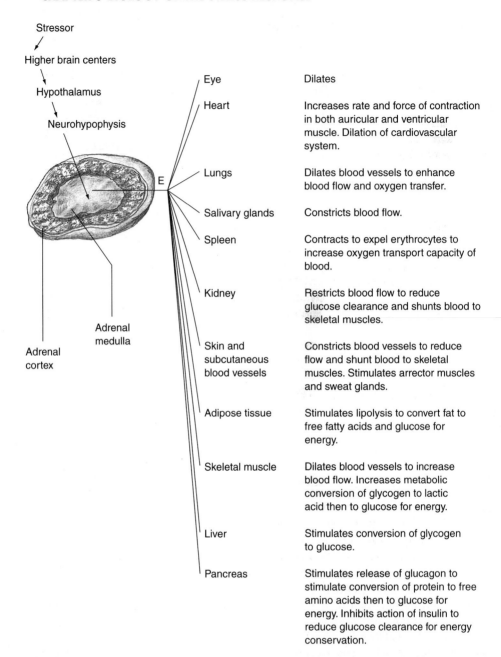

Stressor

Higher brain centers

Hypothalamus

Neurohypophysis

Adrenal medulla

Adrenal cortex

E

Organ	Effect
Eye	Dilates
Heart	Increases rate and force of contraction in both auricular and ventricular muscle. Dilation of cardiovascular system.
Lungs	Dilates blood vessels to enhance blood flow and oxygen transfer.
Salivary glands	Constricts blood flow.
Spleen	Contracts to expel erythrocytes to increase oxygen transport capacity of blood.
Kidney	Restricts blood flow to reduce glucose clearance and shunts blood to skeletal muscles.
Skin and subcutaneous blood vessels	Constricts blood vessels to reduce flow and shunt blood to skeletal muscles. Stimulates arrector muscles and sweat glands.
Adipose tissue	Stimulates lipolysis to convert fat to free fatty acids and glucose for energy.
Skeletal muscle	Dilates blood vessels to increase blood flow. Increases metabolic conversion of glycogen to lactic acid then to glucose for energy.
Liver	Stimulates conversion of glycogen to glucose.
Pancreas	Stimulates release of glucagon to stimulate conversion of protein to free amino acids then to glucose for energy. Inhibits action of insulin to reduce glucose clearance for energy conservation.

FIGURE 3.2 Broad effects of epinephrine, a part of the fight-flight syndrome in response to a stressor (*E* represents epinephrine produced by the adrenal medulla)

preparing the animal to meet emergencies that are mediated by receptors existing in effector tissues and stimulated by this hormone and norepinephrine. Effects are increased metabolic rate; increased heart rate and blood output; increased blood flow to the brain, heart, and muscles; and reduced blood flow to the skin by shunting of blood circulation patterns and vasoconstriction in body surface tissues and visceral regions. These effects are obviously preparing the animal for action. Epinephrine also initiates conversion of liver and muscle glycogen to glucose and may initiate gluconeogenesis to generate more readily available energy from protein and fat. Figure 3.3 illustrates the metabolic effects of the SA stress response. This gluconeogenic action is later supplemented by similar effects of glucocorticoids from the adrenal cortex in the longer-term stress-response system. In general, the gluconeogenic effect of epinephrine is faster than that possible through the glucocorticoid response. However, the net effect is the same metabolically. There is also evidence that a symbiotic relationship exists between the medulla and cortex (i.e., one enhances the action of the other).

The origin and effect of the hormones involved in the SA system are as follows:

Origin	*Neurotransmitter (chemical messenger)*	*Effect*
Preganglionic nerve cell	Acetylcholine	Neural transmission to cells of the adrenal medulla; secretion of epinephrine results
Adrenal medulla	Epinephrine; norepinephrine	Broad system tissue responses (e.g., increased heart rate, blood flow, and metabolic rate)
Postganglionic nerve cell	Norepinephrine	Neural transmission to effector tissues such as smooth muscle in blood vessels of skeletal muscle; the tissue responds by dilation of the vessel to increase blood flow

Hypothalamic-Pituitary-Adrenocortical System

The hypothalamic-pituitary-adrenocortical (HPA) stress-response system (fig. 3.1), referred to as the general adaptation syndrome, was conceptualized by Hans Selye. The response represents a longer-term, sustained response to stressors as compared with the SA system discussed earlier. Effects are heavily oriented toward metabolic changes to strengthen the animal's ability to cope with stressors. Higher brain centers including the cerebral cortex serve as cognitive sensors, and thalamic and hypothalamic centers capable of detecting changes in the internal environment of the body cause the endocrine cells of the hypothalamus to initiate physiological actions that may be without cognitive influence to respond to stressful conditions. The response in this system is geared to stimulating metabolic activity to provide the energy resources to meet the emergency over an extended period. The hypothalamus, following the initial stimulus, responds by releasing CRH. This peptide stimulates the anterior hypophysis (anterior pituitary gland) to release adrenocorticotropic hormone (ACTH). This

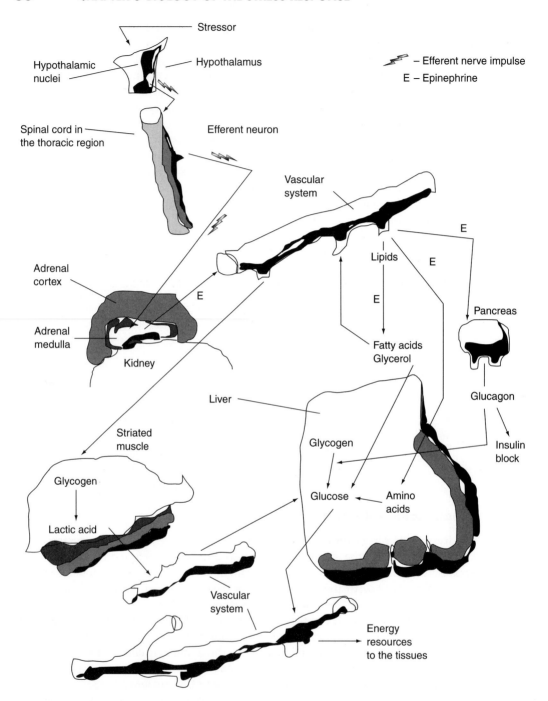

FIGURE 3.3 Innervation and metabolic influences of the sympathetic nervous system in the sympathoadrenal stress response

hormone activates the adrenal cortex to produce several hormones essential in the stress response.

The major adrenal cortical hormones are cortisol, corticosterone, and aldosterone. Cortisol and corticosterone are glucocorticoids that mobilize liver and muscle glycogen stores by conversion to glucose (glycolysis) for use as an energy source (fig. 3.4 illustrates the broad metabolic effects resulting from the HPA stress response). Glucocorticoids, through gluconeogenesis, will also mobilize energy from protein by conversion to amino acids (proteolysis) then to glucose following deamination. Lipolysis also makes energy available from free fatty acids and from potential conversion of glycerol to glucose. The enhancement in carbohydrate reserves fostered by glucocorticoids is a mechanism to increase available energy. If glucocorticoids are not present when an animal is severely stressed, there is no sustained metabolic response and the animal will be less able to cope with the stressor. Such animals will survive only in protected environments. Aldosterone is a hormone referred to as a mineralocorticoid and is involved in the regulation of mineral balance in the body. It is thus involved in the maintenance of blood composition, blood volume, and blood pressure. In these processes, a major function of aldosterone relates to electrolyte balance of body fluids through sodium homeostasis.

The origin and effect of the hormones involved in the HPA system are as follows:

Origin	Hormones	Effect
Hypothalamus	CRH	ACTH secretion by the anterior pituitary
Anterior pituitary	ACTH	Production of adrenal steroids by the adrenal cortex (adrenal steroidogenesis)
Adrenal cortex	Cortisol and corticosterone	Carbohydrate metabolism
	Aldosterone	Sodium balance

HOMEOSTASIS AND REGULATORY SYSTEMS IN THE STRESS RESPONSE

Regulation of Body Temperature

Homeothermy represents a form of homeostasis involving temperature and refers to regulatory mechanisms to maintain a condition of relatively constant body temperature. However, variations in temperature may exist in different parts of the animal's body, and diurnal variation in temperature is normal. Mechanisms involved in thermal homeostasis are depicted in figure 3.5.

Heat or cold stress results when extremes in environmental temperature challenge the animal's system to maintain a relatively constant temperature. Animals that are able to respond by maintaining body temperature within a narrow range are designated homeotherms. Extended temperature-related stress can exceed the ability to maintain homeothermy, and death will result when the response systems are depleted.

Management of animals to ensure minimal negative impacts of temperature-related stressors is a major issue in selection of animals that can successfully adapt to a given environment, proper nutrition, facility design, and related cost-benefit ratios.

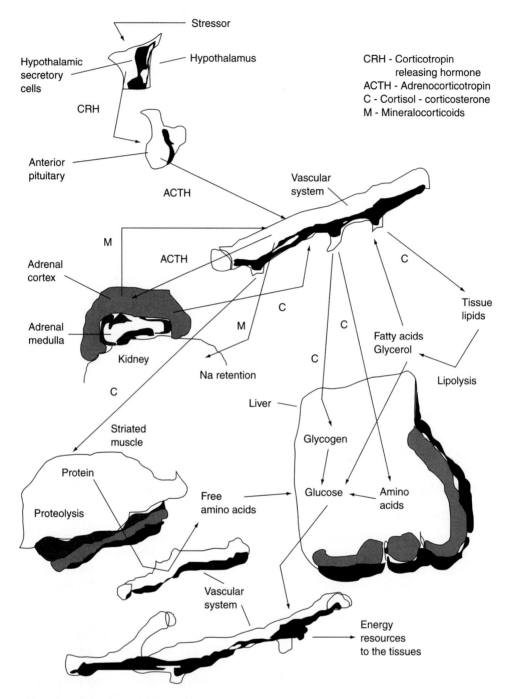

FIGURE 3.4 Metabolic influences of the hypothalamic-pituitary-adrenal stress response

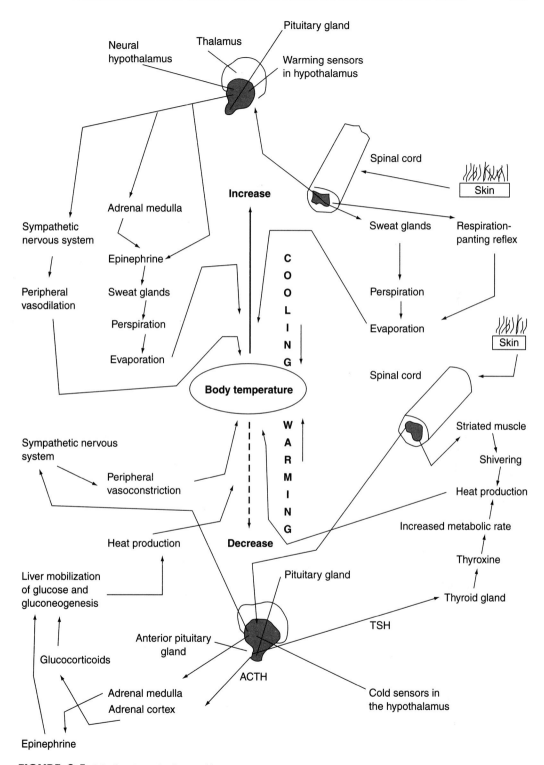

FIGURE 3.5 Mechanisms in thermal homeostasis

Cold Stress Terminology

The following terminology and related concepts (NRC, 1981a) dealing with thermal zones are useful in understanding variations in environmental demands. These concepts will be applied later in part 4 to design of the physical environment.

Environmental Heat Demand   Environmental heat demand is a characteristic of the environment that alters heat loss (e.g., at a given environmental temperature, increasing air movement will increase the demand and appropriate thermoregulatory processes of the animal will respond).

Effective Ambient Temperature   Effective ambient temperature (EAT) describes the animal's thermal environment that considers all influences related to heat demand.

Lower Critical Temperature   Lower critical temperature (LCT) is the point in effective ambient temperature below which an animal must increase its rate of metabolic heat production above the basal range to maintain homeothermy. Processes related to conservation of heat including vasoconstriction in the periphery, piloerection, and behavioral adjustments to reduce heat loss from body surfaces are at a maximum at this point. LCT occurs at the low end of the thermoneutral zone (TNZ).

Thermoneutral Zone   The thermoneutral zone (TNZ) is the range of effective ambient temperature in which the heat from normal maintenance and productive functions of the animal in nonstressful situations offsets the heat loss to the environment without requiring an increase in the rate of metabolic heat production. This zone or range represents that between LCT and upper critical temperature (UCT); it may also be referred to as the thermal comfort zone or zone of thermal neutrality.

Cold Stress Intensity Zones

Cool Zone   The cool zone is the lower range of the thermoneutral zone, just above the LCT. Physical regulatory processes are initiated. Metabolic rate increases as LCT is approached, increasing heat production. These processes will reflect mild increases in the animal's maintenance requirement for feed energy since animal performance and efficiency of feed conversion are on the verge of reduction. These changes can impact production costs minimally.

Cold Stress Zone   The cold stress zone is below the LCT. Physical measures limiting heat loss move to the maximum, and increased metabolic activity is initiated to provide additional heat. As effective environmental temperature continues to drop, metabolic rate continues to increase to the highest level (summit metabolic rate) that can be maintained for an extended period.

Terminal Cold Zone   Terminal cold zone temperature is below that of the cold stress zone. The animal can compensate to a very limited degree below temperatures of the cold stress zone. Core body temperature falls, hypothermia results and the animal ultimately dies.

Heat Stress Terminology

Thermoneutral Zone   Note the definition under cold stress, since the comfort zone ranges up to the UCT.

Upper Critical Temperature   The UCT is the point above which an animal must engage physiological mechanisms to resist body temperatures rising above normal. These processes are related to cooling effected by evaporation through increased perspiration and respiration and vasodilation in the periphery to enhance heat loss from body surfaces through convection, radiation, and conduction.

Heat Stress Intensity Zones

Warm Zone   This zone represents the upper portion of the thermoneutral zone. Vasodilation and change in effective surface area of the animal's body may occur by shifts in posture. Reduced tissue insulation results. Feed consumption may be reduced.

Heat Stress Zone   Heat-dissipating actions above reductions in tissue insulation and posture adjustments occur (e.g., sweating and panting). Heat production increases due to energy need to dissipate heat, adding to the heat load. The animal will attempt to find protection. Lethargy may result. Feed consumption drops.

Terminal Hot Zone   Body temperature rises above normal because heat loss is less than body heat production and heat gain from the environment and ultimately the animal dies.

Physical Regulation of Body Temperature: Heat Loss

Radiation, convection, and conduction represent major heat loss avenues from the animal's body, but the level will vary with air temperature, humidity, air movement, hair length, and level of heat production.

Voluntary Regulation Mechanisms (Behavioral Responses)

Animals show behavioral responses in seeking shelter, grouping, turning away from wind, and the like. It is not clear that animals voluntarily engage in increased physical activity to generate heat.

Involuntary Regulation

Redistribution of Blood   Vasoconstriction and vasodilation shift blood from the periphery to internal organs and from the internal organs to the periphery.

Variation in Blood Volume   A rise in temperature will cause an increase in blood volume by transfer of fluids from the tissue into the bloodstream. A decrease in temperature will reverse the pattern and decrease blood volume.

Altered Circulation Rate   Heat extremes stimulate cardiac output as a means of increasing heat transfer from the periphery to the environment. Cold extremes may do the same to aid in mobilizing heat production.

Evaporation of Water from Lungs   Increased respiration rate presents more blood to the surface area of the lungs for heat transfer out of the body.

Perspiration   Evaporation losses result from increased water on the skin. Sweating is profuse in horses, more limited in cattle and sheep, and of little importance in swine because the latter species does not have effective sweat glands.

Panting   The panting reflex is characteristic of the farm animal species and pets. It effects cooling by exposing the tongue and mouth parts to the atmosphere, as well as by exchanging expired air that may be warmer than inspired air.

Increase in Temperature of Inspired Air   Increased respiration rates and volume during periods of thermal stress represent an important avenue of heat loss from the body by thermal exchange during breathing.

Loss of Heat Contained in Feces and Urine   Excretions remove heat from the body because they have been maintained at core body temperature before discharge.

Chemical Regulation of Body Temperature through Muscle Metabolism: Heat Production

Muscle Activity through Involuntary Shivering   Involuntary shivering in response to cold increases muscle activity and associated heat production through metabolism of glucose and glycogen to lactic acid. This movement is a response to acute cold stress primarily. (Glucose homeostasis mechanisms are illustrated in figure 3.6 on p. 65).

Muscle Activity through Voluntary Movement   It is suggested that at least some animals may voluntarily increase activity, resulting in heat production. This response is not well established.

Endocrine Metabolic Response   The endocrine metabolic response system is longer term in nature and is mediated by the hypothalamus. Temperature receptors and regulatory centers in this body coordinate the response by output of ACTH by the pituitary, which in turn increases release of corticoids by the adrenal cortex (HPA system). Corticoids aid in mobilizing energy from protein and fat storage areas (gluconeogenesis and lipolysis). The hypothalamus also releases thyroid-stimulating hormone (TSH), which stimulates the thyroid gland to elevate the level of thyroxine to increase basal metabolic rate to meet demands placed on the system by temperature-related stressors.

The sympathetic nervous system is activated by the hypothalamus by stimulating catecholamine production by the adrenal medulla (SA system). Epinephrine and norepinephrine stimulate mobilization of fatty acids as energy sources and stimulate heart rate and cardiac output for cooling effect and the transport of metabolites required in response to cold stress.

The described stress regulatory pathways are the same as those noted previously for stress responses in general. In these cases, thermal receptors are responsible for stimulating the mechanisms involved.

Anatomical Factors in Heat Regulation
Insulation Factors

Hair, Wool, and Feathers Plumage entraps air and provides an insulation layer. Some hair contains vacuoles that further enhance insulation. Piloerection and feather fluffing result from muscle action stimulated by motor nerves. Rain and wind greatly reduce cover protection by compacting hair and feathers.

Skin Thickness Variation in skin thickness has only a minor influence on insulation effect.

Fat Deposits Subcutaneous fat provides significant insulation.

Circulatory Anatomy The magnitude of the animal's tissue insulation is determined largely by the character of the periphery vascular system at a given time. Vasoconstriction increases the insulating effect of surface tissue by altering the thermal gradient between the tissue and the external environment. Such alteration of blood flow to the body surface can increase the insulation value of skin and subcutaneous tissues several fold.

Boundary Insulation In a condition of little or no wind velocity, a thin layer (blanket) of air is maintained around the animal. This air affords some resistance to heat transfer with the environment. Movement of the animal or air largely eliminates this insulation effect. This insulating layer is more important for animals with little or no hair, but wind protection is essential for effectiveness of the layer. Swine are a good example of a species that benefits greatly from this effect if conditions are dry and air movement is not a factor.

Effect of Thermal Environment on Nutrient Requirements of Animals

Improved understanding of interaction between environment and animal requirements has increased management opportunities to establish nutritional needs and to evaluate the associated cost-benefit ratio.

Management does not necessarily seek maximum performance of the animal but a level that may be considered optimal with respect to cost and returns. In many conditions, however, maximum or near-maximum performance rates will be necessitated due to relatively high costs associated with animal maintenance, facility utilization, and other production overhead costs.

Since feed costs typically represent such a large proportion of total production costs, impacts of the environment on feed-nutrient requirements, animal performance, and costs must be considered.

Thermal stressors influence the performance of animals by affecting heat exchange. These influences are reflected in feed intake, weight gains, milk production, level of energy required, and nutrient density requirements for such nutrients as protein and minerals in inverse relationships to feed intake.

Currently used mathematical procedures permit adjustment of dietary requirements to meet needs of some animals under varied conditions of thermal environment (part 4). Such adjustments, however, may not be adequate for all conditions. Calculations based on adjustment equations will arrive at recommended levels for feed nutrients. Further adjustments may be necessary to meet variable specific conditions.

Regulation of Blood Glucose

Carbohydrates provide the major source of energy for fueling the stress response. Thus, the mechanisms regulating the level of glucose in the blood and levels of glycogen in the liver and muscle are critical in ensuring that the animal can meet the demands involved in resisting all stresses leading to a demand for muscle activity and heat generation.

The level of glucose in the blood is maintained within a very limited range. Chemoreceptors in the alpha cells of the pancreatic islets of Langerhans detect reductions in blood glucose level and release glucagon, the hormone that stimulates the conversion of muscle and liver glycogen to glucose. Similarly, beta cells of the pancreatic islets of Langerhans detect increases in blood glucose. These cells release insulin, the hormone that stimulates removal of glucose from the blood by conversion to glycogen. Glycogen then represents a storage form of carbohydrate in both liver and muscle. The general mechanisms involved in glucose homeostasis are shown in figure 3.6.

Regulation of Water Balance

Stress physiology, as a science, places heavy emphasis on effects of climate on the body. Thus, temperature, humidity and wind velocity all place demands on animal systems to cope with environmental impacts. Since temperature regulation is so heavily dependent on water stores and biological factors regulating this component, it is important to consider factors involved in regulating water balance.

Water Balance Terminology

Water balance is a state in which water intake and output are equal, so that neither dehydration nor retention of excess water occurs. Small changes toward dehydration or excess retention trigger major regulating pathways to adjust the body water level to a balanced state.

Water turnover rate is the amount of water passing through the animal per unit of time. Turnover rate equals water intake if the status of hydration within the total body is static (Curtiss, 1983). Environmental influences alter turnover rate and demands for water consumption. If this demand is not met, water balance is influenced.

Sources of Body Water

Oral consumption (water and food) and metabolic water resulting from oxidative processes in metabolism represent the total supply of this critical ingredient in the animal's body.

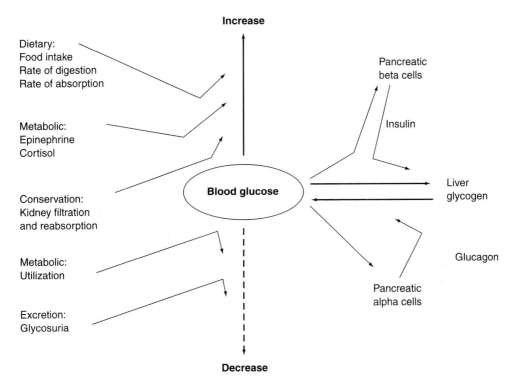

FIGURE 3.6 Mechanisms in glucose homeostasis

Amounts of water produced by metabolism of 100 g of the nutrients are as follows:

Fat	107 g
Starch	55 g
Protein	41 g

Comparative water losses from the body through different routes are as follows:

Urine	60 to 70 percent
Feces	5 to 8 percent
Skin evaporation	15 to 20 percent
Expired air	7 to 10 percent

Regulatory Mechanisms in Maintaining Body Water Balance

Role of the Kidney in Water Balance: Water Conservation The action to shift to a conservation mode results from signals initiated by blood volume receptors in the carotid sinus and osmoreceptors in the hypothalamus. The response is coordinated by the hypothalamus, which ultimately controls water exchange phenomena in the kidney. Mechanisms involved in the regulation of water balance are illustrated in figure 3.7.

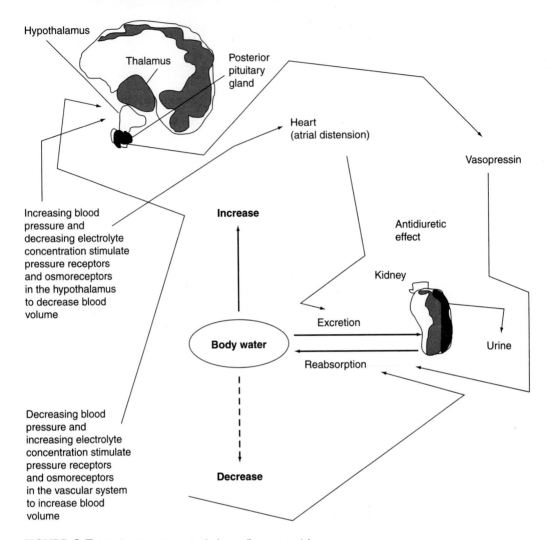

FIGURE 3.7 Mechanisms in water balance (homeostasis)

Endocrine control is affected by aldosterone, secreted by the adrenal cortex, which acts on renal tubules to increase reabsorption of sodium. This reabsorption of sodium is accompanied by an obligatory reabsorption of water. Another endocrine influence results from the release of the antidiuretic hormone (ADH) vasopressin by the pituitary gland, which releases this ADH. Vasopressin acts on renal tubules to increase water reabsorption, again, as a conservation mechanism.

The stimulus for initiating release of aldosterone is low blood volume acting on stretch receptors in the circulatory system. Correspondingly, high blood volume through action on stretch receptors, causes kidney tubules to reduce body water by increasing excretion rate. Pressure receptors in the kidney are also involved in the release of aldosterone.

The stimulus for initiating the release of vasopressin is hypertonic plasma (i.e., when the level of salts is above normal the neurohypophysis, influenced by the hypothalamus, is stimulated to release vasopressin). Similarly, if plasma reaches a hypotonic state, the system shifts to an excretion mode.

Species differ in their ability to concentrate urine and conserve water in hot, dry environments. This characteristic explains the greater capability of sheep to adapt to hot, dry environments. Native desert animals depend on this characteristic ability to conserve water.

Role of the Lungs in Water Balance   Expired air represents an avenue for evaporation and loss of water, as well as heat exchange.

Panting Response   The advent of panting increases the potential heat loss and water loss from the body. Thermal panting in cattle, sheep, and pigs occurs in the area of 36°C skin temperature. This skin temperature occurs when environmental temperature is in the range of 28 to 30°C for pigs and calves and 20 to 25°C for sheep (Curtis, 1983).

Role of the Skin in Water Balance: Evaporation   The skin, through sweating, serves as a heat exchanger. Evaporation provides for a highly variable level of water loss depending on the environment and associated activation of the sweating response.

Great species differences exist in level of water evaporation that can occur under heat stress. Cattle, for example, show many times the maximum loss for sheep. As discussed earlier, swine do not have effective sweat glands.

Role of the Alimentary Canal in Water Balance

Salivary Glands   Dehydration stimulates the salivary glands to reduce output. This reduction in output innervates centers in the brain to initiate a thirst response to correct body water balance, if possible, through rehydration.

Stomach and Small Intestine   Vomiting may represent major losses of water, and diarrhea may impact water balance significantly.

Colon   The colon acts to regulate the water content of fecal materials before excretion. Thus this structure influences water loss and retention. Irritation of the gut, for example, prompts the colon to reduce water absorption to soften the feces. The colon may increase dehydration of the feces as an aid in maintaining water balance.

Rehydration

In general, access to water corrects dehydration; however, certain precautions are necessary in special situations (e.g., water intoxication in poultry and possible founder and colic in horses).

Typical water content of various tissues is as follows:

Tissue	Percentage	Tissue	Percentage
Muscle	75	Adipose tissue	20
Skin	70	Bone	25–30
Connective tissue	60	Kidney	80
Blood plasma	90	Liver	70
Blood cells	65	Nervous tissue	70–85

Regulation of Acid-Base Balance

Environmental stress may place increased demands on pH regulatory mechanisms (e.g., an increased muscle activity increases lactic acid production). An increase in activity increases CO_2 production and challenges the buffering capacity of the blood, resulting in an increase in blood pH. This increase, if severe, may result in acidosis.

Acid-base balance is reflected by a pH parameter of blood and tissues in what is considered a normal range. The bicarbonate buffer system shown below is an example of regulation of pH in body fluids.

Acidosis is a pathological condition resulting from accumulation of acid or loss of base in the body and increase in hydrogen ion concentration (decreased pH).

Alkalosis is a pathological condition resulting from accumulation of base or loss of acid in the body and a decrease in hydrogen ion concentration (increased pH).

Regulation of pH in the Blood

Buffer Systems   The bicarbonate buffer system is shown as follows:

$$CO_2 + H_2O \leftrightarrow H_2CO_3 \leftrightarrow H^+ + HCO_3^-$$

Elimination of Carbon Dioxide by the Lungs   Activity or environmental influences that shift the buffer system to CO_2 decrease the available positive hydrogen ions to buffer (neutralize) negative carbonic acid ions.

Excretion by the Kidneys   High levels of negative ions are excreted by the kidneys in the form of neutral salts of sodium, potassium, magnesium, and calcium. This activity may deplete these electrolytes. A practical effect of imbalances is urinary calculi in some species. (See figure 3.8, mechanisms in sodium homeostasis.)

Regulation of Nitrogen Balance

Nitrogen balance reflects the rate of nitrogen intake and that lost through the urine and feces. Protein and nonprotein nitrogen in foods serve as the source of nitrogen available for use by the body. Nitrogen availability influences functions such as growth, maintenance, hormone synthesis, and antibody development. Excretory functions of the kidney and gut may be influenced by stress. Thus, the retention of nitrogen may be important in maintaining homeostasis, both directly and indirectly.

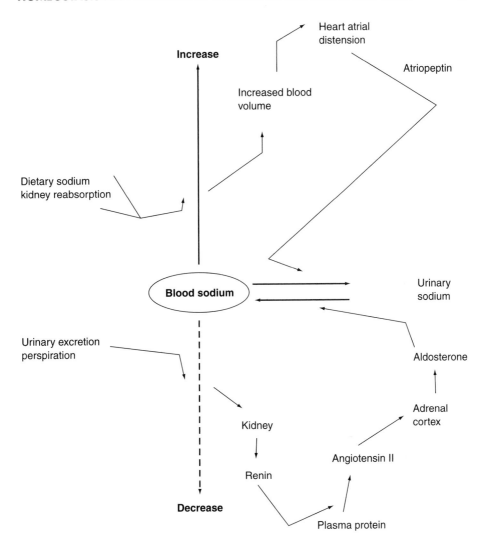

FIGURE 3.8 Mechanisms in sodium homeostasis

Factors Involved in Regulating Nitrogen Balance
Dietary Factors

Protein Quality: Balance of Amino Acids   Animals of different species and at different phases of the life cycle may have different needs relative to kinds and amounts of amino acids required. This variation results in proteins with variable biological values.

Protein Quantity   Animals have a total quantity requirement that must be met in addition to amino acid balance.

Protein Sparing Action of Carbohydrates and Fats   Protein molecules may be broken down and the carbon chain utilized as an energy source. If adequate amounts of energy from sources such as carbohydrates and fats are available, this utilization route is minimized and conserves protein.

Total Feed Intake   There is evidence that the higher total feed intake associated with colder environments will permit lowering the concentration of protein in the diet of some animals. Correspondingly, reduction in feed consumption associated with hot environments may require increased nutrient density in order to meet the nitrogen requirement.

Endocrine Factors in Nitrogen Balance   Adrenal influences on gluconeogenesis (the utilization of amino acids for energy) may deplete nitrogen stores. Pituitary gland influences through somatotropin and growth promoting materials may result in greater synthesis of protein and faster growth. This increase requires higher levels of dietary protein.

Renal Factors in Nitrogen Balance   The kidney is involved in production and excretion of ammonia and as a result may influence the level at which it is available for combining with organic acids, chlorine, and sulfate ions as a means of reducing acidity in acid-base balance phenomena.

Regulation of Calcium Balance

Calcium serves several functions in the body in addition to bone structure. It is critical in the blood clotting process, in muscle contraction, and a variety of cell functions. These examples are sufficient to illustrate the importance of this element in stress responses. Calcium is available from bone stores, as well as normal dietary sources. The mechanisms involved in calcium storage and transfer from bone to blood are referred to as calcium homeostasis and are illustrated generally in figure 3.9. When blood calcium concentration falls, cells in the parathyroid gland sense this shift and produce parathormone, which stimulates release of stored calcium in bone and may also enhance absorption of dietary calcium from the gut. When blood calcium tends to rise above normal levels, parafollicular cells in the thyroid gland produce the hormone calcitonin. This hormone initiates increased calcium deposition thus lowering calcium blood levels.

PATHOLOGICAL RESULTS OF STRESS

Stress adaptation is a critical phenomenon in the general well-being of animals in that life represents a constant series of adjustments to a host of stressors even under normal physiological conditions. In this sense, the adaptation syndrome is operating constantly. Selye concluded early that animals have many stressors operating at all times, but adaptation occurs to counteract such factors. Pathological conditions result from the development of imbalances among stressors and adaptability. Predisposition to pathological conditions is a level of susceptibility based on the animal's sum of hereditary factors and acquired characteristics (environmental factors commonly referred to as influencing the condition).

Integrity of genotype and condition is a precondition for proper adaptability. Disturbance or lowering of either or both may result in failure, malfunction, and damage to organs. Such imbalances may result from sustained response to

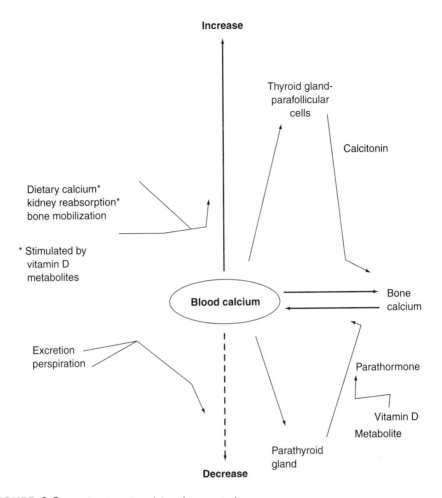

FIGURE 3.9 Mechanisms in calcium homeostasis

stressors or by lower resistance, which may result from a variety of factors including exhaustion of the stress-response systems.

Stress Reactions

Dämmrich (1987) provided a useful perspective by classifying stress reactions as nonspecific and specific. This approach distinguishes environmental influences in terms of general effects on the organism's ability to respond effectively to stress as opposed to those that have damaging pathological influences on specific tissues.

Systemic Reactions

Nonspecific stress reactions were described by Dämmrich as pathological conditions resulting from general deterioration of the animal's ability to cope with stress. The GAS assists the animal in compensating for chronic stress through systems of metabolic reinforcement. Continuation of the stress eventually depletes the elements of this system. Thus, the ability to resist is exhausted by depletion of reserves such as carbohydrates, protein, and fats that sustain the physiological effort. In addition, endocrine resources are depleted. The result is adrenal insufficiency.

Reactions Associated with Damage to Specific Target Organs and Tissues

Damage to specific tissues is usually diagnosed in association with specific environmental causes. Several examples summarized by Dämmrich (1987) are described below.

Gastric Ulcers in Pigs and Calves   Stressors associated with this pathology are typically environmental influences that lead to chronic frustration or social interactions that result in continuing unrest. Thus, behavioral observations are useful in detecting conditions that may be compromising well-being.

Paralytic Myoglobinuria in Horses   This condition, also called Zenkers disease, typically results when exercise is resumed after a period of rest. Acidosis develops from conversion of accumulated glycogen to lactic acid.

Porcine Stress Syndrome (PSS)   This pathology is more common among pigs selected for muscling. It can also occur in other species as well. Predisposition is clearly inherited and is due to a simple recessive gene. Thus, selection to reduce the frequency of the gene is the recommended control.

PSS is an example of lowering the genetic capacity of animals. As a result, even moderate physical activity (e.g., movement, loading, and pre-slaughter management) may result in death. Pale, soft, exuadative pork (PSE) is a related problem in affected animals. Involved in this syndrome is histology of muscle fibers that interferes with oxygen transfer. This possibly results in an energy deficit and is associated with increased lactic acid levels in the tissues. Reduced heart size has also been observed, which may limit normal recovery from lactic acid accumulation. Affected pigs show gradual development of vasodilatation and development of red color first along the underline. This progresses until the flushed appearance is observed along sides and eventually over the back. This reaction is, of course, most obvious in pigs that are white in color. Producers observe these changes when confined pigs are removed from familiar facilities in relocation. Stress susceptible males are likely to show these changes when placed in breeding pens with estrous females. Abnormal calcium metabolism, an additional characteristic, results in a form of tetany that causes continuous muscle contraction. Extreme extension of limbs occurs. The tetany is characterized by a reduction in blood flow to the periphery and related problems of increased lactic acidosis, altered mitochondrial function, myocardial degeneration, and cardiac insufficiency that may result in death. Death may result from the tetanic state of diaphragm muscle tissue. The role of the central nervous system in this syndrome is not clear; however, failure of one or more neurotransmitters may be involved. The described tissue changes are undoubtedly involved in the PSE observed in carcasses of stress-susceptible pigs.

The following specific reactions are observed in cattle in feed lots or upon slaughter.

Rumen Parakeratosis   Rumen parakeratosis was encountered early in the development of the beef cattle feeding industry in the southwest United States as a result of feeding high-concentrate, low-fiber rations. It is characterized by erosion and matting of the papillae making up the lining of the rumen. This dysfunction is partially alleviated by processing feeds to add bulkiness.

Liver Abscesses in Cattle   Liver abscesses may occur in cattle along with rumen parakeratosis. Increased incidence is associated with feeding of high-energy rations. Antibiotics administered as a part of this diet appear to minimize the problem.

Dark Cutters   Animals experiencing stress in shipment or handling before slaughter may produce carcasses with a darker than normal red color of the muscle. This darkness results from depletion of glycogen during the preslaughter period. The muscle is characterized by higher than normal pH, which interferes with the development of the bright red color of postmortem muscle when exposed to usual levels of oxygen in the air.

PERCEPTION AND PSYCHOLOGICAL STRESS

Psychological Stress

Psychological stressors perceived as threats may be equally as important as those of a physical nature in challenging coping mechanisms. Stress resulting from psychological influences, like that of physical stress, is important in any consideration of well-being.

Psychological stressors are sensed by central nervous tissue, as is any source of information. Fear, uncertainty, and social pressures are sensed and processed by cognitive centers. Stress response mechanisms are initiated.

Gradual changes in the environment (e.g., heat or cold) may not initially elicit typical responses to stress due to conditioning. Thus, in some cases, the first stress response may be due to mental awareness or perception that cause the animal to attempt to alleviate the stress by escaping or searching for relief.

Psychological stressors in many cases represent uncertainties presented to the animal. Unsettled conflict or other social interactions leading to discomfort or frustration are clearly recognized as stressors.

Situational uncertainty (described by Warburton in 1987) is the result of many alternative meanings that the input has for the animal, and its available response choices. An animal experiences degrees of uncertainty and thus predictability (i.e., whether an event will happen, what event might occur, and when it will happen) and controllability (i.e., what action can be taken to effectively mediate a challenge).

Control of Stress Responses Related to Perceived Condition

Control of the Psychological Stress Response

Neural and humoral systems associated with physical stressors are also active in responding to psychological stressors (e.g., the involvement of cate-cholamines [adrenaline and noradrenaline] and corticosteroids). The level of corticosteroids in the blood increases in response to a stressor, the output of ACTH by the anterior pituitary is mediated. This lowers the stimulus for the production of corticosteroids. Thus, the typical feedback mechanisms that regulate responses are involved.

Neural pathways in the brain stimulate the hypothalamus to initiate stress responses to situations the animal perceives to be undesirable or threatening. This function is demonstrated experimentally by electrical arousal of cognitive brain centers that in turn prompt corticosteroid release. It also suggests that cognitive evaluation of the animal's condition is important in controlling its response to stressors.

Involvement of the Cerebral Cortex and Limbic Structures in the Stress Response

Hippocampus   The hippocampus is a limbic component of the brain located in the subcortical region; it is likely involved in preparing the cortex for rapid information processing rather than routine storage of information. This function increases the potential effectiveness of responses to threatening stimuli. The hippocampus is also thought to be involved in emotional behaviors. Ladewig et al. (1993) suggest that the hippocampus is involved in how animals perceive spatial environments and their adequacy and in how animals relate their actual situation to expectations. These observations suggest that the hippocampus is probably involved in responses to psychological stressors such as chronic fear and spatial pressures.

Amygdala   The amygdala is also a component of the limbic system located in the subcortical region of the brain. It is thought to be involved in the gradual adaptation of animals to psychological stressors. This adaptability may explain the ability of animals to adjust to some stresses over time that initially may be disturbing to them. The process may be referred to as habituation. Ladewig et al.(1993) suggest that the amygdala plays a key role in aggression and motivation relative to distinguishing between positive stimuli such as rewards and negative stimuli such as aversive treatment.

Cerebral Cortex   Complex events or situations require neural analysis (reasoning and selection of alternative responses). Pain, for example, may stimulate the hypothalamus directly, whereas loud noise may stimulate the hippocampal pathway to respond. Social situations such as space intrusion, dominance, and similar stressors require integration and interpretation to derive the meaning and the related response, if any. The decision to invade the domain of a dominant male requires a considered judgment of benefits versus cost. The same could be true of intrusion upon a marked territory of another animal for feeding.

Coping through Behavioral Responses

Behavior is an alternative means of dealing with stress, as well as interactive with adrenal activity involved in the stress response. Some animals that adopt activities to occupy time have been shown to have lower levels of cortisol. Stereotyped behaviors, for example, may be observed as responses to confined space. The response (resisting) of sows newly placed in crates or those restricted by tethers, while individual sows differ, is gradually reduced with time, and a more passive response over an extended period results. Stereotypies may result.

The development of apathetic behavior in stalled or tethered sows as a coping mechanism appears to hold for most environmental influences except feeding. Group-housed sows are often more responsive to stimuli, which suggests that dullness and apathy are likely related to confinement or a boring environment.

Coping with adverse situations may involve neural gating mechanisms. This process may reduce sensory input or arousal. Opiates, produced by central neural tissue, also contribute to coping ability. The net effect of coping may often require an animal to alter its normal behavior. Some would conclude that this should raise questions about the well-being of the animal and suggests critical need for improving the environment.

IMMUNE SYSTEM RESPONSE TO STRESS

Hypothalamic-Pituitary-Adrenocortical System Influences on the Immune System

The HPA stress response system is involved in a manner that basically suppresses the immune system as a result of chronic stress. The effect results from increased levels of corticosteroids.

Central Nervous System Influence on the Immune System

Psychosocial factors are thought to influence lymphoid tissue. The field of study relating to this area is psychoneuroimmunology.

There is some evidence that lymphocyte-derived neurotransmitters can influence behavior. This relationship suggests a feedback mechanism from the immune system to the central nervous system.

CELLULAR COMPONENTS OF THE IMMUNE SYSTEM

Precursor Lymphoid Stem Cells

Precursor lymphoid stem cells originate in the bone marrow and then mature there (B cells) or in the thymus (T cells).

T (Thymus-Derived) Lymphocytes

T lymphocytes functionally are of two types, effector cells and regulatory cells. Effector (killer cells) directly attack and kill target cells. Regulatory cells enhance cell-mediated immune reactors or suppress them. Those that enhance the system are T-helper cells, and those that have an inhibiting effect are T-suppressor cells. These prevent excess immune function.

B (Bone-Marrow-Derived) Lymphocytes

B lymphocytes are cells that carry surface immunoglobulins that serve as receptors for antigens and are precursors to antibody-forming cells. Interaction of antigen with the corresponding receptor on B cells causes the cell to transform into plasma cells that produce and release specific antibodies.

PREDISPOSITION TO ABNORMAL FUNCTION OR DISEASE BY ALTERATIONS IN EFFECTIVENESS OF RESISTANCE

Glucocorticoid Influences

Increased levels of glucocorticoids in the blood resulting from stress may induce leukopenia (reduction of the number of leukocytes in the blood). Observed impacts are related to suppression of the immune system (i.e., suppression of the system by T lymphocyte cells [T suppressors] that tend to inhibit or slow antibody production).

Alternate Pathways by Which Stress Impacts the Immune System

Examples of influences that may secondarily influence the immune system are nutritional deprivation, altered carbohydrate metabolism, nutrient transport and utilization, mineral metabolism, and changes in body temperature.

Extra-Adrenal Pathways (Neural Influences Generally Exclusive of the Adrenals)

Autonomic Nervous System

The autonomic nervous system provides innervation of tissues such as lymph nodes, thymus, spleen, and blood vessels. Autonomic innervation of the thymus may be involved in the maturation of immune-related cells and soluble factors mediating parts of the immune response.

Central Nervous System

A variety of immunomodulators (e.g., endorphins and enkephalins, opioid agents known to be released in response to stress) are produced under the influence of the central nervous system. Emotional stimuli such as social hierarchy issues have been shown to result in stress responses (e.g., deposing of a leader and avoidance reactions). Mild, short-term stress may enhance immune function, whereas longer-term stress will tend to result in the classical suppression syndrome.

THE LYMPH SYSTEM AS A COMPONENT OF THE IMMUNE SYSTEM

Lymph is the transparent fluid in the lymphatic circulatory system. It is derived from the blood by exchange with the capillaries serving the tissues of the body. It contains a liquid portion and a cellular component made up primarily of lymphocytes. Thus, it is important in delivery of the immune response system to the tissues. The system is made up of a network of vessels that flow to and drain tissue spaces throughout the body.

Composition of lymph is similar to blood plasma with the exception of components held in the blood by vascular membranes. It contains large concentrations of lymphocytes. The function of the lymph system is to receive fluids and lymphoid cellular tissue from the bloodstream (capillaries). Capillaries flow through virtually all tissues of the body and return to the lymphatic vessels and ultimately flow back into the bloodstream by way of the subclavian vein. Thus, the lymph provides a mechanism for cellular nutrition and waste removal from the tissues, as well as delivery of the immune system to tissues.

Lymph nodes are defense structures along the lymphatic system with a primary defense function. An infection in the tissue may cause nearby nodes to inflame, swell, and contain large numbers of phagocytes that attack bacteria and other harmful agents to prevent their entry into the bloodstream and general circulation throughout the body.

Lymphatic tissues provide the origin of cellular components of blood that provide the immune function. Such tissues are found primarily in lymph nodes, spleen, and bone marrow.

BLOOD AND ITS ROLE IN THE STRESS RESPONSE

Blood serves primarily to transport the large numbers of biological materials to and from body tissues. As a result, it mediates virtually all metabolic responses to stressful conditions by delivery of neurohormones, hormones, and nutrients that respectively signal tissues to respond and support their response by dramatic influences important in resisting stressors. In addition, the blood sustains and delivers specialized cells, some providing the immune function and others destroying microorganisms. A major routine as well as emergency function of blood is its role in oxygen transport through function of the red blood cell component (erythrocytes) and the hemoglobin content of these cells.

This system delivers oxygen to the tissues through capillary exchange with tissue fluids and cells and removes the oxidative by-product carbon dioxide and other wastes that are returned to the lungs for discharge in the expired air. Stress increases the demand on all of these functions. Finally, the blood is essential in stress responses because of its role in control of body temperature, acid-base balance, and other characteristics involved in homeostasis. This phenomenon, the collection of metabolic processes challenged by stress, reflects the importance of blood in proper volume with appropriate circulatory rate and with suitable composition in providing critical support to tissues performing stress resistance functions.

Blood Components and Function

Cellular Components

Cells make up just under half of blood under normal conditions and consist of three classes of bodies, erythrocytes (red blood cells), leukocytes (white blood cells), and thrombocytes (platelets).

Erythrocytes   The red blood cells are formed in bone marrow. Cells stored in this tissue, along with those held in the spleen, provide some reserve for supplementing the circulatory level to meet stressful conditions that require increased red blood cell volume and oxygen transport.

The primary function of erythrocytes is the delivery of oxygen to body cells and the removal and transportation of carbon dioxide to the lungs for respiratory elimination.

Platelets   Platelets are formed primarily in bone marrow, although the liver and spleen produce some during fetal development. The primary function of platelets is in the blood clotting process. This function is obviously related to meeting emergencies by conserving blood and in the healing process of damaged tissues by contributing to the stable mass of material at the injury site.

Leukocytes   This classification of blood cell components, the leukocyte, is also referred to as the white blood cell component.

Lymphocytes   Lymphocytes are formed in bone marrow from stem cells and develop as T cells in the thymus or B cells in bone marrow. The primary role of B cells is in antibody formation. The role of T cells is in cytotoxicity and in stimulating other immune cells. Prolonged stress may reduce the effectiveness of an animal's immune system by causing a reduction in the lymphocyte count.

Monocytes   Monocytes are formed in the bone marrow and spleen. Their primary function is destruction of microorganisms and removal of damaged cells in areas of tissue injury through the process of phagocytosis.

Eosinophils   Eosinophils are formed in the bone marrow and may be phagocytic. This cell class increases in response to allergies and allergic reactions. Prolonged stress may result in low levels of eosinophils, which is a condition referred to as eosinopenia. The problem is related to extended function of and hormone levels related to both the HPA and SA stress-response systems.

Basophils   The bone marrow is the site of formation of basophils. The cells are involved in allergic responses and produce heparin (anticoagulant), histamine (vasodilator and gastric secretory stimulator), lysozyme (antibacterial enzyme), and bradykinin (vasodilator).

Neutrophils   Neutrophils are formed in the bone marrow. Their major function is phagocytosis, in which microorganisms and foreign particles are engulfed and destroyed by enzymatic action. Large numbers of these cells congregate at infection sites. They also prompt the central nervous system to increase body temperature (fever) at the onset of an infection. This process is initiated by the formation of pyrogens by the cells. The elevated temperature may enhance the effectiveness of white blood cell function in resisting infection.

Fluid Phase: Plasma   Blood plasma makes up slightly more than half of the total blood volume. It consists of 90% or more water, with the balance consisting of solids including protein, lipids, carbohydrates, minerals, nonprotein nitrogen, enzymes, hormones, and antibodies. It is clear from this partial list of components that the plasma transports a great variety of biological materials to and from the tissues. Materials such as water, oxygen, and minerals are transferred through the selectively permeable membrane of the capillary wall into the tissue fluid. Waste or excess products are transferred back into venous capillaries in the same manner. However, some tissue fluid is picked up by lymphatic capillaries in the tissues to become lymph. This fluid is moved through the lymphatic system of vessels by the physical force of surrounding tissue until it ultimately is returned to the bloodstream by way of the right and left subclavian veins. The lymphatic vessels have one-way valves that prevent backflow, and the fluid is filtered along the way as it flows through lymph nodes.

Blood Composition and Volume

Many components of blood and the volume of blood relative to body weight may be relatively constant. However, these factors vary to some extent among species and with stage of development of individuals. Several comparative parameters for the animal species are shown in tables 3.1, 3.2, and 3.3.

TABLE 3.1   Blood Volume (mL/kg Body Weight)

Cattle	57
Draft horse	71
Light horse	77
Race horse	100+
Pig (newborn)	74–100
Pig (2–12 mo)	50–70
Pig (adult)	35–50
Sheep	55–60
Goat	70
Chicken	80

Source: Swenson, M.J., *Dukes' Physiology of Domestic Animals,* 11th ed. Edited by M. J. Swenson and W. O. Reece (Ithaca, N.Y.: Cornell University Press, 1993), p.45, by permission

TABLE 3.2   Blood Cellular Components

Erythrocytes (million/mm^3 blood)	
Cattle	6–8
Draft horse	7–10
Light horse	9–12
Pig	6–8
Sheep	10–13
Goat	13–14
Chicken	2–4
Leukocytes (thousand/mm^3)	
Cattle	7–10
Horse	8–10
Pig	
Newborn	10–12
1 wk	10–12
2 wk	10–12
6 wk+	15–20
Sheep	7–10
Goat	8–12
Chicken	20–30

Source: Swenson, M.J., *Dukes' Physiology of Domestic Animals,* 11th ed. Edited by M. J. Swenson and W. O. Reece (Ithaca, N.Y.: Cornell University Press, 1993), p.27, 36, by permission

TABLE 3.3   White Blood Cells (Percentage of Total Leukocytes)[a]

	Neutrophil	Lymphocyte	Monocyte	Eosinophil	Basophil
Pig					
Newborn	70	20	5 – 6	2 – 5	<1
1 wk	50	40	5 – 6	2 – 5	<1
2 wk	40	50	5 – 6	2 – 5	<1
6 wk+	30 – 35	55 – 60	5 – 6	2 – 5	<1
Cattle	25 – 30	60 – 65	5 – 6	2 – 5	<1
Horse	50 – 60	30 – 40	5 – 6	2 – 5	<1
Sheep	25 – 30	60 – 65	5 – 6	2 – 5	<1
Goat	35 – 40	50 – 55	5 – 6	2 – 5	<1
Chicken	25 – 30	55 – 60	10	3 – 8	1 – 4

[a]Neutrophils, eosinophils, and basophils are classified as granulocytes; lymphocytes and monocytes are classified as agranulocytes
Source: Swenson, M.J., *Dukes' Physiology of Domestic Animals,* 11th ed. Edited by M.J. Swenson and W.O. Reece (Ithaca, N.Y.: Cornell University Press, 1993), p.36, by permission

PART III

ANIMAL BEHAVIOR AND WELL-BEING

An understanding of animal behavior and factors influencing these phenomena is critical in assessing the needs of animals. Such needs form the basis for planning and providing an environment that ensures the well-being of the animal. One might consider needs relative to the well-being of a group of animals; however, true assessment of well-being requires consideration of the individual, because responses to the environment may be quite variable among those in a group.

Behavior, in many cases, may constitute an important indicator of the animal's success or failure in interacting with its environment. The level of well-being provided by that environment can influence comfort and survival of the animal but may also impact economics of the enterprise as it influences physiological and psychological systems of the animal and, in turn, animal performance. Thus, the individual charged with providing adequate care and a state of well-being must have the ability to interpret behavioral characteristics and patterns that may suggest the welfare status of animals. It is important to keep in mind that the animal's response to its environment may have roots in either or both physical and social characteristics of that environment.

The objective of this section is to consider some important principles of behavior, recognizing that they may serve as benchmarks or barometers, so to speak, of environmental adequacy. To determine environmental adequacy, we must first be able to recognize normal behaviors in order to understand behaviors that indicate an animal is experiencing stress. Stressful environmental demands, whether they have a social or physical origin, may ultimately result in behavioral and other biological responses. Such responses can create social instability in a group of animals, reduced resistance to disease, reduced efficiency, or other negative impacts related to the process of coping.

In recent years considerable emphasis has been given to the concept that design characteristics of the animal's environment should allow expression of normal behaviors. This perspective is often accompanied by the thought that an animal exhibiting aberrant behavior is one experiencing an environment that is less than desirable and it therefore has an unacceptable level of well-being. These views have formed the basis for regulatory action in some countries as they attempt to codify characteristics of animal environments such as type of pen or cage design, space allowances, management practices involving pain, and some handling procedures. The Brambell report issued by the British government (Brambell Commission, 1965) is an example of results of extensive deliberations on issues related to animal well-being that are then provided to governmental agencies for the purpose of guiding the formulation of public policy relating to animal welfare. Although very broad regulatory criteria might require that animals be allowed to accomplish normal behaviors, examples of more specific environmental requirements are those that relate to space allowances and pen or stall configuration that permit the animal to turn around and allow normal lying down, resting, and rising. Others may require that the animals be able to experience grooming behavior characteristic of the species and have the capability of viewing other animals. Animal facilities that prevent one or more such behaviors would therefore be viewed by the involved regulatory agency as compromising the animal's well-being. Proposed and existing regulations must be carefully and continuously reviewed to ensure that they are based on scientific evidence. Time, along with additional research and experience with such regulatory requirements, will hopefully ensure that demands placed on livestock producers and companion animal owners will be carefully based on science.

A major consideration in evaluating the state of well-being of an animal is whether the animal is exhibiting aberrant behavior. Such behaviors may be related to stressful environments and thus are generally considered preventable. The animal demonstrating such behavior may in fact be doing so as a means of coping with a stressful environment. Parameters reflecting stress level in the animal in some cases reflect reduced stress in those individuals engaging in the coping behavior. The fact remains that the animal is using depletable body resources to cope and as a result may well be in a state of reduced well-being.

In considering the effect of stressful environments on behavior, one should be mindful that stress may influence the animal's well-being through either or both physiological and psychological response mechanisms. Often animal enterprise managers and caretakers fail to recognize the significance of psychological stressors that may be related to characteristics of the physical and social environments. It should be kept in mind also that such stressors may cause acute or chronic stress effects. Short-term behavioral responses such as vocalization and activity related to painful experiences are recognized as indicative of stress by most individuals associated with animals. They are often less aware that longer-term stressful conditions related to pressures of the social environment or frustrations related to the physical environment may result in a variety of aberrant behaviors, some of which may be stereotypic in character. Behaviors so classified (e.g., cribbing and bar biting) are referred to as stereotyped behaviors. A number of aberrant behaviors including those classified as stereotypies are reviewed in detail later in this section.

The importance of understanding behavior in developing and maintaining the animal's environment as a part of the management process is illustrated by a few examples. Socially oriented behaviors such as social organization and interactions within a group of animals, aggressive characteristics, dominance, communication, and spacing behaviors are all important in managing and handling animals. These behaviors may determine, singly or collectively, management strategies to minimize the possibility of stressful conditions. Reproductive and maternal behaviors influence reproductive effectiveness dramatically and therefore have a significant influence on the economic status of an enterprise. Feeding and grazing behaviors may influence both animal performance and the efficiency in which animals utilize the feed or grazing resource. Understanding an animal's ability to respond behaviorally to demanding thermal environments can be important to both animal well-being and the cost of producing animal products. These represent but a few examples that readily justify the effort to understand animal behavior, and they illustrate the importance of recognizing these characteristics when designing and maintaining animal management systems to ensure animal well-being.

To fully understand the causes associated with behaviors, one must pursue information provided by several fields of science. The areas of ethology, behavioral genetics, stress physiology and related areas of endocrinology, neurology, biochemistry, and animal psychology are all essential. Scientists working in the field of ethology are providing greatly improved insight into behavioral repertoires of animals and greater understanding of the causes and characteristics of such behavioral phenomena. They, in collaboration with physiologists, are gradually clarifying the complex relationships between stress and behavior and as a result are enhancing our understanding of animal well-being.

ETHOLOGY: THE SCIENCE OF BEHAVIOR

The field of science dealing with behavior is ethology. Hurnik et al. (1995) define *ethology* as simply "the study of animal behavior." Some define *ethology* as the study of behavior in the evolutionary context. Others limit the field to the behavior of animals in the natural setting. Still others include the behavior of both wild and domestic animals in typical environments. In any case, the relationship between behavior and the usual environments in which animals exist represents a critical issue in considering effects on animal well-being and may determine the outcome of management decisions.

Ethologists describe some behaviors as innate or instinctive in nature, reflecting that such behavioral characteristics are the result of evolution of that species. Such behaviors have resulted from natural selection, although they may have been modified by artificial selection over long periods of time in certain subpopulations of a species. An example is the loss of maternal behavior in some dairy cattle as compared with the retention of this behavioral characteristic in beef cattle. In the former case, the common practice in dairying is to separate the calf and mother immediately after birth; therefore, mothering ability has not been a significant factor considered in genetic selection of dairy replacements for many years. In contrast, most beef cattle operations rely on good maternal performance to ensure efficient production, since the ability to effectively produce, protect, and provide for a calf is paramount in the economy of the enterprise. Understanding the genetics of behavior is critical to decisions relating to whether to try to alter these characteristics in a population of animals through selective breeding. Reproductive behaviors are typically characteristic of a given species. These behaviors are established and controlled largely by endocrine function, which is the result of genetics but may be influenced by environmental characteristics such as length of day in some species and also by stressful environments. It should be added, however, that some characteristics of sexual behavior may be enhanced by experience, and likewise previous experience in maternal behavior typically improves the dam's effectiveness in rearing offspring.

Ethologists are vitally interested in behavioral genetics because it forms the basis for understanding evolutionary effects related to behavior. Animal enterprise managers are interested as well in genetic effects on behavior, which must be considered because it relates to both animal performance and well-being. Inherited animal characteristics must be compatible with the environment in which the animal exists, or well-being will likely be compromised. An example is the adaptability of laying hens to the caged environment on which current production practices are based. The fact that laying hens may commonly engage in aberrant behaviors that are normally assumed to be related to the spatial and social environment suggests the presence of both physical and psychological stress. Some genetic lines of layers are more productive under these conditions and reflect stress evaluation parameters indicating that they are better adapted to this system of production.

Although some behavioral characteristics of animals are inherited, others are developed in response to environmental influences. Such characteristics may be described as acquired behaviors and are developed through experience such as observing other animals, associations during development, success or failure in responding to the environment in which they live, exploration of the environment, and training systems designed to produce specific behavioral responses. All of these traits represent learning, which is a major topic in considering animal behavior.

Understanding the relative level of instinctive and acquired influence on a given behavior is very important in animal enterprise management. Achieving or altering a given behavior requires totally different approaches depending on which of the two avenues gives rise to the characteristic. Genetically derived behaviors may require many generations of selective breeding to change. Behaviors influenced primarily by environmental influences must be altered by changes in the animal's environment. The science of ethology provides understanding of both the nature of behaviors and their causes and as a result assists in the design of approaches to influence behavior if it is desirable to do so.

Darwin provided some understanding about the evolution of behavior and the importance of coping behavior in the process, but the field of ethology per se was firmly established in the late 1930s with the publication of the first journal relating to animal behavior in Germany. Alcock (1993) concludes that ethology, as a scientific field, began seriously in the 1950s with the work of three zoologists awarded the Nobel Prize in physiology and medicine in 1973: Karl von Frisch of Austria, Konrad Lorenz of Austria, and Niko Tinbergen of the Netherlands. Tinbergen noted the fact that newly hatched gulls already knew how and thus were programmed to stimulate their parents to feed them. He concluded that these birds were hatched with an innate behavior that required a key external stimulator. Thus, the chick pecks at the red dots on the beaks of its parents, which prompts the parents to regurgitate food previously eaten and allow the offspring to consume it. Such behavior might be classified as feeding behavior on the part of the young and maternal-paternal behavior on the part of the parents, an example of an instinctive behavior (fully functional when first performed) that in this case is essential for an adequate state of well-being for the young.

Tinbergen described spatial learning by wasps that guided the insect back to its home burrow by the use of features of the landscape in the vicinity of the entrance. He also showed the influence of environment on behavior by demonstrating that cliff-nesting gulls showed movement of only the head and neck in submissive behavior to avoid attack by neighboring birds, as compared with their ground-nesting counterparts who moved away from the nest to minimize the possibility of an attack. The behavior developed through natural selection, which resulted in minimizing the possibility of the young cliff dwellers falling from the nest. Von Frisch solved the mystery of communication among honey bees that convey direction and distance to sources of pollen by characteristic dancelike behavior called a waggle dance. He also did pioneering work on visual and color perception of bees. Lorenz demonstrated that imprinting was a powerful form of early behavioral learning by showing that newly hatched ducklings would imprint and identify with him and then ultimately follow him in the same manner as they would normally follow their mother upon leaving the nest. Understanding the importance of innate, fixed-action pattern behavior in survival is yet another major contribution by Lorenz.

Today, applied ethologists are contributing significantly to scientific and educational endeavors relating to practical dimensions of animal well-being. Some examples of contributions include the following:

Evaluating and designing improved methods utilized in livestock production.
Enhancing conditions under which companion animals are maintained.
Increasing awareness of potentially hazardous conditions in some sporting activities involving animals.
Contributing to enlightened public policy development regarding animal care and management.
Improving recognition of the influence of all forms of stress and behavior on the validity of research to better understand the full range of the animal's environmental requirements. Such requirements include physical, nutritional, and social dimensions of the environment.
Designing appropriate environments for animals used in research and educational activities.

BEHAVIOR AND ANIMAL WELL-BEING

The Brambell report (Brambell Commission, 1965) describes *welfare* as a broad term that embraces both the physical and mental well-being of the animal. The Brambell report also comments on the importance of the animal's feelings and behavior as a reflection of welfare. Duncan (1996) suggests that *well-being* must be defined in terms of animal feelings. Although others (McGlone, 1993; Moberg, 1993, 1996) suggest that *well-being* must be defined in measurable terms such as physiological, performance, and health parameters, one must recognize that the numerous aberrant behaviors related to stressors of a psychological nature most likely involve some aspects of how the animal "feels" about its environment. Even though some differences exist among scientists as to how well-being is best measured, there appears to be general agreement that consideration of both physiological and behavioral parameters is likely to lead to the best judgments about the adequacy of the animal's environment. Broom (1986) defines *welfare* by considering the animal's state relative to coping with its environment. Hurnik et al. (1995) describe well-being as a state or condition of physiological harmony between the organism and its surroundings and state that the most reliable indicators of well-being are good health and the manifestation of a normal behavioral repertoire. Lynch et al. (1992) conclude that a problem in defining *welfare* and *well-being* is that consideration must be given to both physical condition and mental state simultaneously and that both are linked and both are considered essential to welfare. It is likely that society as a whole, in considering the issue of animal welfare, would view any form of distress, regardless of origin (i.e., physical or psychological), as representing environmental influences that compromise the well-being of the animal. Certain types of stressful environments are clearly associated with aberrant behaviors, and existence of these behaviors suggests that the animal's well-being is compromised. Thus, the relationships among genetic characteristics of animals and the adequacy of both physical and social environments are indeed important considerations in animal management and facility design.

While much effort is currently devoted to planning and maintaining suitable environments for animals, new technologies will influence how we plan and execute management practices in the future. A prime example is the use of genetic engineering as a fast-track approach to altering animals. Rollin (1995) discusses issues surrounding such practices to ensure that the resulting animals are not at risk in terms of well-being. Major issues are related to the possibility of unexpected genetic effects that may compromise well-being; examples include effects on health and normal function, as well as the ability to adapt to all characteristics of a given environment.

Ethologists emphasize the importance of understanding the causes of behavior in explaining why animals do things and the motivation and drive associated with such behaviors. *Proximate* reasons for behavior and *ultimate* reasons for behavior are distinguished by the respective examples that animals eat because they are hungry and animals eat to survive. In the former case, eating satisfies the immediate demand for food. In the latter case, eating supports survival over the longer term and maintains all functions related to the animal's ability to survive and pass genes on to the next generation.

In the final analysis, how we feel about animal care and well-being will relate to our perception of how animals feel and react to sensations, how they feel pain, our understanding about the cognitive ability of animals or how they perceive their condition, and how we feel about the level of intelligence possessed by animals. We must try diligently to view all of these aspects with a high level of objectivity and urge adequate research to better explain the relationship of the animal to its environment.

Currently, much thought and discussion surround the animal well-being issue. In this process, objectivity is often replaced by emotions, and science-based approaches are frequently lost. There are at least three thought scenarios that may be hazardous in terms of reaching appropriate conclusions about animal well-being.

Anthropomorphism

An anthropomorphic view should be recognized as a potential hazard in making judgments about an animal's condition. The term refers to the assignment of human feelings to those of animals. One may judge an animal to be happy or sad, but these terms are descriptive of human feelings and not necessarily those of animals. Evaluating the comfort level of an animal is an area frequently influenced by judgments made by our own comfort standards. In some cases such judgments may be correct; however, an understanding of the animal's specific environmental needs is essential to making such assessments. An example is the importance of recognizing that species vary greatly in their ability to adapt to and withstand hot and cold environments. As a result, matching animals to a given environment or planning to provide the animal with an environment having characteristics within an acceptable range is critical to animal well-being.

Assumed Teleologic Capability

Teleologic ability of animals should not be assumed. Teleology reflects the process whereby an individual will accomplish something, a behavioral scenario, for example, that will result in an expected outcome. Animals are not as accomplished in this type of endeavor as are humans. For example, an animal may lick a wound, preventing infection. Most likely the animal does so as an innate behavior and does not understand that cleaning the wound may prevent infection.

Reductionism

Reductionism is another potential mistake that may be hazardous in relation to animal well-being. Since one does not understand what makes an animal happy or sad, it might be concluded that such evaluations should be discarded from the logical thought process in favor of assuming that an animal is in a state of well-being if it is not experiencing something obviously aversive such as pain. This reasoning would lead to the conclusion that unless the animal is showing clear responses to pain, it is doing fine. Another reductionist view is that even though an animal is reflecting the presence of pain, this response is innate, without feeling. These views show a lack of understanding of animal response to painful and stressful environments and represent obvious oversimplification.

CHAPTER 4 ▰▰▰▰▰▰

Control of Behavior

Understanding behavior requires insight into the various factors involved in control of behavior in individual animals and how such factors may relate to the behavior of a group of animals. Such understanding is critical in evaluating the animal's response to its environment and the relationship of observed responses or actions to the well-being of individual animals. Genetic influences controlling behavior, including both natural and artificial selection, are important in making judgments about environmental needs and how animals may respond to given environments. Behavioral influences such as learning, stage of the animal's life cycle, and stress are examples of factors that are important in understanding behavior and how animals may be able to cope in the environment provided. Stress may influence behavior in ways that are useful in diagnosing environmental inadequacy. Finally, some understanding of neural and endocrine influences on behavior assists in evaluating the animal's condition. Such physiological phenomena relate to the extent to which an animal may be able to adapt to either acute or chronic demands of the environment.

GENETIC EFFECTS ON BEHAVIOR

Behavior, like all other animal traits, is influenced by both hereditary and environmental factors. Both act in concert to form behavioral characteristics and patterns. The relative influence of each may form the basis for important decisions in managing animal enterprises from the standpoints of both economy and animal well-being. Highly heritable traits are readily altered through selective breeding. Those of lower heritability, should it be desirable to alter or control them, must be managed by planned changes in one or more of the environmental elements associated with the particular system in which the animal is expected to exist. Such elements include factors like physical facilities, diets, the thermal environment, and the broad range of social interactions.

FIGURE 4.1 Innate or inherited action pattern behavior in the pig is reflected by its choosing the most direct route to the mammary area of the dam for nursing; few pigs choose the longer, more hazardous route completely around the back of the dam

Hereditary influences, evaluated through studies in behavioral genetics, are a major consideration in understanding evolution and domestication, whether these processes occur as a result of natural or artificial selection. Figure 4.1 illustrates innate or inherited action pattern behavior in the newborn pig that is reflected by its choosing the most direct route to the mammary area of the dam for nursing.

Environmental influences are evaluated by a range of measures including interactions with other animals and humans, as well as a variety of factors contributing to the resource base supporting the animal. Some of the major genetic factors relating to animal behavior are summarized in the following section.

Effects of Selection and Heredity on Behavior

Behavioral genetics provides important insight into some aspects of animal behavior, which is important in breeding programs designed to alter behavior in subsequent generations. It is important to recognize that environmental influences also affect behavior; thus, an understanding of the relative influence of genetics and environmental factors is essential to developing approaches to changing behavior.

Behavior is controlled by the central nervous system and related endocrine influences. Such systems develop and become active under the control of genes. The degree to which a behavioral trait is inherited can be determined if that characteristic can be measured quantitatively. A given behavioral characteristic represents a complex array of neural and endocrine control mechanisms and will most likely involve multiple gene action, although some behavioral traits have been observed to be due to single gene effects. Nest cleaning in bees is an example (Rothenbuhler, 1964). Genes provide the components important in behavior control. That is, neural, endocrine, and muscular function capability are influenced by gene action. Behaviors result from the various transmissions through neural and endocrine (humoral) mechanisms and the response functions accomplished by specific target tissues responsible for the mechanics of executing the action. The scientific field of behavioral genetics deals with the component of a given behavior that is controlled by inheritance and approaches to altering behavior through selective breeding.

Craig and Muir (1996) reported the success of selective mating in producing stocks of laying hens that exhibit low levels of feather and cannibalistic pecking. This approach may eliminate the need for the practice of debeaking, which is recognized as a painful procedure that results in sensitivity of the remaining tissue for an extended time. The selected stocks were characterized by lower death loss, higher egg production, and improved feathering. Other examples of traits that have responded to selection to enhance a desired characteristic include the following:

Dairy cattle	Temperament and milk letdown
Beef cattle	Mating and maternal behaviors
Pigs	Reduced aggression; mating and maternal behaviors
Sheep	Maternal, mating, and flocking behaviors
Horses	Temperament and trainability
Honey bees	Docility and hygienic behavior
Dogs	Trainability, docility, hunting, and working characteristics
Chickens	Broodiness (characteristics related to hatching and caring for offspring) and cage adaptability (reduced social aggression)

The proportion of the total variation observed in a trait in a population that is caused by differences in genotype is referred to as heritability. This proportion is an estimate and is expressed by the ratio between genotypic variance and phenotypic or total variance. The ratio ranges between zero and one. The utility of heritability estimates is in predicting the response to artificial selection. Traits having values above 0.7 would typically be described as highly heritable; those with values below 0.2 are classified as having low heritability. The animal industry commonly utilizes selection pressure to alter animal characteristics having heritability estimates in the area of 0.25 or higher, and progress is made as a result. To do so in a reasonable amount of time, however, requires substantial selection differential.

The calculation of expected response to selection is as follows: (R) = heritability (h^2) × selection differential (sd), with sd representing the difference between the mean of the selected parents and the population from which they came. The following example estimates the expected change in flight distance of a population resulting from selection of parents with an observed flight distance different from that of the general population.

The hypothetical heritability for flight distance is 0.80 (80 percent). An example mean population flight distance is 12 feet. If we select parents with an average flight distance of 8 feet, what will be the expected response of the offspring of the selected parents?

$$h^2 = 0.8$$

$$sd = 12 - 8 = 4 \text{ feet}$$

$$R = 0.8 \times 4 = 3.2 \text{ feet}$$

The offspring, in this hypothetical example, would be expected to have a flight distance of 8.8 feet or 3.2 feet less than the mean observed for the population.

Estimates of heritability are unique for a given population, environment, and time. A very uniform environment will increase estimated heritability, whereas a uniform genetic makeup will decrease it. Inbreeding tends to decrease estimates of heritability.

Pleiotropy is a term used to reflect that a single gene can affect more than one trait. Pleiotropic effects are the rule rather than the exception in gene effects. For example, an increase in maternal care by sows may also give rise to greater aggressive defense of piglets. *Polygenic* is a term used when multiple alleles at several loci influence one trait.

Evolution and Behavior

As with any trait, natural or artificial selection will influence the frequency of a characteristic in a population that favors those genes for survival.

Genes controlling those traits related to survival and ability to reproduce give rise to population changes that enhance the ability to cope (i.e., survival of the fittest). Evolutionary effects result from gene survival through generations.

Important concepts in evolution of adaptability include the following:

Efficiency of the whole animal in coping (i.e., effective establishment of priorities for use of time and effort in resource use) contributes to survival.
Effectiveness in establishing mechanisms and actions for survival is critical.
Establishment of patterns of behavior that help animals protect, hunt, feed, reproduce, and so forth (i.e., coping is enhanced by effective procedures and in some cases this involves association and cooperation with others) have obvious positive effects on survival.

Some behaviors considered social in nature that are related to survival include the following:

Gathering for protection
Hunting in packs
Animals or birds watching each
    other for guidance to food sources
Development of social dominance

Females tending the young of others
Following the leader
Efforts of separated animals to
    return to the group
Mutual grooming

Domestication and Behavioral Effects

Domestication has obviously occurred over extended periods. General time periods for domestication of some species are thought to be before 10,000 to 12,000 B.C. for dogs; 6,000 to 8,000 B.C. for sheep, goats, and cattle; 2,500 B.C. for swine; and 2,000 to 4,000 B.C. for horses, poultry, and cats. Domestic and wild-type turkeys

FIGURE 4.2 Wild-type and domestic forms of the turkey; both form and function are altered by the process of domestication

shown in figure 4.2 illustrate changes that influence both appearance and ability to function. In this case the domestic bird cannot normally fly. Wide genetic differences in form, however, may not be associated with obvious differences in behavior (fig. 4.3). Domestication affects the quantitative not the qualitative nature of behavior (Price, 1998).

An understanding of relationships between domestication and animal behavior provides important insight into animal-environment interactions. As a result, some issues related to animal well-being may be viewed from a more informed perspective.

The process of domestication may involve marked changes in the animal's environment including the character of relationships with both humans and other animals. Thus, influences on socially oriented behavior become a matter of interest and inquiry as to the effects of domestication and how these may influence considerations related to environmental design to achieve an acceptable level of well-being. For example, a given species may be able to adapt to a higher level of population density than can another. This characteristic will influence individual space allowance specifications in facility design.

Zeuner (1963) describes domestication as having five stages:

Stage 1. Animals associated with humans were allowed to continue to mate with those in the wild. The extent to which selection by humans may have occurred was probably related to compatibility, making the association desirable. Animals that represented a threat to humans or other animals may have been intentionally eliminated.

Stage 2. Association of animals with wild forms was prevented. The animals were captive, although mating may not have been controlled, at least completely, by humans.

Stage 3. Selection was practiced in maintaining animals that were most useful in terms of size, strength, compatibility, and economy.

FIGURE 4.3 Wide genetic differences in form may not be associated with distinct differences in behavior; these two horses, both domesticated, are expected to show similar behaviors (Photograph courtesy of Photo Service, Iowa State University, Ames)

Stage 4. Selection was refined by the recognition that characteristics and performance of animals could be altered to achieve rather specific goals such as meat and milk production. The concept of developing specialized lines, or breeds, probably began to develop during this stage.

Stage 5. Intentional reduction in the population of certain wild forms that are considered undesirable. That is, some wild animals may represent a threat in terms of safety and competition for resources.

Hart (1985) suggests that a sixth phase of domestication could be added because of the highly refined selection for adaptability to the very specialized agricultural environments currently being used in animal production.

Zeuner (1963) also plots an interesting chronological order of domestication, with animals such as the dog, sheep, reindeer, and goat being domesticated during the preagricultural phase; primary food animals such as cattle, buffalo, and pig during the agricultural phase that were described as crop robbers; animals used for transport including horses and camels; those that might serve as pest killers such as the cat; and finally birds and fishes.

The reversal of the domestication process, in a sense, is when animals are allowed to return to some form of the so-called wild state. Such animals are referred to as feral animals and are of interest to ethologists in providing information about both hereditary and environmental influences on behavior. In some cases such animals are studied to better understand the environmental preferences of animals.

Some characteristic results of domestication might be summarized by the following examples:

Increased compatibility with
  humans, other animals within the
  species, and other species
Reduced aggressiveness
Reduced flight distance

Increased flexibility and adaptability to different environments and
  often tolerance of reduced space
Reduced selectivity as to mating
  partners and reduced pair bonding

An extensive review of behavioral aspects of animal domestication by Price (1984) provides an excellent base for enhanced perspectives on relationships between the process and animal behavior. In this review, *domestication* is defined as "that process by which a population of animals becomes adapted to humans and to the captive environment by some combination of genetic changes occurring over generations and environmentally induced developmental events recurring during each generation." Short-term developmental influences such as management practices to which animals adapt may also become influences for genetic change if they are long-term environmental influences. Given the provided definition, one concludes that the process of domestication has effects that are evolutionary (altered gene frequency) resulting in genotypic change based on the various influences of the captive environment.

The assumption that a captive environment is limited to a pen or structural constraints should be avoided. A more appropriate view of this term is to think in terms of a managed environment that may range from very nearly the same as that experienced by the species in the wild to a highly confined, intensive livestock production enterprise.

In the review (Price, 1984), the term *domestic phenotype* was used to describe the animal resulting from domestication. The domestic phenotype reflects an adaptation to humans and to the environment provided. An important question in the domestication process and animal well-being relates to whether a given species or even individual animals within a species are capable of adapting to specific environmental conditions. If not, the likelihood of stress is increased, along with the possibility of associated negative effects (see also Price, 1998).

The level of success in domestication is judged in relation to how effectively a human need is met by the process. Such needs are of course established by humans, and the process is designed to develop animals to meet those needs. One's ability to achieve a level of success in the venture is heavily dependent on factors such as those discussed in the following two sections.

Extremes to Which the Animal Must Adapt

Jewell (1969) and Crawford (1974) suggest that in too many cases we attempt to utilize traditional farm animals in areas to which they are only marginally adapted, and thus we fail or the resulting animals require an artificial environment that may not be cost-effective. It follows that consideration should be given to domestication of native species that are more nearly adapted to the existing or anticipated environment before the domestication process is begun. Figure 4.4 shows two breeds of cattle that have different capabilities for resisting environmental influences. The Zebu is more resistant to insects and heat, whereas the animal of European breeding is better equipped to resist cold.

FIGURE 4.4 Two breeds of cattle that have different capabilities for resisting environmental influences: the Zebu breed (the animal on the left) is more resistant to insects and heat, whereas the Angus breed (on the right) is better equipped to resist cold (Photographs courtesy of J. D. Hudgins, Inc., Hungerford, Texas [left] and the American Angus Association, St. Joseph, Missouri [right])

Genetic Variation Existing in Desirable Traits

Genetic variation within a population provides a greater range of characteristics capable of responding effectively to environmental demands and thus may have a greater likelihood of contributing to the survival of those genes that best meet the demands placed on the animal.

Characteristics Typical of Species Selected for Domestication

Hale (1969) outlined a number of behavioral characteristics that tend to favor or impede domestication. Some of these are summarized below.

Characteristics favoring domestication revolve around behaviors related to social organization. Examples are groupings into flocks or herds and associated establishment and maintenance of dominance orders. Opposing domestication processes are characteristics related to maintaining and defending a territory. Farm animals are not classed as territorial but may identify areas for primary use as a home range. Other animals or groups may occupy the same area even though there may be an order establishing primary use by the groups involved.

Animals that are promiscuous in mating characteristics favor domestication as opposed to those species that tend to form pair bonds relative to mating.

Domestication has been facilitated by the presence of critical periods in development in which offspring bond to their dam. Humans have capitalized on this, allowing young to imprint on a person rather than the dam. Additionally, domestication is favored when females accept foster young.

Species that produce precocial offspring are more adapted to domestication than those that produce altricial young that require intense and extended care for survival. Pigs, calves, foals, and newly hatched chicks and poults are precocial. Except for nursing and protection by the dam in the mammalian species these young can be very independent in terms of survival.

Flight distance characteristics are an important factor in domestication. Farm animal species are characterized as having short flight distance as compared with the species that remain largely undomesticated. One observes, in many cases, animals of species that remain largely undomesticated but are tame and appear to be comfortable in relation to humans. In general, however, such animals are regarded as potentially more volatile or dangerous in associations with humans than those of the domesticated species.

Dietary requirements may influence the domestication process. Farm animal species are flexible in terms of food characteristics. Some are herbivores but do well on a variety of roughage feeds and forages. They can readily accept diets high in concentrates. Ruminants, however, may experience physiological problems when minimum requirements for dietary bulk are ignored. Horses suffer digestive disorders if the level of dietary bulk is inappropriate. Swine are omnivores but do well on varying levels of plant and animal materials. Poultry, like swine, can subsist on a variety of food types but flexibility is limited because of the capacity and character of the digestive system.

Price (1984) summarized the uniqueness of the domestication process as being characterized by six issues.

Artificial Selection
Mating plans are designed and regulated by humans rather than the animals. Any losses in fitness resulting from artificial selection may or may not be compensated for by animal management practices in properly designed management environments.

Reduced or Eliminated Intraspecific Competition
Providing food and controlling the breeding process reduces or eliminates the effects of animal competition on these characteristics, which normally represent two prominent areas for animal competition and related conflict. The net effect may be a comparative reduction in the animal's ability to compete, which must be recognized in certain environments.

Cognitive Mechanisms Are Altered
The human role in the process of domestication serves as a buffer between the animal and its environment. This buffer results in lower cognitive demands on the animal because coping with the environment in providing food, water, and protection, for example, is less complex and demanding.

Quantitative Nature of Behavior Patterns Is Changed
That the quantitative nature of behavior patterns is changed is reflected in an altered response threshold associated with some behaviors. Although the threshold for aggressive behavior is normally increased by domestication, the qualitative dimension of this characteristic may not be altered. For example, dogs may be just as aggressive as their wild counterpart, the wolf, once the threshold for an aggressive response is reached. The threshold for eliciting the response, however, has been increased by domestication. An example is the common livestock production situation where tolerance of spatial intrusion has been raised through

domestication; however, when crowding reaches a threshold point, the animal may fight as viciously as the wild counterpart. Domesticated animals are considered less aggressive and are in fact so inclined in most production environments because fewer aggressive events occur. Concepts based on relationships between quantitative and qualitative aspects of behavior in domesticated species provide added perspective on the implications of space requirements in design of animal facilities.

Reduced Responsiveness to Environmental Change

Reduced responsiveness is seen as an adaptation to living in a so-called safe environment. Such reduction in reaction to environmental change is reflected in factors such as comfortable associations with humans, tolerance of frequent invasions of personal space, reduced flight distance, reduced frequency of aggressive encounters, reduced exploratory activity, and reduced level of fear of potential predators. These responses are associated with both biological and physical environmental influences.

Alterations in the Rate or Extent of Development of Certain Biological Characteristics

Evidence for alteration in the rate of development of certain biological traits in domesticated animals relates to earlier sexual maturity in domestic species. As environments for higher levels of productivity in beef cattle have developed, calving at an earlier age (two versus three years of age at first calving) has become routine except under the most limiting nutritional conditions or where certain domestic genotypes are known to have extended age at sexual maturity. This latter case has been brought to the forefront in recent years as beef cattle breeders have selected for greater mature size that in many cases is related to delayed sexual maturity. This point merely recognizes that in some cases selection occurring during domestication and beyond may delay or accelerate biological development. Hart (1985) raises the question, relative to the extent of biological development, as to the effect of the current use of many highly rated males in artificial insemination rather than natural service on the longer-term maintenance of the species' genetic ability to reproduce by natural means.

Another area of interest related to altered development is the conclusion by some investigators that neoteny (juvenile behavior in adults) is a common component in domestication. Some researchers have suggested that domestication has tended to preserve or extend certain juvenile characteristics in adult animals (Dechambre, 1949; Clutton-Brock, 1981). They base this perception on the fact that a juvenile appearance of the head of an animal has been preferred by breeders and that selection for ease of handling (docility, pliability, and trainability), which is more characteristic of young animals, has been a desirable trait. Price (1978) suggested that rearing animals in isolation away from older, socially dominant individuals may result in a reduction in development of normal adult aggressive tendencies, thereby preserving juvenile behavior. Hart (1985) concluded that the tendency to select for juvenile characteristics in pets was due to the preference for playful behavior. Practical implications of these conclusions in selecting for certain juvenile characteristics in modern livestock production are not completely clear but may partially explain some historical conclusions relative to negative effects of selecting for juvenile characteristics.

LEARNING AND BEHAVIOR

Genetics was mentioned previously as one of the two components that influence the phenotype of animals. The other component is constituted by all of the environmental influences that an animal experiences. Such sensory experiences result in learning. Thus, consideration of learning by animals is an important component in any effort to understand animal behavior. In terms of animal well-being, learning represents an important component in an animal's continuing success or failure in adjusting to and coping with its environment.

All learning involves interaction with a source of information. The information is a component of the total environment in which an animal exists and includes any interaction related to contact with (including observational contact) the physical and social environments and that with humans involved in care and training activities. Such interactions form experiences. Experiences are recorded in the neural memory and will reinforce or modify behavior (see Pinel, 1993). Such memory impressions in neural tissue may support voluntary behavioral actions, as well as some that become sufficiently automatic that they appear to be involuntary in nature or at least require little if any cognitive input.

Some experiences are apparently ignored and result in no perceptible change in behavior. The mechanisms involved in this form of neural gating that determine neural storage or lack thereof are not clearly understood.

The nature of a specific interaction will influence predisposition to learn from that experience (e.g., painful and food-related experiences appear to be of a higher learning priority than some other types of interactions). Individual animals vary in level of predisposition to learning, and learning ability varies among species. Willham et al. (1963, 1964) demonstrated that about 50 percent of the variance, within relatively homogeneous groups of pigs, was attributable to additively genetic causes and that important differences may exist among breeds.

Although learning can be classified in numerous ways, the following categories are relevant to farm animal behavior and related management concerns.

Habituation

Habituation is a progressive decrease in the strength of a behavioral response to a continuing stimulus (Pinel, 1993). Animals "get used to" certain stimuli, and the frequency of startle reactions drop. This characteristic is important in considering animal environments. For example, the noise associated with aircraft, automobiles, or farm equipment appears to cause little if any stress in animals after a period of conditioning. Poultry in large units may panic upon being startled by strange or sudden noises. The hysteria that results can be very damaging in terms of bird injury. Poultry house workers will commonly make a familiar noise or knock in a standard manner before entering a poultry house as a

FIGURE 4.5 The horse has obviously become accustomed, or habituated, to the presence of the large and noisy train in the area

means of prevention of these events. The net result of habituation is a reduction in what would normally be called a fear response. Hart (1985) presents a practical example of this systematic desensitization as being the normal process of conditioning a horse to a saddle and eventually to riding. Animals that learn to reside calmly near an airport or a major highway have habituated to the noise level. Figure 4.5 illustrates a horse that has become accustomed to the presence and noise associated with a nearby train. Figure 4.6 shows a horse saddled and prepared for riding. Both of the illustrations depict forms of habituation.

Sensitization

Sensitization is the general increase in response to a continuing noxious stimulus (Pinel, 1993). Thus, repeated stimulation may not only reduce the threshold of a response but may also increase the magnitude of the response. An example of this characteristic is the increasing aggressiveness or level of escape response to continuing frightening events such as noise or violent threats.

Conditioning: Conditioned Reflex (Classical Conditioning)

The learning, or conditioning, that occurs when a conditioned stimulus (e.g., sound) is paired with an unconditioned stimulus (e.g., food) such that a reflexive response (e.g., salivation) is initiated is referred to as classical conditioning. Hart (1985) suggests that the chief characteristic of classical conditioning is that only responses that are innate or reflexlike can be conditioned. Pavlov's dog is the

FIGURE 4.6 The horse, broken to be saddled and ridden, is gradually habituated to the equipment and rider (Courtesy of the artist, Richard L. Willham, Iowa State University, Ames)

example usually used to illustrate this behavior. The dog salivates when food is presented and consumed. The practice of offering food is accomplished routinely until finally the dog will salivate and go through other motions associated with eating when all activities associated with feeding are accomplished even though no food is present at that point. If the food is not present as a result of these activities, over time the dog will cease to show this behavior. Thus, the conditioning becomes extinct. Such conditioning may also be referred to as a conditioned reflex because the behaviors typically involve systems involving little or no cognitive input. Examples in addition to salivation are milk letdown in dairy cows in response to entry into the milking parlor and some endocrine responses such as insulin, glucagon, and glucocorticoid release related to expected feeding. A practical example in farm animal production is the reaction of animals to neutral stimuli associated with frightening events such as noise of a working chute or the flavor aversion associated with toxic plants (Houpt et al., 1990; Provenza, 1996). In the latter case it is doubtful that the animal can detect a toxic element in grazing but it apparently associates the flavor with previous illness. Animals routinely rise, assemble, and prepare to eat when they hear the noises associated with preparation and delivery of feed. This is a classical conditioned response to neutral stimuli that provide a signal of an expected result. The development of conditioned responses may be referred to as a form of associative learning.

Operant Learning or Conditioning

Operant conditioning, a form of associative learning, is a term used to describe learning related to dimensions of self-control of some aspect of the animal's environment. Examples are operating feeders and waterers. Research shows that animals can learn to push buttons, levers, and the like that turn on lights and heaters. Animals learn to squeeze through fences and open gates or doors to gain a desired goal. In all of these cases animals learn that a reward will result from accomplishment of a task. Thus, they learn to operate something to produce the desired result. If the result fails to materialize, the practice will be discarded over time. This result is referred to as extinction.

Research using operant conditioning to determine the preferences of farm animals has been successful. Livestock have been trained to work for a reward to gain access to water, food, shelter, additional space, and conspecifics. Some of this research (Matthews and Ladewig, 1994) has concentrated on establishing demand curves for these rewards, thereby determining how hard an animal will work for reward. The base assumption is that the harder an animal will work for the reward, the more important is that reward to the welfare of the animal. This application of operant conditioning can be useful in establishing welfare criteria on how to raise livestock. However, the major hurdle for this type of research is determining the differences between an animal's needs and wants. An animal may work very hard for a reward, but its welfare may not be compromised if it does not receive the reward.

Observational Learning

Animals can learn behaviors from other animals. Management systems involving early isolation of animals eliminate this avenue for development of some behavioral functions. In other cases, separation of animals from those demonstrating certain undesirable behaviors may be an appropriate approach to managing behavioral problems. An example of observational learning results when a mature hen is placed with turkey hatchlings to enhance development of feeding behavior. Some stable vices in horses are thought to result from imitation. Recommendations to isolate animals displaying these behaviors are common. Negative aspects of isolation must be considered.

Aversion

Animals learn from negative experiences and may simply choose to avoid contact with certain animals or things. This aversive behavior may represent the submissive component of agonistic behavior in establishment and maintenance of a dominance order. Animals learn to avoid areas that may be inhabited by predators. They also learn to avoid electric fences. Grazing animals may readily learn and remember the location of undesirable feedstuffs of low quality or those having toxic properties.

Shaping

Shaping is a form of learning that may involve waiting until an animal performs a specific behavior. Then the reward is given and the animal, over time and with repetition of the sequence, learns to accomplish the desired behavior on command. A more common dimension of shaping involves progressive development of the desired behavior such as training a dog to roll over. Development of complex behavioral sequences by animals typically observed in the circus ring may involve shaping techniques. The bear that rides the bicycle may first be rewarded for sitting on the bike, then for placing feet on the pedals, then for movement of the pedals and associated movement of the bike, and so forth. Dog and horse acts involving mounting and riding by the dog require shaping of two animals simultaneously. In such scenarios the animals are actually trained to do a series of simpler responses that combined make impressive behavioral repertoires. Shaping techniques are commonly used to housebreak dogs, one such method is referred to as the newspaper method.

Prenatal and Neonatal Learning

Learning during very early development of animals is an important consideration in terms of animal well-being and economics. Effective maternal, fetal, and neonatal behaviors often depend on the preparation that occurs to ensure appropriate bonding between mother and offspring. Without such preparation, acceptance of the newborn by the dam may not occur, and failure of identification between the two may result in inadequate care of the offspring.

Prenatal Learning

The fetal brain reaches a stage of rapid development well in advance of parturition. In pigs, for example, near peak development as a percentage of fetal weight is approached at least five weeks before birth, with rapid development occurring well in advance of this time. The interaction between dam and fetus may provide early learning experiences that are important to survival following birth. There is evidence that animals may learn vocal characteristics of their dam during the prenatal learning period. Stresses on the dam during pregnancy may have effects on the fetus that are observed later as emotional and social abnormalities. Genetic strains may vary in such influences. The mechanisms involved in these effects are not well understood; however, the observations emphasize the importance of management in ensuring environments that minimize stress. Some suggest that a fetus may learn other sensory characteristics such as odors and tastes; however, such influences are not well established.

Neonatal Learning

Neonate is a descriptive term referring to an animal during a brief period following birth. The length of this period may vary in definition but will generally be considered as the first few weeks or the first month following parturition.

Infant handling or other manipulations such as mild stress have been shown to have favorable effects on such factors as social development and development of organs, immune function, and even growth characteristics. These effects, although demonstrated with laboratory animals, have not been adequately evaluated in farm animal species. Such positive effects are considered to

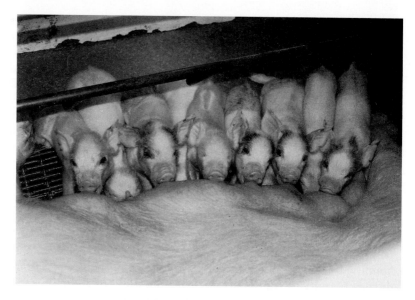

FIGURE 4.7 Pigs form a teat order (a form of dominance order) the first few days after birth by fighting for the most productive teats

be less likely in the precocial species, which suggests that these effects could be less pronounced in farm animals. All of the mentioned factors, however, are of extreme importance in farm animal performance and well-being and should be researched extensively for possible positive or negative effects of neonatal learning in animal management systems.

Learning in the neonate to enhance innate behaviors involved in nursing and following, as well as those that relate to identity of the dam and other offspring, is widely recognized as rapid and efficient. Figure 4.7 illustrates early learning as pigs fight to establish teat order, a form of dominance. Early weaning (at one to three weeks of age) in swine management systems makes early, effective use of housing and feeding equipment essential. Thus, design and adequacy of space in such equipment are critical issues, as are dietary formulation and management. Another example of even earlier weaning is the management of dairy calves, which typically may be separated from the dam and provided milk or a milk replacer diet soon after birth or after a day or two at the latest. Learning to drink from buckets or mechanical nursing equipment is an immediate necessity.

Imprinting and bonding are important in the care and survival of offspring and as a result are important considerations in animal management. Thus, immediate association of dam and offspring after parturition, usually reflected as measures to prevent separation, is a common management goal. In instances where cross-fostering of offspring is desirable, efforts are made to disguise odors and appearances to encourage acceptance by the foster mother. In this case, early association with the birth mother will usually make cross-fostering more difficult. Figure 4.8 illustrates a foal following the dam, a behavior that is

FIGURE 4.8 The foal follows the dam beginning one hour or less following birth; early bonding of the dam and offspring is critical in this following species, where the innate behavior is to follow the mother

initiated within the first hour following birth. In a following species such as the horse, it is critical that the dam and offspring avoid separation. Very early bonding is essential for survival.

COMPARATIVE LEARNING ABILITIES OF SPECIES AND INDIVIDUAL ANIMALS

The interest level in comparative learning ability among species is usually high, but more important in farm animal management is variation in abilities among animals within a species. Well-defined patterns reflect the types of things the various species may learn with reasonable ease or at all. Training techniques vary also among species. Animal caretakers observe such differences; however, there is little information available in terms of heritability of such traits and whether artificial selection is a useful approach to altering the frequency of desirable learning characteristics in the farm species. Issues involved in comparative learning abilities among several species of domestic animals are presented in detail by Hart (1985), along with some interesting examples of experiences in training some species to perform specific types of activities.

Many different comparative measures have been made in attempts to evaluate the relative intelligence of various species. Typical measures have been brain weight, brain volume, brain weight relative to body weight, brain volume relative to body weight, learning rate, the number of trials required to result in operant conditioning, delayed discrimination responses to evaluate effective memory, learning avoidance responses, various approaches to maze negotiation learning, and choice tests. These measures have not provided conclusive answers to the common question about relative species intelligence, yet efforts to rank the species for this characteristic continue.

Individual animals vary in level of predisposition to learning and learning ability varies among species. As an example, Kilgore (1987) ranked several species relative to learning abilities by evaluating their ability to negotiate a maze. Children ranked first, hens last and those ranking intermediate were dogs, cows, pigs, sheep, and goats. Studies of this type are designed to test specific abilities that may relate to learning, memory or other characteristics. Other studies have demonstrated clearly that individuals of the farm animal species utilize memory in discriminating among environmental choices even when sequential choices must be made to achieve a goal.

Although attempts to quantitatively assess intelligence and predisposition to learning are interesting, they are of minimum utility in animal management. The important issue in terms of well-being and animal performance is whether an individual animal can accommodate and make adequate use of the environment provided. Farm animals are generally very capable of learning how to accomplish relatively complicated tasks such as opening feeder covers and various types of feeding gates and identifying which electronic feeding gate or stall is assigned to their individual, routine use. They quickly learn location of utilities such as feed and water. Grazing studies reflect a high level of selective grazing related to nutrient value of the forage and avoidance of toxic plants. The important management consideration is whether all animals are able to accommodate to a given environment and meet the demands of that environment by learning to make effective use of it. Thus, animal well-being as it relates to learning is a matter of sufficient observation by the manager to ensure that the environment provided does not exceed the learning capability of the individual animals. Figure 4.9 illustrates an easy-to-use self-

FIGURE 4.9 A common self-feeder design, with covers for protecting the feed, that requires the pigs to learn to open the lid; most do so readily

feeder that protects the feed with a cover when animals are not feeding. Animals quickly learn to raise these covers to eat. Few fail to do so when the equipment is simple to operate.

MOTIVATION AND BEHAVIOR

Motivation is a process within the brain that determines which and when behaviors will occur (Fraser and Broom, 1990). Goal orientation of animals forms the basis for motivation (Toates, 1987). Motivation is a function of internal stimuli such as physiological parameters and interactions with external stimuli such as those of the various dimensions of the animal's total environment. It is useful to think of motivation resulting from priority establishment among competing goals. An understanding of animal motivation and behavior allows one to more clearly assess the likelihood that certain behaviors will occur. Such assessment is important in effective management and in the design of systems used in all aspects of the livestock industry because ability to achieve important goals is related to well-being.

The fact that the range of possible behaviors is prioritized and that such priorities may be changed by the animal at a given time leads to much inquiry into causes of such changes. Factors resulting in changes in behavior or patterns of behavior may sometimes be referred to as causal factors. Understanding the nature of motivation and factors related to the level of motivation of animals is important in effective management and handling of animals and is, of course, a critical issue in animal training. Recognizing motivational effects of an animal's encounters with both physical and social environments aids in understanding matters related to animal welfare and can assist in understanding the needs of animals.

If motivation is described in its relationship to priority establishment for behavior at a given period of time, then efforts to explain factors that are related to initiation, maintenance, and termination of such behaviors represent major scientific challenges. A variety of factors are clearly involved in establishing the level of motivation associated with a behavior. Some of these factors are described below.

Influences Related to Homeostatic Functions

Physiological systems related to normal function are clearly involved in behaviors such as feeding, drinking, and behavioral characteristics related to regulation of body temperature. These behavioral responses involve both neural and endocrine systems. Responses to external threats and the endocrine systems involved in stress resistance provide good examples of priority motivation. Koolhaas et al. (1997) suggest that homeostatic needs are a major motivation relative to social aggression.

Influences Related to Reproduction

Stages of the reproductive cycle involve major hormonal influences on behavioral priorities. Thus, motivation is influenced dramatically and is clearly related to preservation of genes. Motivational level may change abruptly during mating, as evidenced by a typical refractory period in males following ejaculation. An animal's motivation toward sexual behavior and care of offspring are important in numerous decisions involved in good management of animals.

Influence of Social Characteristics

Social characteristics such as flocking behavior in sheep represent a basis for strong motivation to maintain close association with the flock. Isolation in this case can result in a significant stress level for the individual. Extended isolation, in general, is associated with a variety of aberrant behaviors. Animals within a group are also strongly driven to establish a dominance order. Dominant relationships are major issues in the management of animal enterprises.

Influence of Learning and Training on Behavior

In general, learning and training will shape an animal's behavior to satisfy a basic drive. Experience can result in changes in the behavioral priorities of animals. Experience also increases the ability of animals to perceive the environmental condition in which they exist and to evaluate its possible effects. It allows development of avoidance behaviors based on environmental feedback and may assist the animal in anticipating possible threats such as predation or potential physical injury from things routinely encountered in their environment. Predictive capabilities develop through experience, and animals may have greater levels of stress in environments that hold unpredictable negative threats. Learning also results in certain expectations about the adequacy of the environment. Once such expectations exist, lack of routine, absence of some dimension of the environment, or failure to achieve some expected result causes frustration. This situation may be described as mismatch between the expected and the actual. Thus, learning may create a demand for an environmental characteristic that, if not satisfied, may result in negative effects on the animal.

Hughes and Duncan (1988) provide an interesting concept in explaining the relationship between motivation and the occurrence of aberrant behaviors typically described as stereotypies or repetitive behaviors. These often reflect repeated, incomplete, or fragmented components of normal behaviors. The concept suggests a closed-loop phenomenon in which animals are trapped in an environment that will not permit normal performance or completion of sequential appetitive and consummatory behaviors. Thus, the motivational goal is not met by completing the normal response. Normally the exercise of a consummatory behavior has a negative feedback effect on motivation. As an example, initial sucking behavior of young, which is an appetitive behavior, appears to increase the motivation for this activity toward nursing or drinking of milk, the associated consummatory behavior. Even if milk is consumed by drinking quickly, the animal is likely to engage in nursing behavior by substituting the sheath, ear, or other parts of conspecifics for satisfaction. In this case, even though the food is adequate, the animal will continue this behavior until satiated. The behavior is terminated only when the animal reaches satiation. Other incomplete behaviors such as those that substitute for nest building in hens and sows are examples where even though no nest building materials exist, the animal may go through vacuum behavior patterns similar to those used in nest building. Such relationships underscore the importance of planning animal facilities and systems to avoid the negative effects of an inability to perform some behaviors.

Some ethologists suggest that a more complete understanding of the causes of behavior will eliminate the need for the rather vague term *motivation*, which some conclude is a crutch term describing a collection of phenomena that are not well understood.

TRAINING

Training animals obviously involves motivation. Motivation can be influenced externally through positive and negative reinforcements and punishment. Reinforcements represent the primary approaches to getting animals to do what is desired on command.

Positive reinforcements are rewards administered in response to an animal's accomplishment of a desired act. Such rewards are commonly food items or demonstrated affection and acceptance such as petting and verbalizations that some animals find satisfying. Behaviorists classify rewards as primary and secondary; examples are food and music, respectively. The animal likes food and learns to associate the pleasant experience of food with music in this case. At some point the primary reward may be removed, and the animal will respond to the secondary reward. Rewards are essential in training, and trainers utilize a variety of reward schedules or systems to achieve certain results.

Negative reinforcement involves the withdrawal of some factor that the animal finds objectionable. An aversive stimulus is applied until the animal responds as desired; then the stimulus is removed or terminated as a reward. A shock collar activated by remote control or by underground cable is an example, as is the bit used to control horses. The bit presses on the tongue and jaw until the animal responds, and the pressure is released as a reward. The typical use of a whip or bat in racing is also a form of negative reinforcement. Figure 4.10 illustrates riding equipment, the bridle bit, that serves as an instrument for negative reinforcement. Pressure is released when the animal responds in the desired way, such as changing direction of travel as directed by the rider.

Punishment represents an aversive stimulus applied when an animal is observed doing something that the trainer wishes to prevent. The aversive treatment must be applied immediately or the animal may not associate the punishment with the act. It is easy to confuse punishment and negative reinforcement since the same kinds of acts may be used for both. The bit in the horse's mouth, for example, is used routinely as a negative reinforcer in that removal of the pressure to one side results in the horse turning its head and moving in a specific direction. The desired behavior is the directional movement. Sharp pulling on the reins may also be used as punishment if the horse is doing something undesirable.

FIGURE 4.10 The bit, a part of the bridle, is an instrument used for negative reinforcement

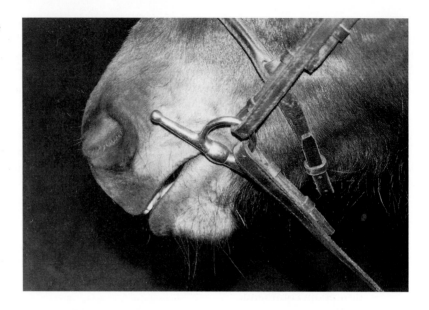

Some examples of reinforcement schedules or systems outlined by Houpt (1991)[1] and Hart (1985) are discussed in the following sections.

Reinforcement Schedules in Animal Training

Fixed-Ratio Reinforcement

A reward is given after a certain number of responses in fixed-ratio reinforcement. The higher the fixed ratio between the number of responses and the reward, the faster the animal will respond. The higher the ratio, the longer the animal will retain the behavior before it becomes extinct when the reward is not given (Houpt, 1991). If the ratio is one to one the reinforcement is referred to as continuous (Hart, 1985).

Fixed-Interval Reinforcement

The animal is rewarded for a response that occurs after a certain period of time has elapsed since the last reward. The response will slow after the reward is given and then increase as the time for the next reward approaches. Thus the animal's behavior relates to the expectation that something will happen about the same time each day, for example. This system assumes that animals have a good sense of timing, which in fact they do (Houpt, 1991).

Progressive Ratio Reinforcement

Progressive ratio reinforcement is a system whereby the animal must respond one time for the first reward and then two times for the second, three times for the third, and so on (Houpt, 1991).

[1]Descriptions of reinforcement schedules are from Houpt, K., *Domestic Animal Behavior,* 2d ed. (Ames, Ia.: Iowa State University Press, 1991), p. 234, 235, by permission

Variable-Interval Reinforcement

In the variable-interval reinforcement system a reward schedule is developed in which the reward is given at variable times following the accomplishment of a desired act (e.g., after one minute, after five minutes, after three minutes, or after seven minutes). Obviously this system makes it very difficult for the animal to predict the timing of the reward, although it knows one will be provided at some point. Behaviors reinforced in this way are more resistant to being extinguished (Houpt, 1991).

An extensive body of published work exists on training procedures for companion animals, horses, and various performing animals. Sources noted above should be consulted.

STAGE OF THE LIFE CYCLE AND BEHAVIOR

Some basic characteristics of behavior related to stage of development are summarized in the following sections. A more detailed discussion of each is presented later under types of behavior.

Behavior of the Fetus

A detailed review of part of the body of data treating fetal behavior is presented by Fraser and Broom (1990). Characteristic movements and positioning during pregnancy, especially those of positioning before the birth process, provide examples of the importance of these behavioral patterns in fetal development, function, and survival during the birth process.

Behavior of the Newborn

Priority behavior is dedicated to survival in a relatively hostile environment. The risk of not receiving adequate nutrition can be high, and vulnerability to predators may be great. An example of a routine behavior pattern of the newborn is the movement of piglets directly to the teat area of the dam by the shortest route. This behavior is an example of one that is termed innate, or genetically fixed.

Animals with advanced systems at birth are termed *precocial.* The farm animal species are considered precocial since they are mobile, are born with their eyes open, and require limited parental care other than providing food and protection. Species born with less well-developed functional systems are referred to as *altricial.* Examples of species classed as altricial are humans, dogs, cats, and monkeys. Newborn of these species require high levels of parental care for extended periods.

Behavior of Neonates and Juveniles

Behavioral development is related to exploration and acquiring the ability to cope in a totally unfamiliar world, physically and socially. Generational diversity in the group influences behavioral development and may relate to the development of abilities associated with exploitation of the set of resources available.

Rate and efficiency of development of functional behavior systems vary with the system. As examples, feeding-related behavior is developed rapidly, whereas social behaviors are developed much more slowly.

STRESS AND BEHAVIOR

Behavior is generally defined as some action or pattern of action in response to a stimulus. Other attempts to define the term suggest that behavior is an action resulting from interaction with the animal's environment. Thus, behavioral and physiological responses to stressful stimuli occur. Both are involved in coping and ultimately in determining whether an animal is disadvantaged; they may even determine survival. Both are important in evolutionary processes. Elimination of animals that cannot successfully cope with the environment in favor of those that can represents survival of the fittest and moves the population to a higher level of adaptation to a particular set of environmental characteristics. Figure 4.11 illustrates effects of boredom or a barren environment. The pig exhibiting bar-biting behavior may reduce or eliminate this response if, for example, limited amounts of straw are provided for the animal to manipulate.

Genetic Effects on Behavioral Responses to Stress

Evolution results in long-term adaptation of animals physiologically and behaviorally to certain environments. Survival of the fittest is the net effect. Shorter-term genetic effects may be produced by artificial selection, practiced in planned breeding

FIGURE 4.11 The bar-biting behavior is associated with a barren and restricted environment, including feed restriction

programs. An example of evolutionary effects in the adaptation of farm animals are cattle of Zebu breeding that evolved for many generations to be more resistant to heat and some insects. This adaptation has prepared them for hot, tropical climates. In the shorter term, selective breeding has been practiced in an attempt to render the animal more docile behaviorally, since some lines of these animals were typically difficult to handle. Thus the threshold for aggression may have been increased in a relatively shorter time than the development of effective heat and insect resistance. During this period it was important to retain the animals' climatic adaptability characteristics.

In an evolutionary sense, if animals cannot successfully cope with an environment they fail reproductively, and the population with coping deficiencies disappears. In the case of artificial selection, animals that do not perform as expected are usually eliminated by human decision to cull the individual, which prevents future reproduction by that animal. Thus artificial selection becomes a part of the evolutionary process. Because animals are not selected for all possible traits relating to successful interaction with the environment, natural selection continues at some level. Regardless of whether selection occurs naturally or by human design, the result is that changes in neural-physiological processes controlling both behavior and internal biological phenomena occur in animals.

Domesticated species, for the most part, have been selected over many years, intentionally or otherwise, to be more docile and more comfortable in association with humans and other animal species. Consequently, the level of stress associated with being near people has been greatly reduced. It still exists, however, as demonstrated by domestic animals that grow up in a wild state without learning that humans may pose little or no threat. Such animals will have a greatly increased flight distance because they may view humans and some other animals as predators.

Because the state of an animal at any given time represents both genetic and environmental influences, enlightened animal management must understand principles of heritability. The component of a phenotypic variable that is inherited or the component that results from environmental influences is expressed as a fraction of total variation. For example, the variation in aggressive character is thought to be about 0.20 (20 percent) heritable and 0.80 (80 percent) due to environmental influences. This low level of heritability would suggest that progress through selective breeding for docility would be slow but possible. If this heritability estimate were 80 percent (genetic and environmental influences reversed in this example), progress in reducing aggressiveness by selective breeding could be rapid assuming a useful selection differential.

Environmental Effects on Behavioral Responses to Stress

The environment in which an animal exists places demands that require the exercise of coping mechanisms. Developmental (ontogenetic) processes influence how animals are able to cope with various environmental demands. If these demands are novel and new, stress may be more likely. As animals learn to cope, the threshold level for triggering a stress response may be altered.

Learning

Learning prepares animals for later events. Social skills are examples of learning. If animals are deprived of learning in early life, they tend to be less explorative, have lower learning capacity, and have greatly diminished capacity to detect and deal with contingencies. This situation leads to serious coping problems in later life and therefore has special implications in livestock production-management systems that differ greatly from the optimum. The work of Harlow and Harlow (1962) with neonatal monkeys provides a classic example of impaired social and psychological development when young animals are deprived of maternal interaction.

Types of learning experiences that animals encounter may have different effects as to the level of stress associated with them. For example, animals may be able to predict certain types of events but be unable to exercise any control. A challenge by a dominant conspecific or appearance of a predator (at certain locations or at certain times) are examples of only predictable events. Controllable situations, with learning, pose little or no uncertainty. Examples of controllable factors are knowledge of where to go to get food or water and that successful dominance eliminates significant risk from a competing conspecific. Thus, the animal can learn to cope by controlling or by avoiding the threat in some cases (see Toates, 1987).

Changes in the levels of predictability and controllability represent potential stress conditions because of the associated uncertainty.

Behavioral Constraints

There is evidence that animals are motivated to accomplish tasks and achieve certain end results. Consequently, they are described as being goal oriented. Animals will try to learn the process for achieving certain goals. If achievement of the goal is repeatedly thwarted, frustration results and the condition is stressful. Such stress may result in behaviors that are different than those normally observed for such animals. Examples of such behaviors are redirected, displacement, or highly repetitive stereotyped activities that will be discussed later under types of behavior.

As mentioned earlier, uncertainty can be a stressor for animals. Decreases in either predictability or controllability present the animal with a confusing situation, referred to as the *mismatch concept*. Based on the fact that animals have certain expectations, mismatch results from the failure of expected events to occur. Animals apparently have some inherent or learned standards by which to judge their circumstances. Examples are spatial requirements and environmental temperature comfort zones. If such standards are not met by the environment, the mismatch occurs as a stressor. In some cases, learning will modify expectations over time and may reduce the stress potential.

Conflict

Another environmental effect commonly leading to stress in animals is conflict. Such effects may appear in typical relationships such as maintenance or acquisition of a dominant position, territorial defense, learning difficulty (complexity of the task exceeds the ability of the individual), and lack of success.

Unresolved conflict may be a chronic stressor that results in a variety of behavioral problems such as redirected aggression and some stereotyped behaviors.

The preceding points demonstrate some relationships between stress and behavior. It is important to recognize that the neural and metabolic effects of those stressors causing such behavioral patterns are influencing the same neural and metabolic processes previously discussed as the SA and HPA stress-response systems. For example, the release of increasing amounts of ACTH associated with stressful environments prompts increased corticosteroid production by the adrenal cortex, and if the stress is sustained beyond the animal's ability to respond satisfactorily, the system will be exhausted and the animal will likely suffer important negative effects. An example of an important negative influence in animal production is that extended high levels of corticosteroids may inhibit gonadotropin effects and result in reproductive failure. A second example is the negative effect of sustained high levels of corticosteriods on immune function. These effects of chronic stress have long been recognized and may have extreme consequences for the economics of livestock enterprises.

Examples of Behavior Associated with Stressful Environments

A variety of behaviors may be associated with environments that cause the animal to become frustrated in accomplishing normal activities. An animal's behaviors are typically goal oriented, and aberrant behaviors may result from the inability to achieve related goals. Some behaviors are associated with aversive stimulation and may involve the discharge of some form of aggression. Examples of some behaviors related to stressful conditions follow.

Redirected activities apparently provide some form of important behavioral outlet. Such behaviors are appropriate for the context but directed toward an inappropriate stimulus. Feather pecking is an example of redirection of actions normally associated with food pecking.

Displacement behaviors appear irrelevant to the situation. Some may be performed incompletely. Sham chewing in sows is thought to result from incomplete satisfaction of the animal's innate drive for the normally longer periods involved in foraging and consuming food. Ranting and biting by confined sows may be due to an inability to accomplish nest building, which is a normal activity before parturition. Nest building is a strong, innate (fixed-pattern) behavior associated with preparation for the birth process and the related hormonal changes that subsequently occur.

Stereotypies are behaviors that show repetitive patterns of activity such as route tracing in caged animals. Some animals confined in zoos demonstrate this behavior. Stallions may exhibit route tracing in paddocks (fig. 4.12); bar biting by sows is an additional example. Such behaviors may appear to have no obvious function and to be out of context in a given environmental situation. A specific definition of *stereotypies* is difficult because such behaviors do not have clearly delineated borders (Mason, 1993). However, the consensual definition of *stereotypies* in the ethological and animal welfare literature suggests behaviors that are unvarying and repetitive patterns that have no obvious goal or function (Mason, 1993). Perhaps more important than definition is that occurrence of stereotypies reflects

FIGURE 4.12
Route tracing, a stereotypic behavior, is evidenced by the path followed repeatedly by this stallion

serious shortcomings in the environment, including housing during development, although the effect of environment is also strongly influenced by the coping capabilities of the individuals involved (Wiepkema, 1993).

Common Behavioral Signs of the Presence of Stress

Numerous behaviors reflecting stress conditions are easily detected. Examples include the following:

Reduced appetite	Dullness
Unusual stance	Self-isolation
Irregular gait or unusual movement	Elevated respiration rate
Reproductive failure	Restlessness
Increased aggression	Unusual level of vocalization

EMOTIONS AND BEHAVIOR

The question of existence of emotions in animals is one commonly debated. That some responses of animals appear to be so similar to those of humans suggests the possible existence of animal emotions such as pleasure and fear (see Duncan, 1995).

One rationalization favoring the existence of emotions in animals might be that all vertebrates have great similarities in anatomy and function in a variety of tissues and systems involved in stress-related phenomena. Therefore, one might conclude that animals, like humans, have emotions, and thus emotional reactions would be a part of the animals' behavioral repertoire.

Price (1992) cautions against anthropomorphism or ascribing human traits and feelings to animals in assessing emotional criteria. For example, judging that a dog is happy by the fact that it jumps, licks, and dashes about is a human perspective of happiness and not necessarily the dog's view. Some conclude that the dog's performance, in this case, is one that it has learned and will prompt a desirable response on the part of the people with whom it associates. The action may result in some reward such as petting, feeding, playing, or other actions from which the animal derives pleasure.

HORMONES AND CONTROL OF BEHAVIOR

A few examples provide an introduction to the concept of hormonal involvement in behavior. More specific details on some of these relationships are reviewed elsewhere.

Hormone	*Related behavior*
Testosterone	Reproductive behavior and aggression in males
Estrogen	Estrus-related behaviors
Epinephrine	Fight-flight syndrome characteristics
Oxytocin	Maternal behavior in sheep
Prolactin	Maternal behavior in birds

INNERVATION OF BEHAVIORAL STRESS RESPONSES

Concepts relative to the control of behavior must include influences of both genetics and the animal's environment. Both play interactive roles; however, some behaviors are influenced more by one or the other of these factors. Some term behaviors assumed to be largely due to genetics as innate. Environmental influences are experiential and represent the result of learning. Some ethologists speak of genetic influences as "nature" and environmental influences as "nurture." Thus the comparative terminology often becomes genetic and environmental or nature and nurture in speaking of these relative influences on behavior.

Neural development supporting innate behaviors, those fully functional at birth, necessitate the development of related neurons, synapses and tracts of the central nervous system *in utero.* Such systems develop in the uterine environment with limited influence by the external environment. Exceptions are effects of sounds and some biochemical components of the dam's blood, both of which may be important later in parent-offspring bonding. Dam-fetal stress may also influence neural development *in utero.* Systems supporting behaviors that develop following birth are influenced by the myriad of environmental experiences that result in learning. This results in the formation of neurons, interneurons and synapses supporting behaviors developing in the neonate and later in life. Pinel (1993) provides a discussion on the formation of neural anatomy relative to behavioral development and function.

Another important concept in understanding the development of behavior is the recognition that certain behaviors and behavioral patterns develop as a result of the growth and completion of extensive neuron connections and complexes during the critical period for that development. If these complexes are not developed during this period, they may not occur normally and the associated behavior may not develop, at least not in the normal way. The unused neurons degenerate. This phenomenon provides a partial explanation as to why young animals that are reared in isolation may not develop normal socially oriented behaviors.

Behaviors thought to be largely instinctive, some of which may be referred to as fixed-action patterns, are generally characteristic of a species. Other behaviors may appear to have highly repetitive, patterned characteristics but may actually be acquired through environmental influences. Thus, an understanding of the animal's ethogram or specific behavioral development patterns is helpful in making such distinctions.

Innervation Routes and Characteristics

The route of behavioral innervation lies first in the capability of receptors to convert physical and chemical stimuli to neural energy and transmit these impressions to the central nervous system. These stimuli may initiate reflex responses, requiring transmission to and from the spinal nerve tracts without involvement of the brain in the stimulation of effector tissues such as muscles or sweat glands. Such mechanisms are referred to as spinal reflexes. Other responses require the involvement of the brain to coordinate an appropriate response by the tissues involved. Some responses may involve the limbic system, whereas others may include influences by the cerebral cortex in the generation of the necessary response. Regardless of the response avenue, however, the receptor triggers a neural response that is based on neurotransmitters and related motor or neuroendocrine function. All behavior occurs as a result of physiological phenomena, even though the stimulus may be of a psychological origin, or a mental perception of condition. Such perceptions may give rise to responses to environmental inadequacies of a social or physical nature, explaining the similarity of neuroendocrine stress responses to a wide variety of stressors. Action patterns that have been observed when the organism's brain connections have been severed may result from a phenomenon referred to as central pattern generation. Behavioral control of such action patterns is assumed to generate from the development of neural systems (neuron connections) outside of the brain that are more complex than the usual spinal reflex responses. Such phenomena have been observed in the praying mantis and result in mating behavior even though the male's brain is removed and eaten by the female. It is not clear whether such control exists in mammals.

Behavioral Response Initiation Avenues

A response to an environmental event requires that receptors react to such events by initiating appropriate behavioral actions. Behavioral responses to external events require innervation through receptors involved in the animal's communications systems. Animal management often utilizes such capability in devising animal systems for production and handling. Specific sensory capabilities are reviewed in a later chapter.

Visual
Field of vision is well established for the various species. Less concrete evidence is available regarding the level of resolution and color perception. It is clear, however,

that many species have anatomical structures in the eye similar to those of humans to accommodate color vision (Jacobs, 1993). Visual communication is clearly a factor in many aspects of animal behavior (e.g., leadership, grouping, and displays related to reproduction).

Auditory

Auditory stimuli are evidenced by common signals (e.g., alarm signals, nursing signals, and distress signals) between mother and offspring.

Misconceptions exist in relation to playing music or having a radio on in animal quarters. Rather than providing a relaxing atmosphere, positive results may be related to reducing the surprise of other sounds that occur in the area.

Extremely loud sounds over an extended period may reduce animal performance. Lesser sounds such as plane flight and sonic booms appear to be within a range to allow easy adaptation (habituation).

Species differences in sensitivity to frequency level are established. In general, animals have sensitivity levels far above those of humans (Hefner and Hefner, 1992).

Attempts are made to classify sounds of various species. In some cases patterns are well established, but efforts to determine some type of specific language are as yet unrewarded.

Olfactory

The ability of animals to communicate through odor marking (e.g., territorial marking by urine or glandular secretions and depositing urine as a reflection of stage of the reproductive cycle) is well established.

Odor is a factor in individual animal recognition and is thus heavily involved in several aspects of social behavior among animals. Birds, in general, appear to make little use of smell in behavioral patterns.

Because the mentioned routes involved in initiating behavioral responses represent communication media (i.e., sights, sounds, and odors), these characteristics are commonly used in animal management and some sporting activities involving animals. Examples of such practices include the following:

Use of boar odor to identify sows in estrus (an attraction syndrome and an immobile response characteristic only of those in estrus).

Use of boar odor to indicate the presence of boars to develop receptivity in sows.

Use of animal secretory and excretory odors in trapping.

Presence of a strange male to stimulate and synchronize estrus in some breeds of sheep.

Use of odor of deceased offspring or odor of the new lamb on young to be cross-fostered.

Dramatic displays in wild birds and mammals in reproductive attraction.

Marking trees or other objects as a means of conveying danger or social attraction.

Calling sounds used in hunting some birds and animals.

Predator odor as a means of avoidance response.

Commands between humans and other animals (works both ways: humans respond to certain signals sent by pets, for example).
Caged decoys for attracting fish.

Homeostatic Influences

Examples of homeostatic influences on behavior are body temperature and a variety of blood chemistry parameters. To prevent imbalances, animals will attempt to respond voluntarily to challenges by seeking shade, windbreaks, water, feed, or other life support needs. Thus, some homeostatic influences translate to voluntary behavioral responses that make important contributions to the animal's well-being.

Homeostatic influences are mediated by a variety of sensors located in both the central and peripheral nervous systems. For example, thermoreceptors in the hypothalamus detect changes in blood temperature, osmoreceptors monitor osmotic pressure, and pressure receptors monitor blood pressure. Such receptors alert response systems to attempt to restore normal parameters to maintain function and life.

Social-Psychological Influences

Animals have certain tolerances in terms of spatial intrusion, competition, boredom, and dominance, for example. If this tolerance level is reached for extended periods of time, the challenge serves as innervation for the development of stress responses including disorders having a psychological basis.

NEUROCHEMICAL CONTROL OF BEHAVIOR

The preceding discussion outlines the roles of genetics and the environment in which the animal exists as factors in the development and execution of the animal's behavior. The field of neurobiology considers the neurochemical control of behavior by treating neural activity that occurs in response to a stimulus. Included are the mechanisms by which the neural activity develops and is transmitted by neurochemical pathways to effector tissues involved in the response. The important basic information involves receptors that acknowledge and mediate the signals and the anatomical components involved, pathways of neural signals, and various transmitters involved in the response that represent the chemical means of transfer between neural cells and target muscle cells (motor responses by smooth or striated muscle) or secretory cells (humoral responses). The science relating neurobiology to behavior is sometimes referred to as *neuroethology*.

It is conceivable that virtually all components of the central nervous system are involved in behavioral responses because behavior involves both cognitive and noncognitive processes. The cerebral cortex receives the variety of sensory stimuli arising from olfactory, tactile, visual, and auditory receptors. This body serves a cognitive role in analysis of inputs and alternative responses and directs neural communication with other parts of the central nervous system for responses.

The group of central nervous system structures referred to as the limbic system contributes to behavioral responses. This system is the central nervous system component of the autonomic nervous system and is therefore heavily involved in home-ostatic regulatory processes, as well as several behavioral responses demonstrated by animals. This system includes portions of the cerebral cortex, as well as the thalamus, hypothalamus, hippocampus, and amygdala and portions of the balance of the brain-stem. Motivated behaviors such as feeding, sexual behavior, fleeing, and aggression are influenced to a great degree by the limbic system. Emotional behaviors and gen-eral feelings of well-being involve the hippocampus and amygdala, as well as cogni-tive centers. Feeding behavior is controlled by feeding centers in the hypothalamus.

Body movements, which are also components of behavior patterns, are con-trolled to a great extent by the large nuclei of the brain referred to as the basal gan-glia, the cerebellum, and the cortical motor centers.

Neural signals are moved from higher centers of the central nervous system through the various spinal cord tracts to other neurons and to effector tissues by efferent neurons. Neural impulses returning from the periphery to the central ner-vous system are conducted by afferent neurons.

All nerve impulses involved in the response systems are transmitted from one nerve cell to another and to effector tissues by chemicals at synaptic junctions with the target cells. The chemical compounds making or influencing these transfers are referred to as neurotransmitters. The nerve cell generally is the site of synthesis and release of neurotransmitters, which are described in the following sections.

Acetylcholine

Acetylcholine is produced by preganglionic sympathetic neurons and both pre- and postganglionic cells of the parasympathetic nervous system. It is present in both central and peripheral neural tissue.

Norepinephrine

Norepinephrine is produced by postganglionic cells of the sympathetic nervous system and by the adrenal medulla. It is also produced in central neural tissue.

Epinephrine

Epinephrine is produced predominantly by the adrenal medulla and also in central neural tissue. Effects are discussed as a component of the SA stress response system.

Serotonin

Serotonin is produced by the Raphe nuclei and possibly by other central neural tis-sue. It is present in the pineal gland as an intermediate in the synthesis of melatonin.

Dopamine

Dopamine is produced in the area of the substantia nigra and other central neural tissues. Dopamine is an intermediate in the synthesis of epinephrine and norepinephrine.

Amino Acids

Aspartic acid, glycine, gamma amino butyric acid (GABA), and glutamic acid may function as neurotransmitters in the brain or endocrine tissue located away from the central nervous system.

Opoid Peptides

Some peptides, referred to as neuropeptides, act as neuromodulators and therefore may be involved in behavioral responses. Beta-endorphin, dynorphin, and the enkephalins influence opiate receptors in the central nervous system and as a result may cause an analgesic effect by blocking or partially blocking pain-related neural signals from reaching centers recognizing and reacting to such stimuli.

Neurohormones

Neuropeptides are produced and found widely in brain tissue and generally in the circulatory system. Oxytocin, the pituitary peptide that stimulates the release of milk (also commonly referred to as milk letdown) and is associated with maternal behavior, is also classed as a neuropeptide, as is cholecystokinin. The latter, gut peptide, has been typically considered a hormone (produced by the duodenum) involved in feeding behavior. These two examples indicate that continuing research alters views as to function and influence of various chemical messengers.

CHAPTER 5

Types of Animal Behavior

SOCIALLY ORIENTED BEHAVIORS AND CHARACTERISTICS

Social Organization

In many cases considerations in managing animals correctly give primary attention to the individual animal. Social behavior such as organizational structure and other relationships among animals and with humans are frequently either not considered or given little attention in developing facilities and management systems. Current production practices tend to group animals into large herds or flocks. These groups are usually quite uniform, consisting of similar-sized animals in same-sex groups. Groups are purposely formed in this manner to reduce sexual activity and to allow animals to be more competitive in attaining food. However, groupings of large size may increase aggressive interactions because it is more difficult for the animals to establish a stable dominance hierarchy.

Farm animal species tend to be social; therefore, such relationships are important issues in planning to meet the needs of both animals and workers associated with them. These species, in general, tend to form groupings referred to as herds, flocks, or bands. When great amounts of space are available, animals tend to form subgroups, each having an established social order. The area occupied by a group may be referred to as *home range*. This area is fully explored for the resources that are used to maintain the group and to a great extent represents the area used by the group unless some event or lack of resources stimulates a relocation.

High population density in grouping results in frequent space violation of animals, and dominant-subordinate interactions help to lend stability in the group. For this to occur, animals must recognize individuals and recall social (dominance order) position and previous encounters that have established the existing dominance order. Social dominance can be an important management problem in high-density systems because a few dominant individuals may control the feed, water, and available space. Mixing of unfamiliar animals always results in the reestablishment of the dominance order with associated agnostic interactions; thus, it is obvious that frequent mixing and remixing will result in negative effects on animal performance.

An important dimension of group behavior is communication among the individuals. Typical communication characteristics for the major species are established. Sheep, cattle, and horses maintain visual contact and will on occasion communicate by sound. Frequency of vocal communication by these species is variable, depending on environmental influences such as those associated with fear, separation from a group, or maternal and reproductive behaviors. Swine try to maintain visual contact and are in constant auditory contact by a variety of grunts. Body contact appears to be a more important component of communication for swine than for cattle (with the exception of Zebu) and sheep. Communications may relate to interactions among individuals, group activities, or warnings of danger. Responses to warnings have some degree of species specificity. For example, upon disturbance sheep will form a tight group, then flee. Cattle and pigs are more likely to scatter and move to escape individually or in small groups. Horses may group or flee individually and with vocalization.

Types of Animal Grouping

Aggregation

Aggregation describes a group of animals that come together because individuals are attracted to the same thing. Assembling at watering sites and salt licks are examples.

Society

Societies vary in their cohesiveness but are the result of animals functioning together, which enhances the effectiveness of the group. Hunting in packs and watching for predators, and associated warning of the group, are examples of benefits of cooperation. Some level of cohesion exists in maintaining the group, but the group may change depending on needs. Individuals tend to remain with the group, and the group resists entry of other individuals even though they may not be conspicuously different. Certain activities may be shared by individuals even to the extent of specialization. Examples are leadership of movement, initiation of activities, and imposing a level of control as to what a group does. Animals serving as leaders, initiators, and controllers are not necessarily the same individuals, however.

The existence of a social grouping does not eliminate competition among individual animals, and a typical hierarchy or dominance order exists. This dominance order contributes to stability of the group and may also control relative access to resources if they are of a controllable nature.

General Types of Interactions among Animals

Animal interactions, like those of humans, may have characteristic purposes such as cooperation, selfishness, spite, or altruism. Other purposes may relate to control of other animals, control of space and other resources, and maintenance of social stability by minimizing negative effects of aggression among animals in a group. Some typical interactions among animals are discussed in the following sections.

Space-Oriented Relationships

Personal space requirements reflect characteristics of a given species and the stage in the individual's life cycle or purpose at a particular time. Animals apparently view space as an important resource and will compete readily for this component of environmental resources when limited. This consideration is an important dimension of designing the physical and social environments of animals. A more detailed discussion of this topic appears later in relation to spacing behavior.

Leadership and Control

Influence over other animals is reflected in behaviors that provide leadership, exert dominance, or otherwise control the behavior of others. Initiation of activities (e.g., grazing, resting, and signaling of danger) may, however, be stimulated by a leader that is not necessarily the dominant individual in the group. Control of activities may be exerted by animals other than the leader. Individuals within the group may be influenced by a controller that signals approval by following the leader in starting an activity or in beginning the relocation of a group of animals.

Facilitation

A common behavior among animals, facilitation takes many forms but generally enhances the welfare of the group. A herd's move to improved grazing benefits all individuals, particularly those lower in the dominance order. Such movement probably resulted from the fact that some animals in the group were effective in exploring for better resources or more experienced as to location of a better environment at a given time. In animal management, attention to this characteristic may be an approach to enhanced performance. Animals may learn to be more effective in a group environment and as a result benefit from such behavior. An example is that a group of animals in a defensive behavioral pattern protects some individuals from exposure to an attacker. Another example is that greater feed consumption may occur among individuals in a group of steers in the feedlot than is typically true if animals are fed individually. Figure 5.1 illustrates feeding competition among pigs. In contrast to the greater feed consumption of group-fed steers, research has shown (Gonyou et al., 1992) that individually housed swine gain weight more rapidly than group-housed swine. However, in studies noting an increased gain for individually housed swine, the pens are open such that the individual may see and hear its neighbor eating. It is believed that the lower rate of gain for group-housed swine is because of aggressive interactions. Therefore, if swine were housed in groups with much more available pen and feeder space, differences in rate of gain may disappear.

Cooperation

Animal behavior frequently has a cooperative character in which collaboration enhances the probability of success in activities. Resisting predators, mutual grooming, and the opposing stance of horses, in which tail switching aids in fly removal from the face of another, are examples. This stance in horses may also be observed

FIGURE 5.1 Pigs eating from the feeder may stimulate other pigs to approach and eat from the feeder; however, aggressive interactions may limit weight gain for the group

in mutual grooming (fig. 5.2). Cows commonly serve as baby-sitters for a group of calves while the other mothers range over a wider area to graze (fig. 5.3).

Association

Associations among individual animals develop as a part of their routine behavioral repertoire. Pairing of individuals is observed in many groups of animals and is one form of association. Another example is the formation of bachelor groups of males. These examples suggest that some animals simply develop preferences for close association. Animals in a particular group develop bonds and will limit admission of unfamiliar animals or groups of animals. Such behavior to limit admission requires effective individual recognition. This ability is well developed in the farm animal species. Limitations that exist in confined systems make avoidance by the submissive individual difficult. Effective management of livestock enterprises requires attention to group size as it relates to the ability of animals to maintain recognition of group members and the established social order. Evidence as to optimum group size in such situations is limited, but some research and practical observations suggest that effective recognition prevails in groups of swine on the order of twenty to forty individuals and possibly up to one hundred, in the case of cattle. However, cattle in dairies and feedlots are commonly maintained in groups larger than these with manageable levels of aggression when introduction of new animals is minimized. Group size is tending to increase in swine operations as a result of management procedures that prevent mixing of unfamiliar pigs. Group size exceeding the animal's ability to recognize individuals may lead to an increased intensity of agonistic behavior and as a result group instability. Such

FIGURE 5.2 Horses engaging in mutual grooming in which each individual is scratching the back of the other; a similar stance is seen when horses stand parallel so that switching the tail of one brushes flies from the face of the other

FIGURE 5.3 A form of cooperation is shown by the cow tending a group of calves as a calf-sitter while other cows graze over a wider area

instability may increase injury level in the group and have negative impacts on the performance and economics of the enterprise, as well as reduced animal well-being.

Altruism

Altruism represents acts that may enhance the fitness of the animal to which the behavior is directed and may reduce the fitness of the contributing individual. Maternal behavior designed to protect the offspring is an example. Sows may also attempt to protect litters other than their own, and stallions will place themselves in danger to protect the band of mares.

Selfishness

An animal may engage in behavior that reduces the fitness of another by restricting access to important resources. The successful animal, in exercising such restriction, may enhance its own fitness as a result. Limiting access to feed and water is an example of such behavior that warrants effective management to minimize negative effects. Proper design of facilities and equipment is often critical in preventing negative effects of restricted access to resources.

Establishment of a Social Order

Animals within a group form a dominance order that once established provides a high level of social stability. This stability limits potential costs of fighting in terms of injury and inefficiency in resource use. The establishment of the order results from learning in agonistic behavioral encounters (aggression and submission) and the animal's ability to remember and recognize the identity of others. Figure 5.4 reflects a series of agonistic encounters in cattle that results in the establishment of a dominance order (Hafez and Bouissou, 1975). The hierarchy is established while a group is being formed or when new animals are introduced into a group. The order remains fairly stable thereafter, although there is a continuing level of testing among individuals and over time the order may change as a result.

A major result of a dominance order is the establishment of individual priority access to resources available to a group of animals. In some cases groups of animals will establish a dominance order and related priority access to resources that are controllable, such as a watering pond. This characteristic among individuals presents commonly encountered management problems in terms of uniformity of access and in turn uniformity of performance among individuals within the group. The management goal, then, is to utilize systems that provide security of position for feeding and watering and for adequate resting space. This encourages the highest level of social stability and minimizes waste in energy and other losses associated with excessive agonistic behavior. A settled social environment contributes to well-being.

In general, dominance encounters among individuals in a group consist of signals or threats of aggression, so that actual aggressive attack and fighting are kept

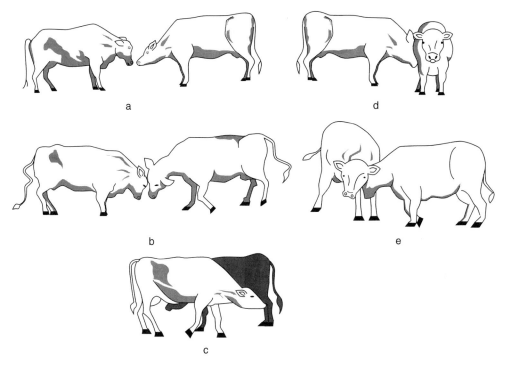

FIGURE 5.4 Agonistic encounters (aggressive and submissive) between cows that will result in the establishment of a dominance order. (*a*) Cows meeting after an active approach. (*b*) Physical combat or fight—the cows push against each other head to head, each striving for a flank position. (*c*) The clinch—one contestant of an evenly matched pair slips alongside the other; the head of the former is pushed between the hind leg and the udder of the latter. In unusually prolonged contests, the combatants rest briefly in the clinch between bouts of head sparring. (*d*) Flank attack—the animal that gains a flank position is at a decided advantage over the other. The flanked animal either submits, flees, or strives to regain the head-to-head position. (*e*) The butt—a dominant animal directs an attack against the neck, shoulders, flank, or rump of the subordinate, which in turn submits and avoids the aggressor. (*Source:* Hafez, E.S.E., and M.F. Bouissou, The behavior of cattle. In *The Behavior of Domestic Animals* [London: Bailliere Tindall, 3d ed., 1975], p.227, by permission)

to a minimum. Thus, routine testing of a relatively mild nature occurs frequently and maintains the order until it is altered by removal or incapacitation of a dominant individual or until a subordinate makes a successful challenge.

Types of Social Hierarchies or Peck Order

Dominance hierarchy is a rank order of individuals in a social unit. Hierarchies may be unidirectional or bidirectional. In groups of farm animals, complex directional arrangements are the norm. Figure 5.5 from Lynch et al. (1992) illustrates an example of dominance hierarchies.

Unidirectional (Linear) Order

In a linear order aggression goes in one direction only to a subordinate individual. Most animal hierarchies are more complex than a unidirectional order.

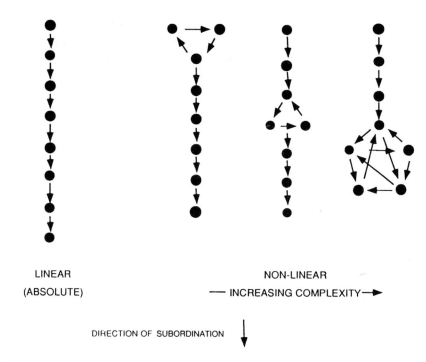

LINEAR

(ABSOLUTE)

NON-LINEAR

— INCREASING COMPLEXITY ➡

DIRECTION OF SUBORDINATION ↓

FIGURE 5.5 Simple hierarchies in groups of animals (*Source:* Lynch, J.J., G.N. Hinch, and D.B. Adams, *The Behavior of Sheep* [Wallingford, UK: CAB International, 1992], p.65, by permission)

Bidirectional Order

In bidirectional order, a subordinate animal directs aggression toward a dominant animal occasionally as a means of testing the resolve of the latter. The dominant animal responds in a manner to retain its position or shows some indication that its position may be in question in which case the challenge will continue.

Complex Orders

Complex systems are commonly described as triangular, square, parallel within a unidirectional system, and despotic. Visualizing these orders suggests the complexity of hierarchical arrangements characteristic of farm animals. The important thing in management, however, is to observe animals adequately to determine whether individuals are disadvantaged by the dominance order that is developing or exists in the group.

During formation of a hierarchy, fighting normally occurs. With time, threat and avoidance will replace fighting; thus, animals convert from demanding physical encounters to those of relatively low physical demands in terms of energy and hazardous exposure.

Increasing group size usually extends the time required for formation of the order. The longer period is due to the increasing difficulty animals have in recognizing other individuals in the group, the complexity of the order being formed, and the number of interactions involved. Recognizing the complexity of working through this combination of factors and the limited amount of time required in most animal groupings to establish a hierarchy reinforces respect for the cognitive ability of animals.

Tests to Assess Hierarchy

Paired Tests Two animals are placed in a competitive situation, and it is determined which obtains control of the test resource. All pairs in a test group must be tested. For example, if five animals are to be evaluated, ten tests are required. The number of tests required is calculated as follows:

$$X\left(\frac{x-1}{2}\right)$$

Group Tests All animals in a group are allowed to compete for the test resource. A commonly used test resource is food. Another approach is to simply watch for agonistic encounters during the animal's normal activity within a group and record the results.

Effects of Dominance

If resources are limited, dominant animals continue to do well while subordinates do less well. This is particularly true if the resources involved are defendable (e.g., sows are able to defend a feed trough but cattle cannot defend an entire pasture).

If a resource is not limited or defendable, dominance will have little effect. The stress of maintaining the social hierarchy does exist, however. The middle positions in a dominance hierarchy may be most stressful, because the individual may be attempting to move up in the order but must also be concerned with challenges by those immediately below it in the system.

General Characteristics Related to Establishment of Dominance among Animals

Males are usually dominant to females and both are dominant to juvenile animals.

Certain strains of animals may tend to be dominant to others.

Older animals are usually dominant to the younger until age impairs physical capability.

Larger animals are generally dominant; however, if the animal learns to be subordinate to a larger individual this position may continue even though relative size is reversed due to growth.

Stronger animals are dominant.

Animals with specific display features, such as horns, are dominant.

Injections of androgens, if limiting, typically increase the status of an animal, such as castrated males.

An animal in familiar surroundings tends to be dominant over new arrivals.

Animals among already familiar animals may dominate a newcomer.
Animals with a winning attitude, from having just won a previous fight, tend to have a favorable situation for achieving dominance.

Communication and Sensory Capabilities

Communications are critical to animals' social behavior, as well as their ability to maintain a positive relationship with the environment in which they exist. Thus, consideration of communications characteristics and sensory capabilities of the species is important in understanding animal behavior and the relationships these may have to management.

All animals communicate by some combination of visual, auditory, and olfactory (chemical transmission) means and through physical contact. Such communication is critical in the survival of individuals and the species because it relates to protection, reproduction, maternal behavior, and learning.

Heffner and Heffner (1992) provide useful auditory characteristic comparisons of the various farm animal species and humans in the audiograms shown in figure 5.6. These comparisons reflect limits in terms of low and high frequencies, along with best frequency and best sensitivity for each. In general, these authors conclude that the hoofed animals have hearing sensitivities very nearly that of humans at lower frequencies but significantly greater at higher frequencies. Best sensitivity for humans is about -10 dB, and best frequency is about 4 kHz. The high-frequency limit is about 17 kHz for humans, whereas high-frequency limits for each of the farm animal species are considerably higher. The low-frequency limit is about 32 Hz for humans, whereas respective values for horses, swine, and cattle are 60, 55, and 25 Hz. Low-frequency limits for sheep and goats appear to be in the area of 125 and 70 Hz, respectively. Low-frequency limit is the lowest frequency sensed at 60 dB. The high-frequency limit is the highest frequency sensed at 60 dB. The best level frequency represents that graphed at the lowest decibel threshold. Figure 5.6 shows the auditory characteristics of some farm animals, as estimated from Heffner and Heffner (1992). The auditory ranges quoted are from the low-frequency limit to the high-frequency limit, with each determined at 60 dB.

Visual capability of animals varies based on the anatomical structure of the eye and eye placement which determines the degree of binocular vision. Horses, cattle, sheep, and chickens have wide eye placement which limits binocular vision. Thus, they may have difficulty seeing strange objects at ground level in their pathway. Eyes of swine are more forward, providing better binocular vision; however, some genetic lines may have sight restricted by large, drooping ears.

Swine, cattle, sheep, and horses have both rods and cones (duplex retina), while chickens have only cones. Piggins (1992), in a review article, suggests the following functions of the two anatomical components of the eye: Rods can signal both the presence of very low light intensities and differences between such intensities, while cones do so less well. At higher levels of intensity, rod action becomes relatively insensitive because of the number of rods that have already become activated and no longer signal such differences, while cones, being less sensitive, con-

FIGURE 5.6 Audiograms for the horse, pig, sheep, cattle, and goats (solid lines); human audiogram is shown in dashed lines (*Source:* Hefner, H.E., and R.S. Heffner, Auditory perception. In *Farm Animals and the Environment.* Edited by C. Phillips and D. Piggins [Wallingford, UK: CAB International, 1992], p. 165, by permission)

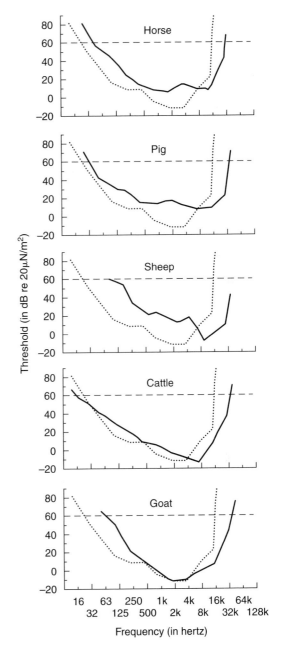

tinue to do so. Thus, rods are associated with high light sensitivity, and cones with relatively moderate to low light sensitivity.

The presence of both rods and cones in the same retina gives rise to the best of visual worlds: high sensitivity and high acuity.

The preceding suggests that chickens do not see as well in low light as the other species. This finding may explain why low light level in chicken housing is effective in minimizing some undesirable behaviors such as damaging pecking among the birds.

The presence of color vision in animals is a topic of common interest. Piggins (1992) concluded that the farm animal species have the anatomical structures present to see color (see also Jacobs, 1993) and cautioned against assuming that animals see and perceive color as is characteristic of human sensory and cognitive systems even though similar anatomical structures exist in a given species. Caution was also urged in assuming animals use color perception in similar ways as do humans. This presents an interesting question relative to experiments to determine an animal's use of color perception. Researchers in this field make every effort to design tests in which color is the only changing variable. This assumes that the animal perceives all variables in the same way as humans do. While this may seem a reasonable assumption it may also be hazardous.

The following discussion summarizes sensory characteristics of some farm animal species. Auditory ranges presented are estimates from figure 5.6.

Pigs

Auditory   Pigs rely heavily on vocal signals. Auditory range in pigs is from about 55 Hz to 40 kHz, with a region of best sensitivity from about 500 Hz to 16 kHz. The best frequency appears to be about 8 kHz.

Common auditory signals described by Houpt (1991) are grunts, barks, squeals, and screams, all varying in character. Longer grunts typically occur as a response to familiar sounds or while rooting. Short grunts are typical in excited pigs. Grunt frequency increases with increasing levels of excitement. Squeals and screams replace grunting during periods of increasing fear, such as while under attack or when caught and held by a caretaker. Longer grunts are associated with a satisfying environment. Sounds similar to a bark are associated with surprise and expressions of dominance. Pigs that are separated from their group vocalize by squealing until they rejoin this or another group of pigs. Sows give a series of short grunts that may be followed by repeated "whuff" sounds as attack warnings. Sows have a characteristic nursing and milk letdown vocalization pattern. At nursing time, short sustained grunts begin, and the frequency increases as the pigs prepare to nurse by manipulating the udder area. These sounds continue to increase in frequency until milk release. Although the total nursing period may continue for one or more minutes before milk ejection, the actual milk consumption period lasts only about 10 seconds. This time is followed by several minutes of continued nursing behavior, however. Efforts to unravel vocalizations of swine to establish some semblance of language patterns have been numerous but inconclusive.

Olfactory   Olfactory stimuli are important in identifying offspring by the dam and individual pigs in a group. Pigs carefully investigate any other pigs introduced into the group by nosing them, emphasizing the underline and the rear quarters. This behavior is important in ultimately identifying pigs in relation to the social order in the group. Females identify males by characteristic odor. Males determine mating receptivity of the sow through use of the vomeronasal organ, as well as by observing behavior of the estrus female. Piglets can identify their mother by odor within a few hours after birth. Pigs, like other domestic animals, are macrosmatic (i.e., they have highly developed olfactory capability). Humans are classed as microsmatic (i.e., they have poorly developed olfactory capability).

Visual Eyesight is well developed in pigs, including the presence of cones and rods, which suggests the capability for color vision. In general, swine appear to depend more on auditory and olfactory capabilities in communicating than on visual contact. Sight may be restricted depending on the breed, by floppy ears. Light wavelengths to which pigs are sensitive are slightly lower than those for humans.

Pigs originally were basically brushy forest dwellers and nocturnal feeders. They do not have discernible day-night preferences but periods of activity are influenced by demanding environments.

The species readily develops conditioned responses to nonvisual cues for food. Noises associated with feed delivery are examples of such stimuli.

The boar uses visual and olfactory clues to assess the reproductive state of the female. A standing behavioral response of the female to pressure placed on the back by the boar's headresting represents a signal reflecting sexual receptivity. Pressure applied to the back by a caretaker pressing on the area is used to assess sexual receptivity for mating management (fig. 5.13).

A curled tail position is suggested as a useful guide to assess a pig's general well-being. A drooping tail may be associated with a reduced state of well-being. The significance of this in animal communication is not clear.

Cattle

Auditory Sounds and how they are heard and perceived are of importance in communication both within and among species and clearly important between dam and offspring. Auditory range in cattle is from about 25 Hz to 35 kHz, with best frequency at about 8 kHz.

Attempts have been made to express cattle calls phonetically but are difficult to describe to one not familiar with such sounds. Calls are exchanged between dam and offspring, probably as identity or simply association vocalizations. Cattle such as dairy cows and feedlot steers will commonly emit calls while waiting to be fed. Cattle may give more intense calls when demands for feed are stronger, as is characteristic of animals experiencing limited feed environments such as low levels of feeding during the winter under range conditions. Frustrating or stressful situations (e.g., when a cow is isolated) may result in longer, more intense calls. A hungry calf will give frequent intense calls. Bulls tend to emit a deep roaring sound during agonistic behavior. They may issue a loud call changing dramatically from low to high pitch as the sound progresses. The latter appears to be a call to exchange location information. It is unclear whether such calls are associated with territorial behavior; however, casual observation of how bulls move about at times during this process is suggestive of some relationship or at least may represent an effort to warn potential competitors.

Olfactory   Olfactory capability is especially important in bonding and related identity between dam and offspring and in mating behavior. Specific estrus odors are processed by the male's vomeronasal organ as an indication of reproductive status. The source may be urine or the genital region of the female with which the bull is associating at the time.

Odor associated with forage is likely unimportant for cattle grazing a uniform pasture with a limited range of botanical species but more important in feeding areas where selective grazing is important. Cattle may use olfaction to avoid toxic plants, but the relative importance of smell, taste, and visual detection is unclear.

Visual   Vision appears to be the dominant sense used by cattle. The species has rods and cones in the eye, which suggests the possibility of color vision.

Cattle have limited power of lens accommodation and by virtue of eye placement have restricted close-up binocular vision. This feature makes them very cautious in examining things at ground level immediately ahead and represents a factor in driving and handling cattle if unfamiliar objects are in the movement pathway.

Sheep

Auditory   Vocalizations between the ewe and lamb are important in bonding, in continuing identification, and in finding each other in groups of animals. When the dam and lamb are separated, typically both animals attempt to vocalize until together again. Adult sheep use vocalizations in group socialization and probably in maintaining contact with those that may be preferred associates. Rams produce a deep, characteristic sound as they approach the female. Increased vocalization may occur with activity level in a group, and larger flocks may reflect a higher vocalization level than smaller flocks. Auditory range in sheep is from frequencies of about 125 Hz to 40 kHz, with best frequency at about 10 kHz.

A high correlation exists between vocalization and activity level. Size of a flock influences level of vocalization. Larger flocks are noisy, probably relating to efforts to maintain contact.

Olfactory   Sheep identify individuals by means of smell. Odor appears to be critical in the identification of lambs by their dams, although there is evidence that visual recognition is important also.

Rams evaluate mating receptivity of the female by smelling urine and the genital area. The common response is flehmen, which is more likely when the female is in estrus.

Visual   Sheep are mainly active in the daytime, and members of the flock appear to maintain contact with each other through vision. Sheep stamp a foreleg as a visual display in aggressive behavior. Color vision may be similar to that in cattle; however, sheep are grazers and browsers and therefore may be more adept at selecting forage by color.

Horses

Houpt (1991) presents an extensive summary of communications among horses including the following vocalizations.

Auditory   Vocal communication appears to be important in maintaining herd cohesion. Some typical vocalizations (Houpt, 1991)[1] include the following:

Neigh or whinny: greetings or separation calls that usually elicit a vocal reply.
Nicker: caregiving, care-soliciting call (mare to her foal upon reunion). The call will typically elicit a vocal reply.
Short snort: an alarm call in horses. Prolonged snorting and sneezing appear to be frustration calls given when horses are restrained from galloping or forced to work.
Squeals: high-pitched sounds that occur when two strange horses meet or when horses have been separated for some time. The squeal is a defensive greeting and is typical during the formation of dominance hierarchy. Squeals are also heard in response to pain.
Roar: a high-amplitude vocalization of a stallion normally directed toward a mare.

The auditory range in horses is from frequencies of about 60 Hz to 32 kHz, with best frequency at about 1 kHz.

Olfactory   As with other species, olfaction is important in reproductive behavior. The stallion, after placing his lips in the urine of an estrus mare, will show the flehmen response.

Horses use manure deposits of their own or that of other horses for geographical marking (note figure 5.9).

Piles of manure may be used to mark an area occupied by bands of horses. Males may dung in response to piles of dung they have investigated. The horse's olfactory capability results in discrimination against consuming grass contaminated with fecal material although other species show the same characteristic.

Visual   Horses have a wide visual field with only slight head turning because of the lateral placement of their eyes. Binocular vision is limited immediately in front of the head, however.

Horses tend to position their ears in the same direction they are attempting to direct attention (figure 5.33). Individual ears directed to front and rear, for example, may be associated with attention to objects ahead and the rider behind. Ears placed tightly back and against the neck usually indicate dissatisfaction with surroundings and possibly aggressive behavior.

Horses have cones and rods in the eye, which suggests color vision capability. Their ability to distinguish among certain colors is not clearly established.

Goats

Auditory   An alarm call by the buck is a spittinglike call. Correspondingly, it will show a jerking action of the head along with foreleg stamping. Males vocalize frequently during courtship. Does and kids vocalize frequently during active grazing

[1]Houpt, K.A., *Domestic Animal Behavior* (Ames Ia: Iowa State University Press, 1991), p. 21, by permission

together. Different alarm calls of does causes kids to hide in the undergrowth or from their lying-out position. The kids' alarm calls are high-pitched wails.

The auditory range in goats is from frequencies of about 70 Hz to 35 kHz, with best frequency at about 2 kHz.

Olfactory   The flehmen response is frequently seen in male goats and may be shown in nonsexual as well as sexual situations. Goats use olfaction as a part of bonding and in individual identity of offspring and associates.

Visual   Color vision is likely in this species, because both cones and rods are present in the eye.

Chickens

Auditory   Chickens do not have earlobes, but they have well-developed ears. Range of calls is from 250 Hz (the broody hen cluck) to about 3 kHz (distress call). Frequency hearing in chickens ranges from 60 Hz to 12 kHz, with an optimum range from about 80 Hz to 2 kHz (Hou et al., 1973).

Olfactory   There is little information on olfactory capacities in chickens; however, it is suggested that the smell of blood can be detected and may be involved in attraction to areas of others associated with damaging pecking behaviors. The sense of smell in chickens is considered to be poorly developed in general.

Visual   Chick embryos respond to light as early as 17 days after the start of incubation.

The visual field is about 300 degrees, thus this species has a limited binocular field. The acuity in sight (sharpness) is high, and distance vision is good. Chickens have purely cone retina (Piggins, 1992). They are assumed to see color but have limited ability to adjust to low light levels.

Experimentally, chickens have shown an ability to discriminate between squares and triangles and red and black dots. Chickens prefer to peck at round rather than flat objects. They also appear to show some preference for orange-red color over green. Newly hatched chicks, however, may prefer to peck at blue colors.

AGONISTIC BEHAVIOR (AGGRESSION AND SUBMISSION)

Aggression is typically defined as a physical act or threat of action by one individual that potentially reduces the freedom or fitness of another. Aggressive activity by an animal along with submissive behavior of another is referred to as agonistic behavior. Thus, agonistic behavior includes both aggression and submission between or among animals.

Examples of aggressive behavior toward other animals include biting, butting, kicking, pawing, and stomping. Horses utilize rearing and pawing as offensive maneuvers. Cattle use bunting and butting. Swine slash with their teeth and bite. These same aggressive behaviors may be utilized against humans, and extreme caution is encouraged in the interest of personal safety when around farm animals of all types.

A number of factors may give rise to conflict between animals. Competition for food, rest areas, territory, and mating partners are common sources of conflict. Others are fear, frustration, pain, crowding and lack of opportunity to escape, protection of offspring, and intrusion of an animal's individual space. Finally, conflict arises in the establishment and maintenance of the dominance order. The behavior that results from conflict in the establishment of a dominance order is a critical matter in maintaining social stability in a given animal population and in the evolutionary process of procreation by the fittest. Attacks on humans may commonly arise from fear or protection of offspring but also from the perception that the person entering an animal's space may represent a challenge to the established dominance order for that animal.

Agonistic behavior among animals within a species is termed conspecific agonistic behavior; that involving another species is termed interspecific agonistic behavior. Neural injuries or abnormalities may be involved in unusual levels of aggression toward any subject, in which case the level of aggression would be considered an aberrant behavior.

Agonistic behavior between two animals will lead to one or more activities defined as threat, attack, submission, and flight (figure 5.7 illustrates agonistic encounters between boars; note also figure 5.4 for similar encounters in cattle). Offensive, threatening behavior suggests that an attack may be forthcoming. These behaviors are apparently designed to make the animal appear large and formidable. Examples are stomping, pawing, piloerection, bowing of the neck, standing

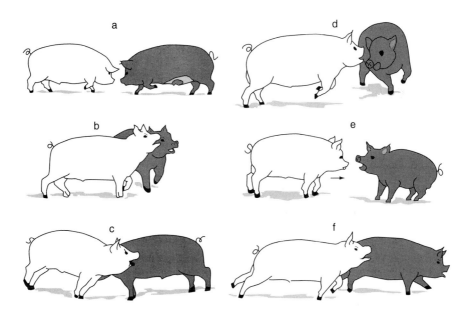

FIGURE 5.7 Agonistic behavior in boars (*a*) initial encounter; (*b*) strutting; (*c*) shoulder-to-shoulder contact and slashing; (*d*) perpendicular biting attack; (*e*) submission of pig on the right; (*f*) flight (*Source:* Hafez, E. S. E., and J. P. Signoret, The behavior of swine. In *Behavior of Domestic Animals.* Edited by E.S.E. Hafez [Baltimore, Md: Williams & Wilkins, 2d ed, 1969], p. 382, by permission)

tall, growling, showing teeth, salivating or foaming at the mouth, and bulging eyes. Submissive behavior reflects that an animal will not attack. Examples of submissive behavior are lying down, crawling on the belly, turning away, diverting eye contact, turning belly up (dogs may include urinating, which is a submissive activity retained from early life of the pup when the dam prevented urine and feces in the nest by licking), and lying motionless as if dead or totally incapacitated. Convincing actions of a submissive nature usually prevent attack.

Fighting behavior among animals is most commonly a test of strength rather than a process for injury or death. In general, when an animal reflects submission and a willingness to accept a lower social position in the dominance order, fighting ceases. Agonistic behavior between bulls, for example, is often a pushing and shoving match rather than a vicious fight. This level of strength challenge is usually sufficient for the routine maintenance of dominance. Figure 5.8 illustrates subtle agonistic encounters among horses (Hart, 1985).

In conflicts between horned and polled (hornless) animals, the horned animals usually predominate unless there is an extreme difference in size. Some conclude that horns are of greater significance in the evolutionary process (because of the importance in the dominance order) than in protection from predators.

Of general interest are concerns related to situations involving domestic animal attacks on humans. Such considerations involve safety issues that are critical in the management of relationships with animals. Such attacks most likely involve fear, stress, territorial protection, or protection of offspring, although they may result from the animal's perception that the intrusion may represent a threat to a dominant position. It is readily apparent that a number of factors normally leading to agonistic behavior can evoke attacks on humans. When animals are experiencing conditions reflecting these factors, even greater caution is in order when entering the animal's space. Keen observation of animals to detect or anticipate such factors is a critical component of maintaining a safe environment for both humans and animals.

Types of Aggression

Fear Related

Fear-related aggressive behavior is characteristic of confined or cornered animals and is usually preceded by movement to try to escape. Threats by conspecifics, predators, or humans are common stimuli.

Pain, Space, and Annoyance Related

Aggression may be associated with illness, age, a persistent environmental annoyance, physical limitations, or injury. Nielsen et al. (1997) reported an inverse relationship between aggression and bedded resting area up to $2.7 \, \text{m}^2$ for 400 kg heifers.

Frustration-Boredom Related

Frustration may lead to aggression in any animal. Sterile and crowded environments may also be frustrating. Blackshaw et al. (1997) reported toys for pigs reduced aggression but did not enhance growth rate.

FIGURE 5.8 Agonistic encounters among horses. Submission stance on left (*Source:* Hart, B. L., *The Behavior of Domestic Animals* [New York: W. H. Freeman, 1985], p. 40, by permission)

Resource Defense

Defense occurs only in an area in which an animal has established itself and which it defends against intruders. Farm animals are not typically territorial in terms of social behavior but will tend to form dominance orders that may preclude access to feed, water, and other utilities rather than a territory as such. Koolhaas et al. (1997) emphasized the significance of homeostatic motivation in social aggression.

Maternal Defense of Offspring

Maternal aggression is a form elicited by the proximity of some agent that is threatening to a female's young. It is a component of maternal behavior and will most likely disappear when the hormone profile changes from that typical of a dam caring for an offspring.

Sex Related

Sex-related aggression is typical of males in competition for females and in controlling females. This behavior is at its highest intensity when the female is in estrus. Sex-related aggression is assumed to be elicited by the same stimuli as is mating behavior. Testosterone is clearly the primary factor in such aggressive behavior in males. Females attack other females, but the form is related largely to normal establishment and maintenance of a social hierarchy.

Genetic and Hormonal Basis of Aggression

Aggressiveness is heritable. In chickens for example, the heritability of aggressiveness is estimated at about 0.22 (22 percent of the variation is due to genetics). Several studies with cattle show heritability of temperament to be moderate, ranging from 0.44 to 0.53 (Dickson et al., 1970; Stricklin et al., 1980). Selection for docility has been commonly practiced by some livestock producers, but change is effected only over many generations due to the low level of heritability.

The primary hormone associated with aggression is testosterone. Males are more aggressive than females, and aggression is reduced by castration. This reduction is gradual and may be partially or fully restored by androgen treatment. There is some evidence that adrenal hormone levels may be higher in more aggressive animals.

The amygdala, hypothalamus, and other areas of the central nervous system are all involved in the neural control of aggression. Surgical techniques have been used in attempts to alter or remove such tissues in the central nervous system of highly aggressive pets. Success has been limited and variable.

Typical Aggressive Behaviors in Farm Animals

Swine

Boars slash at their opponents' shoulders with their tusks and bite at the front legs, neck, and ears. Sows and young pigs bite. Rather than slashing, their bites are directed to the head, neck, shoulders, and ears. Swine do a great deal of pushing with the body and grunting as a form of aggressive testing. A major concern is the intense aggression that occurs when unacquainted pigs are mixed together. This fighting involves biting the opponent's flank and biting of shoulders and neck associated with frontal and parallel attacks. The fight usually ends as the subordinate animal gives way and is chased by the dominant opponent. In white or light-colored pigs, evidence of such bouts is obvious as multiple red streaks (subcutaneous hemorrhaging) on the skin. Serious injury may occur, which encourages management to minimize fighting.

Maternal aggression occurs and may be directed to other animals or humans; however, it is not a problem in many production systems because sows are confined in stalls or crates.

Newborn pigs are very aggressive in competing for forward nipples and as a result establish a dominance order in the first few days following birth. These bouts are especially damaging to littermates because of the pigs' very sharp teeth. Producers usually clip these teeth at birth to minimize such injury to other pigs and also to the sow's nipples. Dominance order established at this early time will remain largely in place as long as the litter remains together.

Tail biting is a common vice among young pigs and usually starts as a part of exploratory behavior; however, it can go to the point of major tissue damage and death once pigs are attracted to the blood that results from the activity. Crowding appears to be associated with increased incidence of this problem.

Early fighting among regrouped pigs may be reduced by areas in the pen in which a pig can hide its head (McGlone and Curtis, 1985). The most typical management technique, however, is to minimize mixing of unfamiliar animals once a dominance order is established in a group.

Cattle

Bulls usually present a side view to the opponent as the first stance to demonstrate size. They will almost always paw with the front hooves and may dig the ground with horns, if present; polled bulls may produce a similar move with the head. Vocalization will be low pitched and might be described as similar to a growl. Snorting is also common. The two will enter the contest by butting head-on, which is accompanied by strong pushing and shoving. The ultimate goal appears to be to move into a position to butt the opponent in the side, shoulder, or flank area. Exposing the side in this fashion by the subordinate suggests submission, and escape will follow if the dominant animal continues aggressive butting advances (figure 5.4 illustrates some agonistic interactions among cattle).

Cows usually approach each other in a manner to reflect their expected dominance. Head position of the dominant animal is perpendicular with the ground. The opponent extending the neck to position the head more parallel to the ground reflects submission and fighting will probably not occur. If submission is not exhibited by one of the animals, the opponents will butt by pushing and trying to slide alongside the other to gain a position in which the head is in the flank and then possibly forced between the hind legs of the opponent. Thus, the opponent may be immobilized by lifting the hind legs sufficiently to prevent normal movement. Butting the opponent in the side, flank, or shoulder area appears to be the goal in achieving final dominance. Refusal to submit is reflected by an attempt to regain a head-on butting position and resuming the scenario. Submission involves an appropriate head position, side exposure, and finally escape.

Maternal aggression can be characterized by strong attacks on other animals or humans. Cows are very likely to attack dogs. Possibility of human injury by butting or kicking should be avoided by exercising extreme care in entering confined spaces and in attempting to assist a newborn calf to nurse, if the dam is not restrained.

Cows vary greatly in predisposition to attack humans in these circumstances, but if an attack occurs it can be severe.

Unlike in pigs, there is little likelihood that young calves will compete for social dominance at a level that results in injury because they are not competing for nursing position.

Severe aggression normally occurs when bulls are mixed or being transported to slaughter. Mixing of unfamiliar cows or steers is not as likely to result in severe aggressive behavior but adds an element of instability and testing; this element can increase the stress level and potential negative effects on carcass value associated with an undesirable dark color in the beef caused by preslaughter activity level depleting muscle glycogen. Such animals are referred to as dark cutters. When bulls are mixed in the feedlot, fighting will continue for one to two days and can be at a strenuous level. Mixing steers or heifers in the feedlot is of lesser consequence, but it is most desirable to allow a group to stabilize socially and to avoid new introductions. In general, animals should be kept in their pen groups during transportation, marketing, and the period before slaughter. This management practice is difficult under some marketing scenarios, but group size can usually be coordinated with truck capacity. For example, most cattle trucks hold forty to fifty head of beef cattle in the weight range of 1,200 to 1,400 pounds. Some trucks can be easily divided to accommodate grouping of the animals.

Sheep
Sheep show potential aggression by stamping, kicking of foreleg, and appearance of readying for a butting attack. Butts may be aimed at the side or rump of the opponent, and head-on clashes occur between rams. A clash is followed by a period of still stance, apparently to judge whether one is moving into a submissive stance. If not, another clash will occur.

Goats
Goats commonly rear up as an initial action of aggression before an actual butting attack. This action is followed by butting, which is often a head-on clash. They jab each other in the side and abdomen with horn tips. The presence of horns is clearly an advantage in establishing dominance.

Horses
Aggressive posturing in the establishment of dominance may involve attempts to chase on the part of the aggressor and fleeing by the subordinate. Agonistic encounters between stallions involve a series of behaviors that allow individuals to size up their opponents before any aggressive acts. Initially, stallions will approach with an arched neck and elevated tail and move using an animated gait. The two stallions will first trot alongside each other at a distance, and then each will approach and initiate nasal-to-nasal contact. At this point, the stallions are most likely to squeal and exhibit a foreleg kick. The stallions will then move to nearby positions, where one will defecate; the second stallion sniffs the fecal pile and proceeds to defecate directly on top of it (fig. 5.9). These preliminary stages

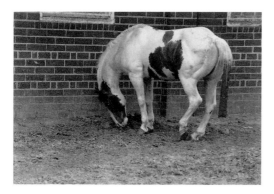 

FIGURE 5.9 Formation of fecal piles is characteristic of stallions; even captive stallions, housed singularly, will form fecal piles, as do their feral counterparts

may be repeated several times. Typically, both stallions will then move off to their respective herds. The last stallion that defecated on the fecal pile is the dominant animal. If the dominance hierarchy is not well established between the two stallions, neck wrestling and bites to the neck are the next level of aggressive contact. Such encounters will then progress to kicking and rearing. Front legs are used to strike the opponent. Severe encounters between two stallions that possess harems are unusual. Mares do not normally rear and paw but may duel by kicking with rear legs from rump-to-rump positions.

Maternal aggression can be strong. Threat of attack involves extended neck and head, ears back, and a quick advance to animals or humans.

Foals, like calves, do not usually represent a threat to each other in establishing and maintaining a social order.

Threatening Behavior as a Component of Agonistic Behavior

Threats reflect the animal's intention and readiness to fight. Frequent threats are important in maintaining the dominance order in groups. Symbolic fights involve movements used in fights but without aggressive physical contact. A fighting posture is assumed, but the animal stops before making contact. This behavior occurs commonly in all of the farm animal species. Snapping and sharp vocalization are characteristic threats by pigs. Head (or horn) threats are common in cattle, sheep, and goats. Ears held tight to neck, pointing toward the rear and advancing body movements are common threat behaviors in horses. In sheep, a kick with a stiff foreleg along with sharp downward movement of the head, stiff walk, and raised head are common threatening behaviors. Pushing and shoving, vocalizations in various forms, and characteristic postures are normal threatening activities in most species. Such threats aid in the establishment and maintenance of social order and group stability without the more severe results associated with fighting. Specific examples of both offensive and defensive aspects of threat behaviors are provided in the earlier discussion of agonistic exchanges for the different species.

SPACING BEHAVIOR

The behavior of animals relative to the space they occupy and distance relationships with other animals, and even humans in some cases, is normally referred to as spacing behavior, or spatial behavior. The characteristics of spacing behavior have both social and physical dimensions. A social dimension is that animals have some characteristic tendencies to position themselves at certain distances from other animals. Factors affecting such spacing may relate to innate species characteristics, dominance position in a hierarchy, fear of other animals, and sensory capabilities that allow them to maintain contact with other animals in a group as desired. A physical dimension in spacing behavior relates to meeting the animal's needs for space to move about and accomplish consummatory behaviors such as feeding and reproduction. Space requirements may relate to some homeostatic demands such as air movement for cooling or huddling to conserve body heat. Thus, environmental design must consider meeting space needs for sustaining life (homeostatic needs), a reasonable level of comfort, efficiency of resource use, and a normal behavioral repertoire if a state of animal well-being is to exist. Commonly used terms that refer to characteristics of an animal's spatial requirements and behavior are discussed in the following section.

Terminology Related to Spacing Behavior

Individual distance is the distance an animal attempts to maintain between itself and other animals.

Group distance is the distance a group of animals attempts to maintain between it and other groups.

Personal or individual space refers to the space occupied by an animal.

Head space refers to the area that an animal attempts to maintain around its head in crowded conditions. The animal vigorously attempts to maintain this space when crowding occurs.

Social distance refers to the maximum distance animals will allow in terms of separation from a group or another individual. This characteristic is related to the desire to maintain social contact and probably security.

Flight distance is the distance an animal attempts to maintain between itself and other individuals such as those that threaten well-being or may offer something the animal assumes to be undesirable. Such distance may be based on previous experience or fear of the unknown. Specifically, it is the minimum distance to which the animal will allow another individual (animal or human) to approach without movement to escape, a prominent characteristic of spacing behavior when handling systems are considered. Farm animals generally are capable of adapting to short flight distances with appropriate experience. Figure 5.10 reflects spacing terms used to describe some aspects of spacing behavior (Lynch et al., 1992).

Home range refers to an area selected and used routinely by a group of animals. It is fully explored and will be changed or expanded when resources become limited. Otherwise, the area may be maintained as the range of a group for an indefinite time. Groups share resources by occupying a home range with overlapping boundaries. With the exception of mature bulls, farm animals are not territorial in terms of social behavior; thus, overlapping home ranges are

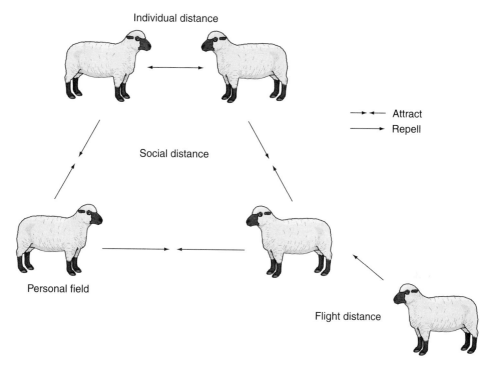

Individual distance

→ ◄── Attract

───→ Repell

Social distance

Personal field

Flight distance

FIGURE 5.10 Spacing behavior terms using sheep as an example species (*Source:* Lynch J. J., G. N. Hinch, and D. B. Adams, *The Behavior of Sheep* [Wallingford, UK: CAB International, 1992], p. 71, by permission)

common, and resources such as grazing and water are shared. This overlap provides priority access to utilities that are limited or defensible. Groups may typically share a home range. More than one harem of mares and foals, along with the stallion, may share a home range. More than a single farm animal species may also share a home range area. Swine typically form compound social groups (a few adult females, more than one juvenile male, and offspring exist together) that may occupy a home range shared with other groups. Adult boars are solitary and enter the maternal groups when sows are in estrus. In all of the cases where groups share a home range, in whole or in part, a group dominance order normally prevails in terms of access to limited resource components of the environment.

Territorial refers to a behavioral characteristic in which animals define and in some cases mark an area and defend it from use by other animals. Normally, farm animal species are not territorial. In contrast, dogs protect a territory, which in many cases is a family residence and the grounds it occupies.

Space Needs in Relation to Production Systems in Animal Agriculture

Over the last several decades, animal agriculture has moved heavily to intensive systems of production. While extensive systems such as ranching or use of large pastures in farming areas still exist, a much higher proportion of total animal agriculture is maintained under intensive systems than in the first half of the century. Efforts have been made to design intensive facilities that accommodate animal needs and meet requirements of efficient animal production. A considerable amount of research has been done and is continuing to enhance system and facility design to achieve these objectives. In some cases, at least, too little attention has been given to behavioral needs of animals. Failing to meet these needs may compromise animal well-being as a result of stress associated with such deficiencies.

Characteristic spatial and social behaviors and associated needs are important considerations in the design of environments. An increasing amount of research-based information is becoming available to provide better service to the animal industry relative to such designs. Specific space requirements are illustrated in part 4; however, one must be mindful that new information may alter such recommendations. Thus, the services of qualified system and facility design professionals who are familiar with current recommendations should be involved in planning for changes in existing facilities or the design of new ones.

Increasing use of intensive systems in animal agriculture and the development of a variety of confinement systems and even very intensive grazing systems place great emphasis on optimizing cost per unit of output from such systems. Such emphasis increases the likelihood that individual animal well-being will be sacrificed in favor of enterprise efficiency. Major factors in the cost per unit of output are involved in providing the space occupied by animals and the cost of supporting it in terms of utilities to assure effective function. Restricted space for animals may violate behavioral needs, resulting in a variety of negative responses. Many of the aberrant behaviors discussed later in this section are related to inadequate quantity or quality of the space occupied by animals. Thus, understanding some of the important space-related behaviors of animals is important in developing good management strategies. The view that something in the environment is inadequate, socially or physically, when aberrant behaviors occur provides strong motivation to demonstrate effective stewardship for the animals with which we are associated.

Planning for an animal's space needs considering function and well-being is important in facility design. A given animal needs space to accommodate normal body functions such as lying down, rising, and resting. Species characteristics and space requirements related to these activities differ. Animals need space to move about and access the utilities that support the function intended for a production system or simply for maintaining normal behavioral repertoires such as grooming or social behaviors. Adequate space to allow avoiding dominant animals and escape from aggressive attacks is important for normal behavior and comfort and in maintaining social stability in a group. The ability to view other animals can be important in some cases and will determine design characteristics of animal facilities. Partition designs that allow animals to visualize one another are economical and may have positive effects in some cases. The dutch door characteristically used in many horse stalls provides a design that enhances the occupant's perception of space and provides an enriched environment by simply allowing the individual to see other animals, caretakers, and activities taking place in the stable.

At least some species prefer to spend most of their time at or near the outside perimeter of pens. Pen configurations such as circular or rectangular pens provide different relative amounts of linear outside partition or fence to floor space. Rectangular designs of varying dimensions to allow similar floor space can provide different amounts of linear perimeter distance. Square, rectangular, and triangular structures have the disadvantage of having corners, where an animal may be trapped by an attacker. These areas should be blocked by fixing boards or other solid construction material across the corner, preventing an animal from being caught there.

In many animal facilities population density is great. Farm animal species are flexible in terms of adapting to different levels of space allowance, which is a characteristic of species that are typically domesticated. Increasing density to the point of crowding will likely have negative effects such as increased aggression. Chronic annoyance associated with violations of individual space can be a strong psychological stressor.

Space restrictions limiting access to feed and water can have severe negative effects on the animal's performance and well-being. In general, the most severe competition for space occurs in areas around feeders and waterers. Thus, location and distribution of these utilities are critical to the effectiveness of the space provided. Location of this equipment should not result in pockets where subordinates can be trapped in the normal process of agonistic behavior.

PLAY BEHAVIOR

Extensive research has examined play in animals. Fagen (1981) presents an extensive review. Play has beneficial effects, including those related to strength and skill, which assist in behavioral and physical development. It is critical for animals to be able to play for normal development of cognitive skills necessary for behavioral adaptability and versatility (Fagen, 1981). Most play behaviors include behavioral repertoires that are used in reproduction, agonistic interactions, and predator-prey conflicts. Isolation during the development processes may result in behavioral abnormalities and physical limitations. Thus, play is related to the social development of the young animal and its well-being, possibly over the lifetime of the individual.

Characteristics of Play

Behavioral characteristics in play activity have been extensively documented by Fraser and Broom (1990) and are summarized and adapted, in part, as follows.[2]

Animals look for play opportunities.
Social inhibitions exist (avoiding injury of a partner in biting, for example).
Play with inanimate objects occurs, which suggests lack of stimulus specificity.
Play is easily interrupted by stronger stimuli.
Stimulation of play may result from signals from others (group effect).
Inventiveness is commonly observed (new actions).
Play activities are often unordered.

[2]Source: Fraser, A. F., and D. M. Broom, *Farm Animal Behavior and Welfare*, 3d ed. (London: Bailliere Tindall, 1990), p. 254, by permission

Repetitious motor patterns are involved.
Continuous return to the stimulus source suggests an orientation factor.
Play normally occurs in relaxed situations.
Exaggerated actions are very characteristic.
There is an appearance of pleasure.
A buildup in level of exaggeration commonly occurs.
Incomplete sequence of actions may occur.
Play partners may commonly change roles.

The extensive list suggests that play behavior and the related practice and exercise have many impacts in the development of behaviors and ultimately in general well-being of the animal. Interactions with young of the opposite sex may influence future function in reproductive behavior of some species. Physical development is enhanced by play activity. Early experiences in interacting with other animals are involved in the development of normal social behaviors as evidenced by the effects of isolation. The extent to which these experiences influence the animal's ability to function normally is clearly a factor to consider in terms of well-being.

Interesting dimensions of play behavior are the species characteristics of initiating play through signaling. Dogs will drop down with front legs extended and rear legs in a position to quickly pounce. Cats will often hump with tail vertical and maintain a perpendicular stance somewhat like initiation of aggressive or defensive behavior but without the aggressive facial expression, hissing, and piloerection. Calves will jump, often with stiff front legs; foals will nod and run; and lambs may jump dramatically and bounce with stiff legs. Such signals are clearly associated with the desire and plan to initiate play.

SELF-CARE BEHAVIORS

Care Related to Homeostatic Influences

Animals accomplish a number of behaviors related to caring for themselves and in attempting to maintain some level of comfort. Examples are seeking shade to effect cooling and shelter to conserve heat. These behaviors have a homeostatic purpose in temperature regulation and minimizing potential thermal stress. Pigs huddle or separate in resting areas depending on respective cold or hot conditions (note fig. 6.2). Some animals may take action to minimize insect damage at a given time such as standing in water or moving away from areas of heavy insect populations. Some behaviors related to rest and sleep are characteristic of a species. Those that bed down may reflect preferences for resting areas for reasons related to thermal protection or areas that offer aid in insulating from cold.

Sanitation

Animals demonstrate some behaviors related to sanitation. Excretory or eliminative behavior is especially important in the management of swine facilities to maintain clean, dry areas for resting. Dunging areas are established by self-training, which appears to be possible due to strong innate behavioral tendencies. Swine in general prefer to defecate and urinate in an area away from the resting area and will do so if the pen design allows enough space. This preference is exhibited from the first few days and throughout life. Baby pigs will begin to eliminate in corners or along the edges of farrowing crate areas. Although cattle are not sensitive about bedding down in an area containing fecal material, cattle, horses, and sheep avoid eating in dunging areas in pastures if possible. These activities are probably related to minimizing transmission of parasites. Horses tend to defecate in concentrated areas of a pasture, which may present grazing management problems because they normally object to grazing in dunging areas. Sanitation is also effected by the cleansing associated with grooming behaviors.

Grooming

Grooming is a high-priority behavioral characteristic and has effects of cleaning and sanitation. It is practiced to some extent by all of the farm animal species and especially by horses, pigs, and cattle. Denial of normal grooming behavior can lead to aberrant behaviors. Grooming serves a cleaning function and may assist in the removal of insects. Self-grooming is referred to as *autogrooming,* and the term *allogrooming* describes grooming of another animal (note fig. 5.2). The latter may be accomplished mutually but is often done by the subordinate animal. Grooming by use of some environmental item such as a post for rubbing and scratching is common. This action achieves treatment of body areas inaccessible to the animal's mouth, tongue, and feet. Rolling on the ground by horses and dust bathing and preening in the case of poultry are forms of grooming behavior. Poultry may typically develop a dust bathing site and return to that area routinely. In extensive areas, horses may also return to specific sites that show effects of continuous use in rolling. Such areas are characterized by presence of sand or dust, both of which are effective in dusting the body surface. Following rolling or dust bathing, the animals shake vigorously to remove the dust.

Rest

All animals show periods during daily rhythmic patterns defined as awake, drowsing, and sleeping. The following sleep periods were reported by Ruckebusch (1972) as a percentage of a 24-hour day for each activity noted for the different species (asleep includes both slow wave sleep and paradoxical sleep).

	Awake	*Drowsiness*	*Asleep*
Horses	80	8	12
Cattle	53	31	16
Sheep	68	17	15
Pigs	47	21	32

Ruckebusch also reported time spent standing and in recumbent positions during a 24-hour day. Values for percentage of time standing were as follows: horses,

92; cows, 41; sheep, 70; and pigs, 21. Values for percentage of time standing during the 10- to 12-hour nighttime period were as follows: horses, 80; cows, 12; sheep, 60; and pigs, 11.

Understanding the normal sleep patterns of livestock is important to animal well-being. Sleep is essential for normal physiological function, and therefore deprivation of sleep can have profound influences on an animal's health.

EXPLORATORY BEHAVIOR

Exploratory behavior is directed to acquiring new information about the environment in which an animal exists. Exploration results in learning and therefore is involved in developmental aspects of animal behavior simply as a matter of providing the necessary experience related to a particular behavior. As animals explore their surroundings and other animals it is obvious that all sensory systems are used.

An early function of exploratory behavior in the newborn relates to the critical need to find food through nursing. In some cases newborn animals will explore other lactating females and may eventually rob milk from alien females.

As discussed in relation to behavior of the newborn, the efficiency of the young in locating the teat area is very high and they do follow an inherited pattern in the search. In older animals, exploratory behavior is related to the location of feed and water resources and the establishment of a mental map of sites and routes to these resources. Such exploration has high priority when animals are introduced into a new area. Exploration will occur in either extensive areas or confined facilities. The size of the area explored depends on boundary restrictions except when the area is very large. In these cases the animal will explore an area of a given size and become very familiar with the existing resources. This area will become what is often termed *home range,* essentially an optional habitat for the particular animal or, more likely, a group of animals. The size depends on the capability of the animals for travel and is related to feed and water resources, as well as topography in some cases.

A practical dimension of exploratory behavior relates to that occurring when animals escape from restrictive boundaries such as fences. Based on expected behavior, the animals set about immediately to explore the new area encountered. If escape involves one or two animals, they may remain reasonably close to the herd from which they have escaped. If a larger group escapes so that distance from other animals is not a factor of concern to them, the distance covered in exploration may be very large as they try to map resources and determine the boundaries of the new area. Thus, exploratory behavior may reduce the ease with which the animals are found and recovered; it also becomes a factor in increasing the liability exposure of the owner related to damage to other areas such as cropland, other animals, and even to humans, who may become involved as a result of accidents by attack or collision with vehicles on public roadways. Another very practical consideration is that animals introduced into a new pasture or paddock will give high priority to exploring the boundaries of the area even before starting to mentally map resources existing within the boundaries. This behavior increases the likelihood of the animals finding openings in the fence due to inadequate maintenance and escape.

Exploratory behavior toward other animals is a routine component of normal behavior and is an intense activity when new animals are introduced to a group. Thus, exploratory activities are related to other behavioral scenarios including the establishment and maintenance of a dominance order. In some cases exploratory behavior may lead to other behaviors with negative effects, such as tail biting in pigs. This result is typically a secondary effect associated with crowding or an environment lacking complexity.

Exploration of novel or complex environments may be a factor in minimizing certain behaviors related to boredom. Animals encountering novel items will explore them extensively. This characteristic forms the basis for attempts to give confined animals objects as playthings to keep them occupied and hopefully reduce the incidence of undesirable behaviors. In many cases the animals will ignore the toys after an initial period of exploration and manipulation. The longer-term value of such practices in farm animal production is not well established.

Other important explorations relate to the location of shelter and hazards. These areas apparently become a part of early and continuing resource mapping.

Factors affecting exploration may involve species differences, as well as differences in individuals within a group. Artificial or natural boundaries or other physical restrictions may limit exploration and be the basis for behaviors related to frustration. This effect suggests the importance of exploratory behavior as a component of the animal's normal behavioral repertoire. Stressful conditions, injury, or cognitive limitations may also be limiting factors in expressing normal exploratory behavior.

REPRODUCTIVE BEHAVIOR

Factors Influencing Reproductive Ability

Attainment of puberty is obviously essential to normal reproductive function. This level of maturity is influenced by age and also by physiological condition. Factors influencing the developmental rate of the animal may delay sexual maturity if such factors are sufficient to have a negative influence on the typical growth curve. An example of the importance of normal rates of growth in reproductive development is the influence of level of nutrition on the development of replacement females. Specific recommendations for optimum feeding programs consider appropriate weight change patterns during development (guidelines are outlined in Part IV). An example of the importance of social development in reproductive ability is the fact that young rams developed in all-male groups are more likely to have mating problems when introduced to females at the beginning of a breeding season.

The development and maintenance of several sensory systems are critical for normal reproductive activity. Visual factors are important in signaling stages of the reproductive cycle in several species. Examples are mounting activity in a group of females. Display activity in poultry and wild animals serves to alert and stimulate potential mating partners. Olfaction is a major functional characteristic in the mating behavior of most species. Pheromones play an important role in olfactory stimulation. An easily observed behavior reflecting olfactory involvement is that referred to as the flehmen response, which is demonstrated by all farm animal mammalian species but most pronounced in cattle, sheep, and horses (note figures 5.11, 5.12, 5.14, and 5.15). Auditory stimuli may also be involved in reproductive activities, depending on the species.

Photoperiodism is a characteristic of some species. The reproductive cycle in these species is regulated by day length or changing day length. Horses, for example, are summer breeders; thus, they start reproductive cycles in the spring. Sheep are fall breeders and thus start cycling in the fall. Regulation of day length is a critical factor in poultry production, and as a result sophisticated light management systems have evolved. Regulation of light to control reproductive function may be referred to as light priming. Examples of photoperiod management systems for poultry, horses, and sheep are outlined in part 4.

Endocrine influences regulate the reproductive process and are therefore essential for proper reproductive function in all phases of the process. As examples, pituitary influences are through follicle-stimulating hormone (FSH), luteinizing hormone (LH), and prolactin. Gonadal influences are through testosterone, estradiol, and progesterone. Influence of the pineal gland on photoperiod characteristics is through the hormone melatonin.

Sexual Behavior

Sexual behavior is affected by gender and hormones, along with developmental and experiential factors, before and after puberty. Genetic sex is dependent on the presence or absence of the Y chromosome in mammals (males carry the Y chromosome). In birds, the female carries a W chromosome, whereas the males do not. During fetal development, the Y chromosome allows the production of testosterone during the sensitive period for masculinization of the brain. (See Gilbert, 1991, for an extensive review of sex differentiation during development.) In birds, the brain develops in a feminine manner if the W chromosome is present. Hormones are responsible for secondary sex characteristics; therefore, castration or antagonizing masculinizing hormones before the critical perinatal period in which these characteristics develop results in feminine mammals and masculine birds (Gilbert, 1991).

Some typically male or female behaviors may be shown by either sex in many species (e.g., mounting other males or females). These behaviors are appropriate and have a functional role (in cows, for example, mounting signals nearby males that they are in estrus). Social mounting in feedlot cattle is sometimes directed to one individual (e.g., the animal attracts mounts from several animals). This activity is known as the buller steer syndrome and is usually observed in cattle feedlot environments. The buller animal (subject of the mounting) may have to be removed from the group to prevent injury or an undesired level of activity. Mounting is a common management problem for feedlot managers and represents behavior related to social order among a portion of the group of steers (Klemm et al., 1983). Some characteristic patterns of mating behavior are illustrated in figures 5.11, 5.12, 5.13, 5.14, and 5.15.

Female Sexual Behavior

Typical Forms of Female Strategy in Mating Behavior

Strategies that may normally be used by females of the farm animal species during estrus include the following:

Females signal receptivity but do not approach the male. In this case, males that see the signal will come to the female for mating. This strategy facilitates mating for the male but may not allow mate choice by the female.

Females seek out the male. This strategy results in a cluster of estrous females around the chosen male. It reduces the effort of the male in locating a female and allows the females to choose their mates.

Females signal and seek out the male. This strategy maximizes the probability of mating. The male will only attempt to mount estrous females, and the females are able to choose their mates.

Components of Female Sexual Behavior

Attractivity   The female attracts the male to it during estrus. To evaluate attractivity of the female, it is necessary to observe the behavior of the male. An example would be to allow males access to stimuli from an estrous female and a nonestrous female. If the male prefers the estrous female, there is evidence of attractivity. Mares urinate in the presence of males.

Proceptivity   The female becomes attracted to males when she is in estrus, which is evaluated by observing the behavior of the female when a male is present. Typical activity includes increased walking by estrous females as they seek the male. Ewes may actively seek out a specific ram.

Receptivity   Receptivity facilitates copulation. It is easiest to distinguish this component of behavior from attractivity and proceptivity by defining it as behavior when the female is in contact with the male and ready for mating. A common method used to evaluate this behavior in sows is by placing hand pressure on the back of the animal. Receptivity is reflected by the female standing still as a response to the pressure (figures 5.12 and 5.13).

Behavior during Estrus   Behavioral characteristics of females in estrus are summarized in table 5.1 (Fraser and Broom, 1990).

Biostimulation Resulting from the Presence of a Male   Introduction of a ram influences the onset of estrus in an ewe flock. As a result, this practice has a synchronizing effect on the reproductive activity of the entire group of females.

Boars provide stimulation of sows through physical contact and olfactory and auditory clues. The presence of a boar stimulates lactating sows to come into estrus.

There is some evidence that the presence of bulls reduces the postpartum period, which is the time from parturition until the female returns to routine estrous cycles. The ability of the male to stimulate estrus in females is dependent

TABLE 5.1   Estrous Behavior of Females

Animal	Typical Estrous Behavior
Horse	Urinating stance repeatedly assumed; tail frequently erected; urine spilled in small amounts; clitoris exposed by prolonged rhythmical contractions of vulva; relaxation of lips of vulva; company of other horses sought; turns hindquarters to stallion and stands stationary
Cow	Restless behavior; raises and twitches tail; arches back and stretches; roams bellowing; mounts or stands to be mounted; vulva sniffed by other cows
Pig	Some restlessness may occur, particularly at night from proestrus into estrus; sow stands for "riding test" (the animal assumes an immobile stance in response to haunch pressure); sow may be ridden by others; some breeds show pricking of the ears
Sheep	May be a short early period of restlessness and courting of ram; in estrus proper, ewe seeks out ram and associates closely with it; may withdraw from flock; remains with ram when flock is driven
Goat	Restless in proestrus; most striking behavior includes repeated bleating and vigorous, rapid tail waving; poor appetite for one day

Source: Fraser, A. F., and D. M. Broom, *Farm Animal Behavior and Welfare,* 3d ed. (London: Bailliere Tindall, 1990), p. 176, by permission

on some combination of visual, olfactory, auditory, and tactile stimuli. Depending on the species, some of these cues are more important than others for bringing females into heat. Pheromones, to a large extent, are known to be very potent initiators of reproductive behavior. The vomeronasal organ is the key organ for detecting these pheromones associated with reproductive status. Figures 5.11, 5.12, 5.13, and 5.14 reflect the typical flehmen behavior for various species.

Male Sexual Behavior

The value of a male in the breeding herd is dependent not only on genetic merit but also on his ability to breed and settle a large number of females in a short period of time. The ability to accomplish these objectives depends on several factors:

Physical capability to produce large numbers of viable sperm. Evaluation techniques include semen evaluation and measurement of scrotal circumference.
Libido or sexual drive, which is described as the level of motivation to mate.
Behavioral constraints, such as injuries or physical defects of the reproductive organs or physical problems that limit activity. Adequacy of facilities to accomodate normal mating behavior is critical.

Performance tests to evaluate breeding capability of males in a commercial situation are termed serving capacity tests (SCT). The tests have proven to be reasonably reliable indicators of the ability of bulls and rams to breed large numbers of females.

All SCTs involve exposing males to females and measuring the frequency of ejaculations. The test is of a uniform duration, and several other variables are controlled. The test should be repeated two to four times. Some operational considerations for bulls and rams follow. However, females well-being should be respected.

Bulls

Bulls should be stimulated by estrous females before the test.

Bulls are usually tested in groups of four to six.

Females should be restrained in stanchions.

Bos indicus breeds will not typically mount a restrained female.

Estrous and nonestrous females will be mounted.

The test is usually 20 to 40 minutes in length.

Number of ejaculations is the most useful information.

Typically bulls testing in the lower 25 percent in a group will not settle very many cows and should be culled.

Rams

Rams are more sensitive to social factors than bulls. Some subordinate rams will perform poorly in a group test but do very well in an individual test.

Use of nonrestrained estrous ewes results in the highest scores.

Tests are usually 20 to 40 minutes in length.

Characteristic Male Behavior Involved in the Mating Process

Mounting behavior develops before puberty and typically persists for the remainder of the animal's life. The propensity for mating behavior is primarily dependent on the production of testosterone by the testes. Considerable variation in this characteristic may exist among individuals within a species.

Flehmen, the olfactory response observed as males examine females in typical mating behavior, is observed in all hoofed animals and some pets. It is very easily observed in cattle, horses, and sheep. Because of the anatomy of the mouth and lips, swine show this behavior much less obviously than cattle, horses, and sheep; therefore, it is sometimes referred to as *gaping*. Common characteristics are extended head and neck and raised upper lip. The process results from a complex physiological and anatomical reaction involving the vomeronasal organ and related olfactory sensors. The vomeronasal organ also has a role in aggressive behavior characteristics of bulls, as evidenced by the fact that cauterizing or plugging the orifice altered such activity (Klemm et al., 1984). The flehman response may be shown by females but is much less pronounced and is normally not observed. Typical flehmen behavior of the various species is shown in figure 5.11 (Hart, 1985).

Demonstrations of strength and power include arched back, curved neck, vocalization, and bulging eyes in bulls.

Other behaviors are forcefully demonstrated and are probably used to indicate strength and dominance. Examples are pawing, bellowing, and horn rubbing.

Turf marking is a term that describes forms of turf or soil disruption by bulls through pawing, rubbing the head on the ground, or plowing with horns if present.

Males often attempt to control the motion of the female by blocking. Nudging and pushing at the side or rear of the female is a common practice in most species and may enhance receptivity. Investigating female genitalia, licking, and smelling urine are typical behaviors in all farm animals as a form of

FIGURE 5.11 Flehmen behavior in some farm animal species (*Source:* Hart, B. L., *The Behavior of Domestic Animals* [New York: W. H. Freeman, 1985], p. 92, by permission)

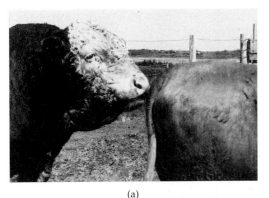

(a)

(b)

(c)

(d)

(e)

FIGURE 5.12 Mating behavior in cattle. The sequence is as follows: (*a*) olfactory evaluation, (*b*) flehmen response, (*c*) back pressing and pushing to determine whether the female is in standing heat, (*d*) mounting and penetration, and (*e*) deep thrusts and ejaculation. (Photographs courtesy of Paul Brackelsberg, Iowa State University, Ames)

exploration and is typically followed by flehmen behavior. Typical back pressing by bulls and boars is shown in figures 5.12 and 5.13.

Intromission tends to be by trial and error in horses and more directly oriented in the other species. Previous experience aids orientation in the mating process. Rearing males in isolation may increase problems in orientation, which are often corrected with experience.

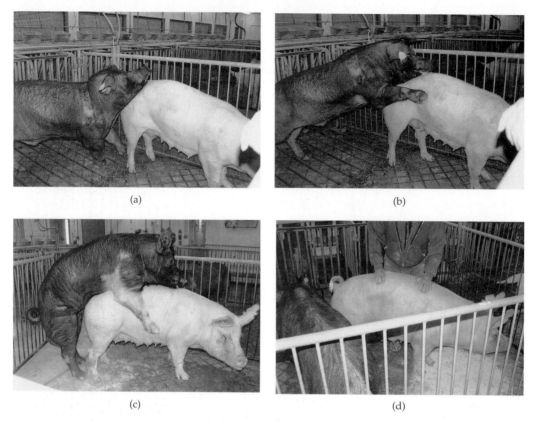

FIGURE 5.13   Mating behavior in swine: (*a*) headresting by the male, a test for immobile response reflecting level of receptivity by the female; (*b*) mounting; (*c*) intromission; (*d*) hand pressure applied by the handler to test the female for the immobile response when artificial insemination is used

Duration of ejaculation is a few seconds in the case of cattle, horses, sheep, and goats and up to 30 minutes in swine.

A refractory period following ejaculation is typical of all animals. Differences exist, however, in the length of the postcopulatory refractory period. Cattle, sheep, and goats are normally capable of more ejaculations per hour than horses and swine. Actual numbers of ejaculations with time is so variable among individuals it is probably inappropriate to estimate a norm.

Approaches to maximizing semen quality for artificial insemination are as follows (Price, 1992; Mader and Price, 1984):

When bulls are collected twice weekly, a false mount before collection increases sperm output by about 50 percent. Two additional false mounts can increase sperm count by another 50 percent.

Restraint of a bull for 10 minutes after exposure to the stimulus animal will increase semen quality at about the same level as false mounts. Such responses may vary with breed and species.

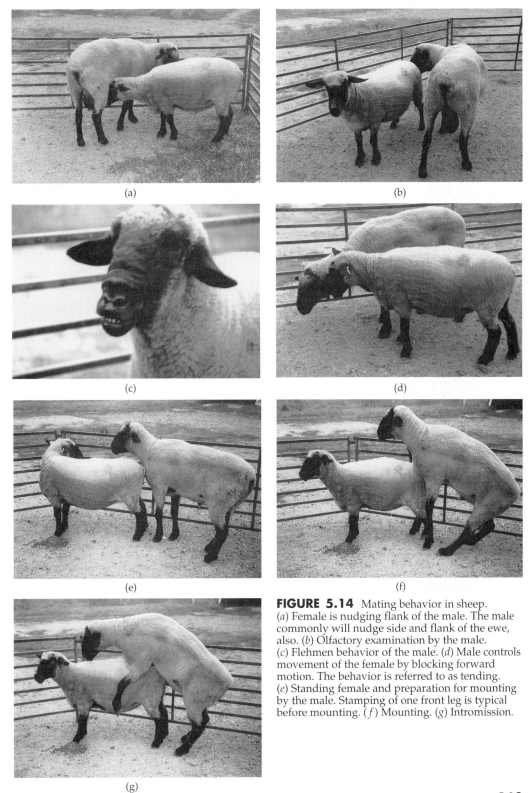

FIGURE 5.14 Mating behavior in sheep.
(*a*) Female is nudging flank of the male. The male commonly will nudge side and flank of the ewe, also. (*b*) Olfactory examination by the male. (*c*) Flehmen behavior of the male. (*d*) Male controls movement of the female by blocking forward motion. The behavior is referred to as tending. (*e*) Standing female and preparation for mounting by the male. Stamping of one front leg is typical before mounting. (*f*) Mounting. (*g*) Intromission.

163

(a)

(b)

(c)

(d)

FIGURE 5.15 Mating behavior in horses. (*a*) Flehmen behavior following the stallion's investigation of the mare's urine. (*b*) Typical nasal contact by mare and stallion. (*c*) Nasal to genital contact. Note typical tail position, winking vulva, and expulsion of urine by the female; this behavioral repertoire is typical of standing heat, a high level of receptivity. (*d*) Intromission. Note typical flagging of the stallion's tail, an indication that ejaculation is occurring. (Photographs courtesy of Brian Nielson, Michigan State University, East Lansing)

Management of semen collection for bulls in artificial insemination studs generally allows two ejaculations per day collected at 3- to 4-day intervals. Collection rates can be greater for some animals.

Restraint of a bull after exposure to the stimulus animal (or sham) and false mounts appear to result in relatively greater improvement in semen quality of those animals showing low quality.

Such stimulation techniques have little or no influence on rams, with the exception of reducing the latency before first mount (Price et al., 1991). Pickett et al. (1977)

reported that observation of other stallions copulating stimulated stallions showing low or abnormal sexual behavior. Hemsworth and Galloway (1979) found that boars that had visual access to copulating boars were slightly faster in reaction to first mount, but latencies and duration of ejaculation were not significantly affected.

FETAL BEHAVIOR

Characteristics of Fetal Behavior

Preparation for Postnatal Environment
Fetuses of the farm animal species show behaviors that indicate development of systems for physical movement and functional sensory capabilities. There is evidence that fetuses can hear. As a result, familiarity with sounds of the dam may develop. Similarly, the fetus may become familiar with tastes and odors through amniotic fluids. Fetal movements include jaw movements related to sucking behavior and extensive body movement. The fetus swallows amniotic fluid at a high rate, which is critical for fetal and amniotic fluid homeostasis and gastrointestinal development (Ross and Nijland, 1997). In sheep, the fetus is capable of discriminating different flavors as early as day 135 of gestation (Robison et al., 1995). The fetus also goes through stages of sleep similar to post-natal activity, and it is thought that REM sleep during development is critical for growth and development of the brain (Richardson, 1992).

Prenatal Exercise
Whole body movements by the fetus are considered important to its health and strength at birth. Postnatal weakness and failure in terms of normal behavior may be correlated with lack of muscle development and tone during the prenatal period.

Preparation for the Birth Process
The sequence of fetal movements, components of fetal behavior before birth, appear to be well established through fluoroscopy. Observations during cesarean surgery also contribute to this body of knowledge about fetal behavior in birth orientation. Typical movements in the process of birth orientation are the following:

Righting (reflecting a change of position).
Extension of forefeet and head toward and into the maternal pelvis.
Rotation of all parts to assume a position for parturition.

PARTURIENT BEHAVIOR

Phases of the Birth Process

Pre-parturient Period
Final fetal positioning and preparation for the birth process involve fetal adjustments for the birth process as outlined previously. Stages of parturient behavior are described in table 5.2 (Fraser and Broom, 1990).

TABLE 5.2   Principal Events in Parturient Behavior

Animal	Immediate Prepartum Period	Parturition	Postpartum Period
Mare	May be shy of interruption Anorexia only evident immediately before foaling Whisking of tail is shown as parturition commences No bed prepared	At first, restlessness and aimless walking; tail swishing, kicking, and pawing at bedding Later, crouching, straddling, and kneeling Finally, mare lies down to strain powerfully and regularly to expel fetus Fetus must rotate from a supine position through 180 degrees for delivery	Frequently remains lying on side for a period of about 20 minutes Does not eat fetal membranes Usually estrus shown by ninth day ("foal heat")
Cow	May separate from herd and seek screened locality Anorexia develops (one day) Restlessness begins (several hours) No bed prepared	Pain and discomfort expressed in very restless behavior (alternately lying and rising) Animal more usually lies during expulsion of fetus Expulsion of fetus slower than in mare	Will eat fetal membranes if fresh, occasionally before uterine separation is complete Maternal responses such as licking calf usually initiated promptly
Sow	Gathers bedding material and attempts to make a nest Becomes restless Lies on side in full extension to farrow	Shows signs of pain Strains when fetuses are in pelvis Tail actions signal each birth The final expulsion of each piglet is usually done with one wave of abdominal straining	Eats fetal membranes and fetal cadavers or parts of them Lies extended on side for long spells, accommodating frequent suckling by litter

Parturition (Birth)

Assistance with the birth process is generally more important for females giving birth to their first offspring. Another factor that contributes to the need for assistance is abnormal presentation of the fetus for the birth process. Specific genetic lines of animals within a species may have a tendency to experience greater difficulty with birthing. *Dystocia* is a term used to refer to difficulties in the birth process.

Post-parturient Period

The post parturient period, an initial period of activity immediately following birth, is extremely important in survival of the newborn. Crucial events that relate

TABLE 5.2   (Continued)

Animal	Immediate Prepartum Period	Parturition	Postpartum Period
Ewe	Interest in other lambs, may separate from flock (66 percent) Most breeds seek some shelter Restless, paws ground	Repeated rising, lying When straining, usually stands; lies during early expulsion	Grooms lamb attentively, licking fluids, nibbling any membranes on lamb Remains at birth site usually
Doe goat	Sluggish walking Becomes restless, agitated	Shows clear signs of pain Repeated straining, occasional bleating	Variable attention to kid Grooms by licking and nibbling If a "leaver," may seem to have weak maternal drive (*Note:* The doe may leave ["leaver"] the kid or may remain ["stayer"] with the offspring; both forms of behavior are considered normal)

Source: Fraser, A. F. and D. M. Broom, *Farm Animal Behavior and Welfare,* 3d ed. (London: Bailliere Tindall, 1990), p. 209, by permission

to management considerations are fetal-dam bonding, early nursing (figs. 5.16 and 5.17), and protection from adverse effects of the new environment and possible injury from other animals. Protection from predators may be an important issue, depending on the location and system of production (note table 5.2).

Timing of Parturition

In general, for the farm animal species, births tend to occur more commonly at night rather than during the day. Management of feeding time has shown some influence on timing of births within the day in the case of cattle. Synchronized births (controlled parturition) are now common as a result of exogenous treatment with hormones or hormonelike compounds to induce parturition.

MATERNAL BEHAVIOR

Pre-parturient care is reflected in choice and preparation of the site for birth. Preparation of the birth site, referred to as nest building in the case of swine, is a strong innate behavior, and environments that do not allow such behavior may predispose some types of aberrant behavior as vacuum activities in this species. In general, females of all farm species try to achieve some separation from the herd in advance of parturition.

FIGURE 5.16 Newborn pigs receive no immediate maternal care, as opposed to that given by cows, ewes, and mares

Post-parturient care emphasizes both care and association for bonding and survival of the young. Important elements of early care after the birth process include the following (note table 5.2):

Early grooming by licking that serves in stimulating the offspring and drying that aids in thermal regulation. Licking of offspring is characteristic of all of the mammalian farm animal species except swine.

Bonding between dam and young including mutual identification to maintain this critical association for survival of the offspring.

Provision for nursing.

Protection from dangers such as predation or attacks by conspecifics forms the basis for maternal aggression.

Observational training in search of food occurs in avian species.

Differences in maternal behavior are closely related to the numbers of offspring and the state of maturation of young at birth.

Monotocous species give birth to one or two young. The young are relatively mature (precocial) at birth. Cattle, sheep, goats, and horses are examples.

Polytocous species give birth to several young. These species, such as dogs and cats, are often relatively immature (altricial) at birth. However, pigs are polytocous, but the offspring are classed as precocial (i.e., more mature at birth). Chicks are highly precocious. Litter-bearing species direct their care toward nest site selection and preparation. They do not need to identify their offspring immediately after birth because the young will be confined to the nest area for a few days. Identification is important, however, before offspring follow the dam out of the nest or other separation occurs.

FIGURE 5.17 Postparturient innate behavior demonstrates the level of dedication to locating the dam's udder and teat selection; this period commences formation of a dominance order for nursing that is normally complete within the first three days following birth. (*a*) Beginning of the search for the dam's udder with umbilical cord still intact; (*b*) progressing toward udder area with umbilical cord remaining intact; (*c*) nursing behavior begins as a somewhat disorganized activity termed "teat sampling" before establishment of nursing dominance order.

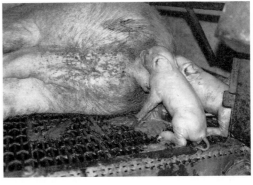

(a)

(b)

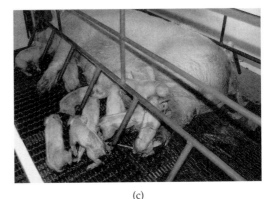

(c)

Pre-parturient Maternal Behavior

Swine select a nest site, isolated from other animals, that may or may not be protected from severe weather conditions. Site selection and nest building usually occur 10 to 20 hours before parturition. Emphasis on site selection and nest building is also common in dogs and cats. The farm animal species often separate themselves during the final 24 hours before birth. Birth site selection may not always take into account protection from severe weather (note table 5.2).

Post-parturient Maternal Care and Neonatal Behavior

Species typically bearing single offspring bond to their offspring at birth. In sheep, females that are soon due to lamb or those that have just lambed may be attracted to birth fluids, which can result in problems with such ewes stealing newborn lambs. Separation of ewes from the flock before parturition is the typical management solution in confinement systems. In systems that allow the ewe to do so, she will usually seek isolation for the birth process.

A typical maternal behavior seen immediately after parturition is licking the offspring extensively, except in the case of swine. Sows do not lick young but, like cattle, do consume the placentae. Piglets usually locate a teat within 30 minutes or less without assistance from the dam. Swine are a nesting species; therefore, the pigs do not follow the dam for the first few days after birth. As a result, bonding occurs over a longer period than is the case with horses (a following species) or cattle (a hiding species). A major problem in the pork industry is that of piglet crushing by the sow. There is interest in evaluating the genetics of maternal behavior in swine to determine whether this problem can be reduced through selective breeding. The work of Minick et al. (1997) comparing Meishan and Yorkshire breeds suggests that differences may exist and that the approach of selecting sows with beneficial maternal ability should be investigated further.

Mares may lie recumbent for 20 to 30 minutes after foaling. The dam then licks the foal during the first hour or two after parturition, and the foal usually begins nursing as soon as trial and error results in the ability to stand. The foal is able to follow the dam within a few hours; therefore, horses are classed as a following species.

Cows and ewes stand within a few minutes after giving birth and begin licking the newborn immediately. The offspring stand usually within the first half hour or less and begin searching to engage in nursing. The dam may appear to assist the calf in locating the udder with appropriate body movements. Cows frequently hide their calves or leave them behind while they leave the immediate area to graze. Therefore, cattle are classed as a hiding species, although newborn calves frequently follow closely during this early period after birth. After a few days it is common for calves to be left in a calf group (kindergarten) that is watched by one cow while the rest graze. Baby-sitting behavior is exhibited more commonly by one or more specific cows in the herd (see figure 5.3).

Ewes usually leave the birth site within 10 hours, and the lamb follows closely until several weeks of age. Sheep are termed a following species.

Bonding

Bonding behavior is the rule in animals just as in humans. An important management tool in livestock production is to ensure bonding of the newborn and its dam. Bonding is the process by which dam and offspring learn to recognize each other; it is referred to simply as parent-offspring bonding as a classification of bonding behavior. Litter-bearing females are already attached to the nest before their young are born, simply because innate behavior places emphasis on birth site selection and nest building. Recognition of individual piglets in a litter, for example, does not develop for several days. This characteristic makes cross-fostering more easily accomplished in this species.

Bonding between dam and offspring occurs within the first few hours following birth in cattle, sheep, and horses. The period for forming a strong bond lasts for only a few hours in ewes and cows. Beyond this critical period (up to about 12 hours), bonding may not occur or may be extremely difficult to accomplish. This characteristic tends to make fostering difficult.

Olfactory identification appears to be the primary avenue for maintenance of dam and offspring association over time, however, auditory and visual characteristics may be important. Price et al. (1984) reported 84 percent successful fostering using odor transfer with stockinettes in lambs.

Animal-human bonding is accomplished in pet and farm animal species. Animals will bond completely to humans in cases of rearing without contact with other animals. In the case of large animal species that, because of increasing size, may easily injure a person and animals that are likely to develop aggressive tendencies, it is important that human dominance be established early to minimize danger as the animal matures. Early and close familiarity with humans is important in management because it minimizes fear and increases the ease with which animals are handled for movement and breeding. A common problem in animal production is fear of humans. Violent fear responses may cause major economic losses in confined birds, with escape efforts resulting in piling and suffocation. This behavioral phenomenon occurs to some extent in swine. Injury of humans associated with animals is also a common result of fear responses by large animals.

Nursing and Providing Food

Teat order, a form of dominance order development, develops among pigs in a litter within 2 to 3 days. The competition is for the more productive forward nipples. Sows nurse their young at intervals usually less than one hour. The nursing sequence consists of a series of piglet squeals, grunts, and udder massage by the pigs. This period is followed by characteristic grunting sounds of the sow just before milk letdown and release. Actual milk release may have a duration of only 10 to 20 seconds. This short time makes preparation for nursing bouts a critical issue for the pigs and is the basis for the characteristic communication behavior by the sow just before milk release. Facility noise may interfere with this effectiveness.

Cattle nurse their offspring at intervals of about one hour initially. The nursing bout lasts about 10 minutes initially, then decreases to half this as the calf develops. Nursing is controlled by the dam, but she responds to demands of the calf to some extent. The ewe may aid the lamb in locating the udder by adjustments in body position and may appear to assist by nudging. Lambs nurse at intervals of the hour or less and for periods that may be as short as one minute as the lamb develops. Twin lambs tend to nurse more frequently than singles. Foals may nurse more frequently during the first few days after birth then at approximately 1-hour intervals after the first few months. Initial nursing bouts may be only one or two minutes. In viewing nursing behavior as a reflection of well-being, it is most critical to observe signs of avoidance by the dam and weakness in the young.

Hens encourage feeding by their offspring by pecking at a food source. The sight and sound of pecking, and the characteristic feeding vocalization of the hen attracts chicks to peck in the same area.

Protection

Offspring are generally vigorously protected against members of the herd during the first few days. Protection from severe weather is through birth site selection and, in the case of pigs, providing warmth in the nesting area. Protection from predators is by flight or through threat or attack.

Learning

What an offspring learns from its dam is important for survival. Some social behaviors are learned by observing the dam. Offspring of dominant dams often become dominant. The dam exposes offspring to the environment, such as sources of feed, water, and salt. Offspring learn to distinguish predators from harmless animals through reactions displayed by the dam and possibly other animals in a group.

JUVENILE BEHAVIOR

Characteristics of Juvenile Behavior

Primary social bonding at the juvenile level appears to be related to those animals with which they have been associated previously, first with the dam and then after a few weeks with peers (Fraser and Broom, 1990) in the group. At weaning, bonding with the mother is altered to a principal association with peers. Peer groups, however, begin formation within the first few weeks postnatally. In pigs, peer groups form within the litter, where grouping normally relates to the prevailing teat order. After weaning, groupings follow species-specific characteristics.

Behavior of early weaned pigs (weaning at 1 to 3 weeks is typical) is different from those nursing for a period of 6 weeks, which was typical several years ago in the swine industry. Early weaned pigs tend to show more behaviors directed at others (e.g., nudging and mouthing the undersides of contemporaries), which are likely related to elimination of the opportunity for nursing at the earlier age. Such behavior may lead to tissue damage in the animals receiving such persistent manipulation (note table 5.3).

Weaning (naturally or imposed by management) normally results in some level of stress. The stress level commonly experienced, however, is mild. If adequate feed is available to the young, adaptation takes place in a matter of 2 or 3 days for most animals. Tranquilizing in the weaning process has shown no consistent value in terms of preventing mild interruptions in performance of the young.

FEEDING BEHAVIOR

Controlling feed intake and the feeding program is a major component in the management of any livestock enterprise. Cost of feed commonly makes up 50 to 70 percent of total production costs. Thus, producers must concentrate on ensuring that the feeding program meets the requirements of animals for the production goals established. Such goals include rate and composition of growth and the various other functions associated with stage of the life cycle (e.g., growth, reproduction, and lactation).

In a production-oriented system, the consumption of nutrients beyond that required for maintenance is usually critical to cost-effectiveness of the feeding program and other inputs in the enterprise. Nutrients above the level required for maintenance are available for production of products or for support of physical activity as might be related to the support of performance of work or athletic activity in performance animals such as race horses.

Factors Influencing Feeding Behavior

Type of Animal

Herbivores and omnivores show feeding patterns related to dietary and species characteristics. Ruminants show unique characteristics related to the digestive process, including extended periods for rumination in which the animal spends both time and energy in regurgitation and chewing of the bolus as part of the routine digestive process. The nature of the diet may exaggerate or minimize differences in feeding behavior of animals within or between species. For example, ruminants and monogastrics fed high-concentrate feeds in a typical finishing program will tend to reflect similar feeding patterns. The feeding pattern for ruminants is quite distinct when consuming high-roughage feeds or pasture forage. Pigs, under conditions requiring foraging as the major method for obtaining food, may spend comparable periods of time in seeking and consuming foods that are typically observed in cattle maintained under range conditions.

Circadian Characteristics

Daily feeding patterns are somewhat characteristic of the species involved but are also affected by environmental conditions such as season and characteristics of the food supply.

Social Factors

Social influences on feeding behavior involve competition and related dominance characteristics in which the feed resource is restricted to varying degrees by the more dominant individuals. Dominance, through restricting access of some animals to feed resources, can be a major management problem. Design of adequate feeding systems including feeder space and operational characteristics is critical. Competition and expression of aggressive behavior increases as feed supplies become more limited. In confinement swine production, for example, limited feeding of sows during gestation is essential to prevent excessive fatness. This creates increased fighting and potential injury. In extensive range or pasture conditions, however, the animal's energetic cost in defending feed resources becomes excessive, and the control by dominant individuals is simply not feasible. Herd influences and leadership by individuals also influence feeding patterns. Groups of herding or flocking animals tend to feed and rest collectively and to move through feeding grounds and to water as groups. Recognizing the relationship between water locations and grazing distribution is important in range management to prevent overgrazing of areas near watering points. Distances and the distribution of watering sites are essential considerations. Visual contact and the related competition (perceived or real) influences motivation for feeding.

Biological Sensors and Neural-Humoral Influences

Feeding centers in the hypothalamus provide both stimulation and inhibition of food consumption. Those located in the lateral hypothalamus are capable of stimulating consumption of food, and those located in the medial hypothalamus are associated with satiety.

Taste and olfactory sensors are clearly involved in the consumption of foods and may represent major management considerations in animal feeding programs. The importance of visual appearance of foods is not clear.

Chemoreceptors reflecting plasma glucose levels and utilization rate are involved in the regulation of consumption, as are sensors reflecting fat storage or level in the body. Control theories based on glucose utilization are referred to as *glucostatic,* and theories based on fat level are termed *lipostatic.* Control based on fat content or body weight is based on the theory (sometimes referred to as the set-point theory) that each individual has a set point or an established value for these characteristics and that biological mechanisms involved in food intake are guided accordingly. As a result, there is a strong tendency for the animal to remain near that set point; even though short-term fluctuations may occur, body weight will return to that level once voluntary food consumption allows. The set-point phenomenon is likely involved in weight and growth curve recovery commonly referred to as compensatory gain. For example, feeder cattle, having experienced a period of reduced gain due to winter temperatures and limited feed, will tend to gain faster than contemporaries wintered at higher levels of energy once both are placed on a full-feed system typical of finishing programs or if given access to excellent grazing.

Humoral influences of hormones such as cholecystokinin, which is released by the duodenal mucosa, and insulin, released by the pancreas, are well established in relation to food intake regulation. Failure of these tissues with respect to humoral function has broad influences on both food intake and utilization. Cholecystokinin is known as a regulator of satiety.

Rate of Ingestion

Rate of ingestion reflects the animal's capability of both finding and consuming food and therefore is related to feeding patterns. The character of the diet frequently is a major determining factor in this characteristic and relates to nutrient density or ratio of fibrous feeds to higher-energy concentrates such as grains.

Rate of Digestion and Passage of Digestive Tract Contents

The time required for food to pass through the digestive tract is a factor determining feeding characteristics since it may influence feeding frequency and duration. Because of the nature of the diet and anatomy of the digestive tract, ruminants experience greater variation in rate of passage of gastrointestinal contents.

Nutrient Density or Dilution

Nutrient density is a term referring to the concentration of a given nutrient in relation to the total weight of a unit of feedstuff. Animals tend to compensate for reduced nutrient density by eating more total feed in order to achieve a given level of a particular nutrient. The compensation level may be limited, however, by the capacity of the digestive tract. Cattle being finished in the feedlot will maintain about the same level of energy intake when consuming feeds ranging from ratios of concentrate to roughage of about 50:50 to 90:10, depending on the quality of the roughage involved. Diluting nutrient density is the basic concept used for limiting energy in feeds and thereby controlling weight and excessive fatness of gestating sows (short feeding periods may result in aberrant behaviors. Note table 5.3). Including higher levels of fiber by adding a source of roughage causes the character of the feed to limit consumption by exceeding the capacity of the sow's digestive tract.

Palatability

Animals will reject off-flavored foods. Common problems are contamination, presence of mold in stored feeds, and spoilage. Method of feed processing and the presence of certain feed ingredients may influence palatability. Cattle, swine, and horses tend to show some preference for sweet flavors. Salt may influence palatability. Salt-deficient animals may select higher-salt feeds; however, this preference may be the result of nutrient need rather than palatability.

Intake Depressants

Factors tending to depress intake include achieving body weight set point, high environmental temperature, high estrogen concentrations related to the estrous cycle, digestive tract fill or distention, character of rumen fermentation, end products present in that organ, and parasitism. Generally, activity depressants administered as therapeutics, such as opiates, barbiturates, and benzodiazepines, also reduce food intake. Although low temperatures may stimulate feed intake to meet requirements of the animal, extremes (e.g., below $-10°C$) may reduce intake as a result of the animal refusing to be exposed to the cold in moving to the feed source or leaving the protection of shelter. The animal may also be reluctant to leave the protection provided by a group of animals that are closely assembled for thermal protection.

Intake Stimulation

Production levels such as high levels of lactation and higher growth rates are normally associated with higher levels of feed intake, given a static maintenance requirement level; thus, in many livestock enterprises, higher levels of feed intake are desirable. Cooler temperatures and particularly cold at the lower critical temperature will tend to increase intake in order to meet energy demands. Factors such as feed additives, implants, or injections that increase level of production may also increase total feed consumption. Efficiency of feed utilization may be maintained or actually improved in such cases by the higher proportion of total feed going into production relative to that supporting maintenance. This efficiency forms the basis for management to achieve high levels of productivity in farm animal enterprises in order to minimize costs per unit of production and in turn the cost of the resulting product. Efforts to develop products that dramatically increase feed intake in farm animals have been extensive and of limited success.

Feeding Interference

A number of factors interfere with normal food consumption patterns. These are important considerations in animal management to ensure dietary adequacy. Examples are climatic influences, presence of predators, abusive competitors, food spoilage, parasite infestations, and disease. From this list, it is obvious that maintaining satisfactory levels of feed intake is a major issue in managing animal enterprises for economy as well as animal well-being.

Food Selectivity

Animals demonstrate an ability to select higher-energy, more nutritious foods. Whether this capability reflects a rationalization of the energetic cost of obtaining and utilizing feeds is not known. Animals have a tendency to consume food at a level to achieve amounts of energy to meet needs for maintenance and production. That level is largely determined by genetic potential, assuming an adequate environment, and will compensate for reductions in nutrient density (fiber level) to the extent that the capacity of the digestive tract will allow. Consequently, high-fiber feeds may limit consumption of adequate nutrients simply because the animal cannot consume enough of the lower-energy-density material to meet the need established for a given level of production.

There is no strong evidence that animals have self-regulatory ingestive systems that give them an adequate intrinsic ability to select the best diet from a cafeteria source of nutrients. Thus, in most animal production systems, the most effective management practice is to formulate balanced diets to meet all known nutrient needs. This formulation does not, however, dictate the use of completely mixed diets. Some effective feeding programs involve separate feeding of primary energy, protein, and mineral components.

Experience

Learning is a factor in feeding behavior because it influences pattern and efficiency in finding and consuming food in many circumstances. Animals learn and follow others in feeding and they learn, through mental mapping, the location of such resources.

GRAZING BEHAVIOR

Many of the previously mentioned factors also influence the behavior of animals subsisting largely on pasture or range forages. Since this area has been studied extensively, it is useful to review some of the animals' characteristics observed in this particular type of animal production system.

Except in cases where pastures are primarily of single species and lush, the act of obtaining adequate nutrients from foraging is a complex and possibly major challenge for animals. Suitable forage must first be located. Some animals exhibit effective memory in terms of returning to certain areas after suitable regrowth has occurred. Following location of a suitable grazing area, selective grazing to pick and consume the most nutritious plants or plant parts is routine. Toxic plants may be refused based on previous negative encounters and possibly by learning from maternal association. As pasture conditions influencing quality of material deteriorates, animals are more likely to consume such dangerous plants, however. One might ponder the level of cognition required to effect good selection of grazing sites and selective consumption (choosing the more nutritious and safe materials).

Factors Affecting Grazing Behavior

Animal Species
Anatomical differences among species influence the way the animals sever plant materials and both the closeness of cut to the soil and the ability to select and sever small parts of plants. Cattle have greater difficulty in selecting plant parts than sheep and horses.

Forage Quality and Quantity
Quality and quantity of forage interact to minimize grazing time and distances covered. Grazing times typically range from about 6 to 8 hours daily on improved pastures to 10 hours or more on more arid ranges.

Plant Population Composition
When possible, animals will select the safest, most nutritious plants and plant parts. Overgrazing and high animal density limit selection.

Time of Day
Typically animals graze longer periods right after dawn and late evening, with a shorter period around midday. Patterns are altered by thermal conditions.

Season and Weather Conditions
Cold weather tends to reduce early and late day grazing periods. Hot weather reduces or eliminates midday grazing.

Insect Annoyance
The presence of a heavy infestation of flies may shorten grazing times and increase time spent standing in water if ponds or streams are available.

Scheduling

Factors such as milking times for dairy cows and goats, riding time for horses, and other management-imposed schedules alter grazing behavior related to time of day.

Efficiency of Harvesting

Some factors influencing harvesting efficiency by the animal include the following:

Stage of plant growth.
Volume of material that must be consumed to meet nutrient requirements.
Physical condition of the grazing animal.
Travel distances required.

Distances obviously influence the amount of time required in feeding patterns. Arid areas typically require animals to travel distances up to 8 or 10 miles daily. More productive pasture areas require much less travel. The distance required to access supplemental nutrients (e.g., salt, water, protein, and energy concentrates) influences required travel distances.

ABERRANT BEHAVIOR

Stereotypies

Stereotypies is the term used to describe stereotyped behaviors. Such behaviors, as a group, are defined slightly differently by ethologists; however, these definitions usually include several common characteristics and suggest generally that these scenarios are movements or actions typically out of the animal's natural repertoire that develop into repetitive movements that are fixed. (See Lawrence and Rushen, 1993, for an extensive review of research related to stereotypies.)

Stereotyped behaviors should be viewed not only as a reaction to stress but as a response that allows the animal to cope with aversive environmental circumstances (Dantzer and Mittleman, 1993).

Some ethologists conclude that causal factors for specific stereotyped behaviors have not been adequately determined. There are, however, some environmental characteristics that are typically associated with such behaviors. Factors such as confinement, social deprivation, and environments described as sterile and of limited complexity are commonly involved. Additionally, restricted feeding (Terlouw et al., 1991; Savory et al., 1992; Rushen, 1984) lack of exercise or limited individual space may be associated with stereotypies (Krzack et al., 1991; Terlouw et al., 1991); however, others report that such conditions do not routinely cause stereotypies. Restricted ability to perform normal exploratory behavior is associated with certain stereotyped behaviors (Stolba et al., 1983). The above situations can be judged as those that frustrate goal achievement.

Variation among individual animals in response to environmental characteristics that appear to be inadequate for some animals underlies the caution related to concluding that specific causal factors explain typical stereotyped behaviors.

In a review of some factors forming the basis for stereotypies, Rushen et al. (1993) provided an approach to considering these behaviors in relation to moti-

FIGURE 5.18 Route tracing in this horse is accompanied by an unusual combination of head slinging and side carriage over the fence that is repeated as the animal moves along in the well-worn path

vational systems. This results in a degree of commonalty of aberrant behaviors relative to associated environmental factors. Motivational states considered were those related to feeding, locomotion, aversion, exploration, mating, and species specific behavioral characteristics. Rushen et al. concluded that the growing evidence that specific motivational states controlling normal behavior are involved in the development of stereotypies provides useful insight as to causal factors.

While further research will define relationships between environmental factors and stereotypies more clearly, there are certain environmental inadequacies that are typically associated with stereotyped behaviors. Several stereotypies are associated with close confinement and restricted movement. *Route tracing*, commonly observed in zoo animals and in confined horses, is an example. Figure 5.18 illustrates route tracing by a stallion. *Stall walking* by stabled horses is a form of route tracing. Caged laying hens may *pace* in stereotypic form as a result of confinement, difficulty in feeding, or inability to engage in normal nest building behavior (Wood-Gush, 1972). Caged hens may also demonstrate *head shaking*, a very quick rotational movement of the head in stereotypic form. Horses confined to box stalls may engage in violent *stall kicking*, *swaying*, *weaving*, or *pawing*. *Windsucking (aerophagia)* and the often associated *cribbing* (figure 5.19) appears to be more common in confined horses. *Eye rolling* is a stereotypies observed in veal calves. This behavior is characterized by extreme rolling of the eyes back into the socket and occurs more commonly in calves maintained in crates. In general, the above stereotyped behaviors are

FIGURE 5.19 Cribbing and wind sucking behavior in the horse (Courtesy of E. O. Price, University of California–Davis; photographed by Lisa Holmes, University of California–Davis)

associated with confinement systems that involve very restricted movement and often an environment that provides little or no opportunity for normal social and environmental interaction.

Some of the more common stereotypies observed in farm animals will be discussed later in relation to environmental associations and some possible preventive approaches. It is important to recognize that there is sufficient evidence to suggest that the presence of stereotypic behavior is a reflection of an inadequate environment. Further, the performance of these behaviors, while recognized as coping mechanisms, must also be viewed as a reflection of suffering that is the basis for their development. Wemelsfelder (1993), in a review, makes a compelling case that animals that exhibit these behaviors do so because they experience frustration and boredom, which lead to depression and possible anxiety. Wemelsfelder concludes that even though animals may maintain homeostasis in impoverished environments, suffering is involved in the development of these behaviors; and, thus, well-being is compromised. Therefore, observation of behaviors like those discussed should prompt efforts by managers of animal enterprises and those involved in animal care to design and provide environments that prevent behaviors that are reflective of inadequate environments.

Stereotyped behaviors do not lend themselves to classification due, in part at least, to individual animal variation in character of the behavior and to inconsistency in response of individuals within a common environment. However, some stereotypies are thought to be related to motivational states. Several of these behaviors are associated with feeding and oral exploration of the environment. *Sham chewing* in swine is illustrated in figure 5.20. This behavior is common among sows

FIGURE 5.20 Sham chewing by a sow

housed in gestation stalls. Oral stereotypies do occur, however, in sows maintained on pasture or in dry lot (Dailey and McGlone, 1997). Lack of environmental stimuli normally associated with foraging and exploration is also associated with this behavior. *Tongue rolling* in calves and cows (figure 5.21); *cribbing* in horses and cattle, *crib whetting* in horses, *wood chewing* in horses and cattle, and *bar biting* in swine and cattle (figure 5.22); and *drinker pressing* in swine (figure 5.23) are aberrant behaviors related to motivation for feeding or oral exploration.

The stereotyped behaviors discussed previously and other aberrant behaviors are described further in tables 5.3, 5.4, 5.5, 5.6 and 5.7. These should be recognized for what they are—abnormal behaviors related to environmental inadequacies. As a result, corrective action is in order. Possible environmental associations with these behaviors are outlined along with environmental adjustments that could be considered as approaches to prevent or correct such behaviors. It should be noted, however, that environmental changes that might have prevented the development of a behavior originally may not be effective as treatment in correcting established stereotypies or other aberrant behaviors.

Other Behaviors Impacting Well-Being or Performance

A number of aberrant behaviors associated with environmental inadequacies (figures 5.23 to 5.28) that result in self directed injury are not typically classified as stereotypies. Examples are polydipsia and flank biting. Additional behaviors, not classed as stereotypies, result in injury to others and are similarly associated with environmental inadequacies. Examples are tail biting in pigs and feather pecking in chickens. Other behaviors that may result in injury are described by Fraser and

FIGURE 5.21   Tongue rolling behavior in a dairy calf (Courtesy of Carolyn Stull, University of California–Davis)

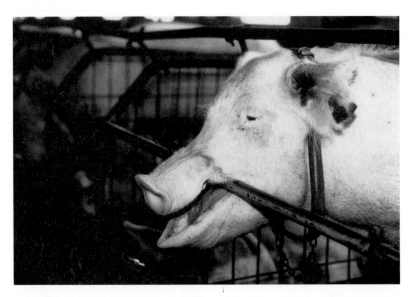

FIGURE 5.22   Bar biting behavior in a sow (Courtesy of E. O. Price, University of California–Davis)

FIGURE 5.23 A sow showing the behavior termed drinker pressing, which is performed for extended periods

FIGURE 5.24 The chicken has been beak trimmed to minimize damage to conspecifics by feather or body pecking; evidence suggests that some genetic lines of birds are less likely to engage in this behavior and, as a result, the need for beak trimming as a routine practice is reduced (Photograph courtesy of George Brant, Iowa State University, Ames)

FIGURE 5.25 Belly nosing in a group of pigs (Photograph courtesy of L. M. Hohenshell, Iowa State University, Ames)

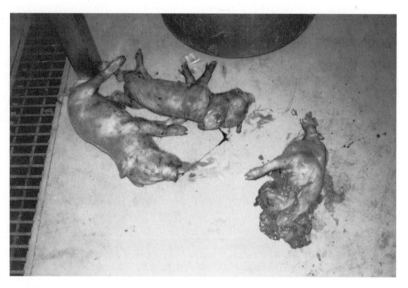

FIGURE 5.26 Newborn pigs killed immediately after birth by the dam; cannibalism followed

FIGURE 5.27 Dog-sitting sow; the behavior may reflect boredom, unsoundness, or inadequate spacing for normal lying or rising

FIGURE 5.28 Tonic immobility in the chicken

Broom (1990) as reactive behaviors. Examples are hysteria in poultry or swine. Still other aberrant behaviors have negative impacts on performance. Incompetent reproductive performance and rejection of offspring are examples. While specific causal factors may not be clearly established for aberrant behaviors, many are associated with environmental characteristics described as confining, isolating, lacking enrichment or complexity, lacking in social contact, or those that preclude a normal behavioral repertoire.

Cause and Effect Relationships in Aberrant Behaviors

Tables 5.3, 5.4, 5.5, 5.6, and 5.7 summarize cause and effect relationships involved in aberrant behaviors for the different species. They also suggest possible avenues to corrective action in altering the characteristics of the animal's environment. Such adjustments may involve one or a combination of changes in physical, dietary, or social components of the environment. It should be noted that a large number of the aberrant behaviors described are related to inadequacies in the physical environment. These behaviors are typically related to confinement, isolation, and boredom, or a sterile environment and suggest need for an improved environment.

TABLE 5.3 **Characteristics, Causes, and Possible Avenues for Prevention of Aberrant Behaviors Associated with Environmental Interactions in Swine**

Problem Behavior	Characteristic Behavior Observed	Probable Cause	Environmental Considerations
Tail biting	Occurs at various weights but frequently when new groups are formed Pigs start this practice by chewing the tail of another; other pigs join in the behavior, complete removal of the tail may occur and biting other parts of the body may follow. Secondary infections are common	Although there may be breed differences in predisposition (Fraser and Broom, 1990), causes appear to be related to crowding for space, feeder, and water access	The common preventive practice is docking of the distal half of the pig's tail Removal of pigs that are subjects or offenders Reduce environmental stressors
Belly nosing	Pigs massage the underline and areas between fore and rear legs of subjects. The activity may continue until injury occurs	Associated with early weaning, sterile environments, and high population density	Later weaning may eliminate the problem Access to straw for manipulation

TABLE 5.3   (Continued)

Problem Behavior	Characteristic Behavior Observed	Probable Cause	Environmental Considerations
Anal massage	Occurs in growing pigs and is characterized by snout rubbing of anal area	Crowded conditions appear to be the primary cause of this behavior	Reducing population density and removing animals showing the behavior Provide straw for oral manipulation
Killing young	Sows may become hyperactive at parturition and crush pigs; cannibalism may follow Neonatal rejection may occur, and maternal aggression may occur; pigs may be killed and eaten In some cases sows simply attack the offspring as any other strange individual Inflamed udders causes some sows to attack nursing pigs	These activities seem to accompany hyperactivity and may be most typical among first-litter females	Farrowing crate systems largely eliminate the problem Providing quiet farrowing quarters for pre-parturient sows and allowing a few days to become adjusted Sows demonstrating a strong likelihood of cannibalism or the initiation of it should be isolated as a control measure Sedation may be used
Sham chewing	Animals chew with no food present; frothing at the mouth is common	Appears primarily in sows stalled alone in a sterile environment, limited feeding, lack of bulk in the diet, and limited eating time (Rushen, 1984)	Access to some straw or shavings may reduce the activity The condition is less likely to occur in group housing systems
Bar biting, tether chewing, crib biting, snout rubbing	Characteristically the animal holds a pen or crate bar in the mouth and moves the mouth back and forth Snout rubbing may be associated	Appears to be primarily associated with maintenance of sows in crates without bedding material	Provide some material such as straw or shavings that provide opportunity for chewing and rooting (an enriched environment)

TABLE 5.3   (Continued)

Problem Behavior	Characteristic Behavior Observed	Probable Cause	Environmental Considerations
Drinker pressing	Constant pressing or manipulation of water nipples; may cause sanitation and waste management problems in the facility	The drinker appears to be the most interesting thing available to occupy the animal's time	Provide an enriched environment
Dog sitting	The animal sits like a dog rather than lying down; this is an abnormal posture for swine and may result in urinary infections	Typically associated with restrained activity in crates	Provide suitable space for normal lying down and rising Floor surface should allow good footing to prevent slipping
Apathetic or unresponsive behavior	Animals appear to be unaware of things going on around them	Associated with confinement for extended periods	Provide an enriched environment

TABLE 5.4   Characteristics, Causes, and Possible Avenues for Prevention of Aberrant Behaviors Associated with Environmental Interactions in Sheep

Problem Behavior	Characteristic Behavior Observed	Probable Cause	Environmental Considerations
Wool pulling and eating	Wool is pulled from other sheep	Associated with crowded housing conditions May be associated with parasite infestation Lower ranking individuals are usually the target (Lynch et al., 1992) Rarely exists when sheep are on pasture (Lynch et al., 1992)	Provide adequate space to avoid crowding Removal of offender
Chewing on newborn	Dam chews on newborn's appendages Death may occur, but severe damage is more likely	Occurs under crowded housing conditions	Provide comfortable lambing quarters and separation from other animals to minimize excitability of the ewe

TABLE 5.4   (Continued)

Problem Behavior	Characteristic Behavior Observed	Probable Cause	Environmental Considerations
	The behavior is usually covert (Fraser and Broom, 1990)		
Stealing young	May occur in ewes during late pregnancy in which lambs attempting to nurse may be accepted	Occurs more in intensive production systems due to separation of offspring and dam Dams adopting alien lambs before parturition may reject their own at birth (Hart, 1985)	Separate lambing pens for effective bonding Maintenance of ewes and new lambs in small groups to minimize separation potential
Neonatal rejection	Refusal to tend newborn lambs; aggression may accompany refusal to allow young to nurse	Cause of fetal rejection is not clear; occurs most commonly after first parturition and after diffcult birth or separation for several hours (Lynch et al., 1992)	Individual penning and care of the ewe and lamb to assure bonding Violent aggression may require sedation or foster rearing

TABLE 5.5   Characteristics, Causes, and Possible Avenues for Prevention of Aberrant Behaviors Associated with Environmental Interactions in Cattle

Problem Behavior	Characteristic Behavior Observed	Probable Cause	Environmental Considerations
Head rubbing	Frequent and extended rubbing of jaw or poll of head area on stationary objects	Related to confinement in small pen or stall	Altered housing arrangement and exercise appear to be the only solution
Eye rolling	Occurs in calves maintained in crates Animal may appear frozen (no movement) except rolling eyes to expose whites of the eye	Appears to be related to confined space	Altered housing arrangements
Tongue rolling	Most common in early weaned calves but may occur in all ages of cattle	Confined housing environments appear to be involved	Supply adequate roughage and a well-balanced diet in total

TABLE 5.5   (Continued)

Problem Behavior	Characteristic Behavior Observed	Probable Cause	Environmental Considerations
	The tongue is extended and the tip rolled back into the mouth The animal may swallow and gulp air	Also related to deprivation of sucking in young calves Possibly related to low roughage in the diet of older animals Occurs most commonly in the grazing species (Mason, 1993)	Since other animals may pick up the practice by observation, visual separation is recommended
Buller syndrome	A buller steer is one that is constantly ridden by others in the group. Occurs commonly in beef cattle feedlots among steers Subjects may be ridden to the point of exhaustion and physical injury Intensity is increased when new animals are introduced into a group or when groups are mixed	Cause is not clear, but increased incidence occurs with some hormone treatments Limited to a few individuals (about 4 percent) (Klemm et al., 1983, 1984) Activity increases with the addition of new animals to a group Crowding stress and group size may be associated factors Bulling may be a part of hierarchy contesting (Klemm et al., 1983, 1984)	Isolation of affected animals for a few days may correct the problem in some animals Movement to an environment of less population density Avoiding introduction of unfamiliar arrivals in established groups
Wood chewing	Chewing on fences and feed bunks in confined conditions	May be associated with crowding Appears to be related to confinement and the limited time required in feed ingestion, which is typical of cattle being finished in the feedlot May occur more among cattle fed very-high-concentrate diets	Altered space allowance and adequate levels of dietary roughage may be helpful, but the behavior may continue to occur under confined conditions even in good nutritional environments

TABLE 5.5 (Continued)

Problem Behavior	Characteristic Behavior Observed	Probable Cause	Environmental Considerations
Hair licking and intersucking in calves	Frequent licking of self or other animals Hair balls in the rumen result In intersucking, calves try to nurse from all appendages; this behavior may occur to the point of injury	Most frequently associated with weaning at or soon after birth Non-nutritive sucking is stimulated by consumption of even small amounts of milk or milk replacer (de Passile et al., 1996)	Use of nursing systems (artificial teat) for feeding milk with replacer Nursing time of about 30 minutes tends to reduce the level of the behavior (Fraser and Broom, 1990)
Intersucking in adults	Most common in dairy cows; may occur in pairs of cows; teat damage may result	Occurs more commonly in larger groupings of cows	Application of devices to the face of the cow to prevent sucking; these devices cause pain to the subject
Neonatal rejection	Refusal to allow the newborn to nurse Aggressive action may also occur General lack of maternal care of the offspring	Cause is unclear; may, in some cases, result from lack of early bonding More common in first-calf heifers, especially in those having experienced dystocia	Separation of dam and offspring from a group and intensive management in aiding the young to nurse Aggressive behavior may require sedation or separation of the offspring to a foster mother or artificial rearing
Milk ejection reflex failure	Failure of cow to release milk in response to milking or nursing stimuli	Disturbances such as noise or shock (Hart, 1985)	Avoid excessive and unusual noise and practices at milking time Ensure absence of stray voltage in milking facilities Avoid management activities that disturb a group of cows such as disruptive treatments, aggressive action of dogs, and introduction of unfamiliar cows

TABLE 5.5   (Continued)

Problem Behavior	Characteristic Behavior Observed	Probable Cause	Environmental Considerations
Stampeding	Violent escape response by a group	Excitement, fear, unfamiliar noises	Avoid unusual noises Special precautions during potential disturbances

TABLE 5.6   Characteristics, Causes, and Possible Avenues for Prevention of Aberrant Behaviors Associated with Environmental Interactions in Horses

Problem Behavior	Characteristic Behavior Observed	Probable Cause	Environmental Considerations
Pacing, stall walking, box walking	Constant pacing around the stall, sometimes to near exhaustion	Confinement and lack of exercise for an extended period	Provide for adequate exercise
Head banging, stall kicking	Violent banging of head on walls of stall, kicking walls or fences	Probably due to restraint and isolation	Provide more space Allow access to pasture or exercise area Avoid conditioned feeding response (Houpt, 1991)
Tail rubbing	Rubbing the rear end against a fence or stall wall to the extent of hair removal	Excess grooming behavior May be related to parasite infestation	Ensure proper parasite control Allow access to other horses for mutual grooming behavior
Windsucking (aerophagia)	Preceded by bowing the neck; the mouth is opened and air is sucked in and swallowed; air may also be expelled with considerable noise May be associated with cribbing	Cause is unclear but may have hereditary predisposition It may also have a social basis but transfer through visual learning from other horses is not established (Houpt, 1991)	Affected animals should not be used for breeding Various mechanical devices such as straps around the neck are available; effectiveness is variable
Cribbing	Grasping and holding a solid object with teeth Wind sucking may be associated	Probably due in part to isolation	Use of cribbing strap Reduce level of confinement

TABLE 5.6   (Continued)

Problem Behavior	Characteristic Behavior Observed	Probable Cause	Environmental Considerations
Tongue rubbing	Frequent and continued rubbing of the tongue on stall walls, feeders, and so forth	Probably due to isolation for extended periods	Allow contact with other horses
Licking	Excessive self-licking	Probably due to isolation for extended periods	Allow contact with other horses
Crib biting, wood chewing	Chewing on boards, pipes, mangers, posts	Probably due to isolation or inactivity	Allow contact with other horses and exercise Stallions chew wood less during breeding season
Flank-biting and related disorders	Biting flanks occurs more commonly in stallions Presence of scar tissue in flank area is indicative if the behavior is chronic	Believed to be due to isolation (Krzack et al., 1991)	Allow contact with other animals and exercise
Polydipsia	Overconsumption of water Excess urination results and may cause sanitation problems (Houpt, 1991)	Believed to be due to isolation when *ad libitum* water is available	Water restriction
Neonatal rejection	A mare's refusal of the newborn foal	Cause is unclear, but the problem is most common in primaparous females	Intensive management to allow newborn to nurse and bond effectively
Chronic standing (motionless)	Lack of movement for extended periods	May be due to injury or inflammation but is also related to isolation	Providing access to other horses in a group environment unless the problem is due to disease or injury

TABLE 5.7   Characteristics, Causes, and Possible Avenues for Prevention of Aberrant Behaviors Associated with Environmental Interactions in Chickens[a]

Problem Behavior	Characteristic Behavior Observed	Probable Cause	Environmental Considerations
Hysteria	Panic reaction, vocalization, and wild flying May be sufficiently violent to cause injury from scratching, bruising, and crowding May be a major problem when gathering broilers for marketing It has been shown (Duncan et al., 1986; Nicol and Saville-Weeks, 1993) that mortality is reduced when machines rather than workers catch poultry	Unexpected or unfamiliar noise Population pressure may increase probability of such reactions Some genetic lines may be more susceptible (Hansen, 1976)	Routine signals such as knocking before entering the housing facility Limited access by strangers, who may unknowingly act to set off a reaction Reducing the number of hens per pen or cage may lower the tendency for birds to react violently Declawing can reduce injury level
Feather pecking	More common in older birds but occurs in both young and adult birds of most species Birds concentrate primarily on ventral areas and the back and vent area This behavior is especially prevalent in intensive management systems (North and Bell, 1990) The behavior may lead to body pecking and cannibalism	Confinement, crowding, and other environmental conditions that create discomfort and high levels of competition for space or feeding Craig (1981) concluded from an extensive review that increased incidence is associated with bright lighting, dense or multiple caging, shorter feeding time associated with pelleted feeds, and larger group sizes (see also Craig and Muir, 1996)	Ensure adequate space and minimize competition in feeding and drinking Because this behavior is thought to be related to exploration and searching, it is sometimes suggested that whole or cracked grains be added to the diet to extend feeding time Beak trimming and declawing to reduce injuries that lead to feather or body pecking are common management practices but may not be allowed in some areas Reduced light levels in the house can be an effective aid

[a] Some characteristics apply to other species of poultry

TABLE 5.7   (Continued)

Problem Behavior	Characteristic Behavior Observed	Probable Cause	Environmental Considerations
Body pecking	Occurs in both young and adult birds and most commonly in intensive management systems Usually starts by birds pecking at ventral areas or back Pecking may also occur about the head and toes Initial injuries may result in cannibalism and the behavior may transfer to others (North and Bell, 1990)	Confinement and stressful environments	Since the two behaviors (feather pecking and body pecking) tend to be related, control measures used to minimize feather pecking are typically utilized, hens per cage should be limited to 8 if possible
Polydipsia	Excessive manipulation of drinkers and consuming excessive amounts of water	Boredom is a primary cause of this behavior	Intermittent water allowance; it should be noted, however, that this practice requires very careful management to ensure adequate water allowances for all birds in a housing system
Head shaking-flicking	Brief, repeated periods of head movement from side to side, dipping, and rotation	Thought to be related to the bird's perception of danger and inability to escape because of confined conditions	Appropriate space allowances Group housing reduces the incidence

BEHAVIORAL CONSIDERATIONS IN ANIMAL HANDLING

Animal enterprises are characterized by many management practices involving movement, capture, and restraint. These practices are associated with efforts related to such things as animal care, facility management and maintenance, marketing of animals, breeding, feeding, and observation. Handling involves interactions among people, animals, and facilities. Understanding animal behavior and related sensory capabilities, as well as adequacy of the handling environment, is important in ensuring safety for both animals and the workers involved. Because handling practices usually involve confinement and restraint, these become important components in designing facilities and systems for both animal care and economic considerations related to stress and injury. (For a comprehensive review of animal handling consideration see Grandin 1988, 1992a, 1992b, 1995.)

Animals learn to accommodate reasonable handling methods and thus are basically trainable to go through working chutes, up ramps, to milking stalls, and through passageways. Those working with animals are well aware that they establish a preference order for certain facilities such as specific stalls or stanchions and will persist in achieving access to these areas.

Handling animals early in life can influence ease of handling later as a result of human-animal association. Such experience establishes expectations that injury will not occur, eliminating fear associations and practices that may typically be involved in the system in which the animal is maintained. Handling effects may be transferred among animals through observational learning.

Intergenerational transfer of handling effects may occur if such effects impose selection pressure. This selection pressure results in a behavioral shift that has occurred in the processes associated with domestication, for example.

Sensory Capacities and Handling

Animals do not see, hear, or smell their environment in the same way as humans. Recognition of this fact and the differences that may exist are important factors in whether various handling methods succeed. Sensory capabilities of animals have been discussed in detail previously in this chapter. Visual capabilities, for example, may explain an animal's response to being approached (figs. 5.29, 5.30, 5.31, and 5.32). Cattle, horses, and sheep have a visual field of nearly 360 degrees with only slight turning of the head. This wide-angle vision, however, limits binocular vision and related depth perception. Thus, it is more likely that strange objects or individuals may appear suddenly and surprise the animal to the extent of creating alarm or at least require that the animal stop and investigate before proceeding.

Farm animals have much greater hearing ability than humans and are more likely to react to strange noises. Consequently, the caretaker learns to move cautiously and quietly around animals. Observing the animal's attention characteristics is useful. The horse's ear orientation in figure 5.33 is an illustration of attention direction. In some cases animals are intentionally alerted by workers making a familiar noise before entering a housing unit to prevent flight or even hysteria.

Most livestock do not like to move into areas of darkness. Thus, lighting is critical to ease of movement in many cases which is probably related to an open grassland evolution. Animals desire consistency in character of footing; for example, they will stop or shy away from the shadow of a post, power line, or guy wire in their pathway and will not normally cross a cattle guard.

The sense of smell is more highly developed in farm animals than in humans. Thus, they are reluctant to approach spaces or objects that have an unfamiliar odor.

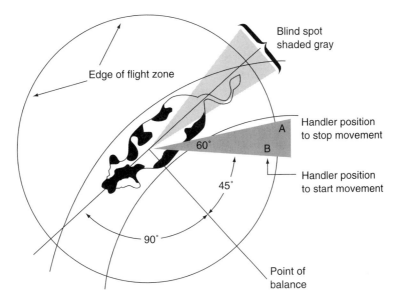

FIGURE 5.29 Flight zone and point of balance characteristics of cattle (*Source:* Grandin, T., Behavioral principles of cattle handling under extreme conditions. In *Livestock Handling and Transport.* Edited by T. Grandin [Wallingford, UK: CAB International, 1995], p. 48, by permission)

The previously mentioned characteristics are examples of a few factors related to principles of animal behavior that may be important considerations in devising safe, efficient handling methods. Understanding these principles allows a reasonably high level of predictability in terms of animal responses to certain stimuli. This predictability, in turn, contributes to successful handling.

Personnel Considerations

A primary goal of the handler must be to prevent unnecessary suffering. Managers must assume responsibility for proper training of those involved in handling to ensure proper care and safety, as well as effective observation and detection of problems related to animal well-being.

The handler must recognize that a bad experience contributes to future problems because the animal is an effective learner from such experiences. Smoothness, quietness in movement, and careful tactile communication facilitate handling and have positive effects. Excessive force usually has negative effects.

Any activity that may produce fear must be minimized. Unfamiliar noise, erratic action, and pain produce fear and future aversive action. Animals respond favorably to routine. Thus, scheduling and consistent procedures contribute to handling ease. Animals are more comfortable with the familiar. Familiar individuals, facilities, surroundings, and routes, make handling and movement practices more successful.

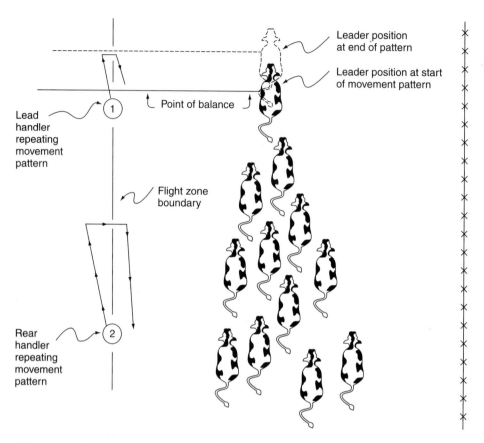

FIGURE 5.30 The appropriate area of activity to be occupied by the person controlling movement of the group (*Source:* Grandin, T., Behavioral principles of cattle handling under extreme conditions. In *Livestock Handling and Transport.* Edited by T. Grandin [Wallingford, UK: CAB International, 1995], p. 50, by permission)

General Handling Considerations

Cattle

Bulls, even if tame and conditioned to routine handling, should be approached with extreme care at all times. Think of escape possibilities before an emergency arises.

Handling equipment, halters, nose leads, and similar control devices should be examined routinely to ensure proper functioning and freedom from sharp edges, lack of slip knots, and properly functioning releases to allow an appropriate response and to prevent possible injury to the animal or workers.

Cattle, like most animals, fight restraint. Thus, minimum restraint should always be a goal to minimize fear and stress. Power-operated squeeze chutes should be evaluated carefully to ensure that excessive force is not involved. Too much force is not uncommon and may result in severe injury to the animal when squeezed.

Attempts to move cattle too rapidly commonly result in excitement and may evoke damaging responses.

Keep in mind flight distance characteristics of the animal in order to assist movement in the desired direction. Recognize that animals have a point-of-balance

FIGURE 5.31
Appropriate area of activity for the person involved in controlling movement of the group of animals (*Source:* Grandin, T., Behavioral principles of cattle handling under extreme conditions. In *Livestock Handling and Transport.* Edited by T. Grandin [Wallingford, UK: CAB International, 1995], p. 52, by permission)

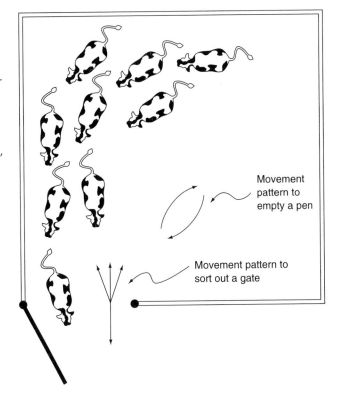

Movement pattern to empty a pen

Movement pattern to sort out a gate

FIGURE 5.32 Flight distance in cattle: when the handler is immediately outside the flight zone, the steer orients toward the handler; as the handler takes only one step into the flight zone, the steer moves away

(a)

(b)

FIGURE 5.33 The horse's attention is directed (a) forward, (b) to the rear, and (c) both forward and to the rear, as shown by orientation of the ears

(c)

characteristic. This characteristic determines the direction the animal is likely to move and can be used to achieve the desired movement (note figure 5.32).

Try to avoid approaching in the blind spot angle to the rear of animals, where they can hear but cannot see (figure 5.29). Such an approach makes escape or defensive action more urgent and usually results in a kicking response.

Avoid placing pressure on an animal by approaching it in an alleyway or corner. Cornering an animal increases the likelihood of it becoming uncontrollable and possibly running by or over you. This response is especially common when an animal is separated and desires to return to a group.

Cattle and sheep tend to circle around the handler to view the person at all times. The circling characteristic in animal movement forms the basis for some handling strategies.

Make every effort to remove shadows and objects from the desired route of movement. Animals will stop to explore and may be diverted by such objects.

In almost all cases, cattle resist moving into a closed area. Thus, the presence of an opening that provides the feeling that they will be able to continue moving is almost essential to gain free movement into a working chute.

Cattle are distracted by being able to see things or activities outside of a chute as they move through it, causing the normal behavioral response of resisting movement. Closed chute sides provide for easier movement as long as the route ahead appears to be open.

Swine

Restraint prompts pigs to be noisy. This response cannot be prevented, and if the handler is sensitive to such noise a protective device should be used.

Extreme care should be taken when working around boars. In cases where boars are fighting, the possibility of injury to the handler is great when attempting to separate the animals. Solid panels are essential to aid in separation and may provide effective protection for the worker.

Pigs should not be supported by their by ears or tail. Hold pigs by supporting the body trunk, head, and neck, as appropriate. Pigs can be held by the hind legs with the anterior part of the body between the holder's legs. This is a common hold for castration. Pigs too heavy to hold by lifting must be held by crates, nose snare, or a rope snubbed around the upper jaw behind the canine teeth.

Pigs should be moved along corridors or alleyways by use of handling panels. Canvas slappers are also used effectively for this purpose. Movement should not be rushed. Pigs should not be struck with hard objects such as canes or poles that may cause pain or brushing.

As with cattle, some appreciation of the animal's response to approach by handlers will aid in an ability to predict the nature of movement and in handling.

Sheep

Sheep have strong flocking instincts and will make great efforts to return to a group. Separation results in fear and excitement and makes handling difficult. Large sheep can easily injure a person as a result of efforts to return to the flock.

Isolation causes a high level of stress. Extreme excitement by one or a few sheep may panic the entire flock.

As with other mature males, rams should be handled with care, especially during the breeding season.

Sheep are commonly caught by hand for management practices. The preferred method of catching is to place the arm around the neck and under the jaw. Avoid choking at the throat. The wool should never be used as a handle for catching or holding. Properly done, catching sheep by the flank until one can hold the head is also effective.

The flocking instinct and the characteristic circling movement of the flock should be recognized in devising handling systems and in actual handling to produce the desired result.

PART IV

DESIGNING THE ANIMAL'S ENVIRONMENT

To ensure well-being, the animal's environment must be adequate in terms of physical, dietary, and social characteristics.

The goal in designing an environment is to provide the resources to meet the various needs of the individual animal or group of animals. The environmental needs of humans associated with animals must also be considered as an integral part of the planning process. Aside from comfort in the workplace, safety and efficiency in caring for animals and maintaining facilities are critical issues.

Planning the environment frequently requires the establishment of priorities among characteristics of an environmental hierarchy as they relate to the well-being of both animals and workers. This task is not unlike planning to accommodate the array of needs that we encounter as humans. One is taught early that basic needs include food and protection from environmental challenges, such as those related to weather. Later, but in more abstract ways, one learns that another area of basic need involves social dimensions of life. Increasing interest in providing completely adequate animal care places more emphasis on meeting such social needs as a component of environmental planning.

Good environmental design forms the basis for an objective approach to animal care. Thus, the term *environmental design* refers to planning and maintenance of the total environment to ensure animal well-being. Elements relating to physical surroundings such as thermal characteristics, space, and lighting are among the important dimensions. An adequate dietary environment must recognize changing needs related to stage of the life cycle of animals, the intended function of the animal, and the variable demands of the physical environment. Social circumstances such as animal interactions or lack thereof must also be considered an integral part of planning for animal well-being. Animal needs, then, translate directly into environmental characteristics that may be described as physical, dietary, and social characteristics. Thus, this section is an effort to summarize criteria that are important in developing the environmental envelope in which animals exist. The review of such information provides some idea of the number and scope of factors that must be considered in the process. The information is instructive also in illustrating similarities and differences that exist among the species considered.

The format used in treating environmental design concepts, as these relate to physical, dietary, and social environments, places emphasis as follows.

First, important basic considerations in developing some major design characteristics are reviewed here and in previous chapters. This brings to the forefront some fundamental scientific principles on which important design concepts are based.

Secondly, the format sets forth numerous examples of specific environmental specifications that are available for application in design of these three dimensions of the animal's environment. Successful design results in environmental characteristics that, collectively, function as an integrated system to meet established needs. A review of such information also provides encouraging evidence that a large and growing body of science-based information allows design of superior systems that do indeed provide for the well-being of animals.

Physical and dietary elements of environmental design are organized and discussed separately for each species, since they have differences that mandate particular environmental specifications. Many biological characteristics are common among species, whereas others may be quite different; as a result, some basic considerations are emphasized more or less depending on the species being considered. Principles that may have common application across species are discussed early in this section and are not repeated for each of the species unless a unique emphasis is appropriate.

Finally, it should be recognized that the examples and guidelines provided serve to illustrate the complexity of designing appropriate animal environments, the numerous issues that must be considered and factors that cause related needs to vary. Professionals who are aware of current information and site specific requirements should be consulted for specific applications.

CHAPTER 6

Design Characteristics of the Physical Environment

The goal in designing an adequate physical environment is to ensure that the envelope in which an animal exists is adequate to meet the needs for physical comfort of the animal, as well as the objectives of the system for which the animal is maintained. These two dimensions of the physical environment are compatible but not mutually exclusive. Often the animal's comfort, security, and general well-being are in the best interests of the economy of the enterprise. We should not assume, however, that maximum rates of production are always associated with a high level of well-being. A high level of production in the individual animal may reflect a high degree of welfare; however, an environment that supports maximum performance may not ensure animal well-being. Thus, in considering the physical environment, the issue of providing for a high level of production by the animals involved always introduces the debate about productivity and well-being of the individual animal and that of the performance and efficiency of the total enterprise. Well-being must be viewed from the standpoint of the individual animal. This view makes the issue much more easily resolved. If individual animal well-being is sacrificed in favor of the economic well-being of the enterprise, most would conclude that the practices involved are not totally acceptable.

This chapter deals with characteristics of the animal's physical environment that are most likely to be associated with well-being and efficiency. In most animal enterprises the thermal environment represents a major component in these considerations. Animals exist in a great range of environments; therefore, the level of protection or modification of the thermal environment is a prominent issue in terms of minimizing stress and costs of production. Numerous other factors are also critical to the well-being of the animal and in some cases to that of the workers occupying spaces in which animals are maintained.

DESIGNING THE PHYSICAL ENVIRONMENT FOR SWINE

The Thermal Environment

The recommended temperatures reflected in the following sections are expressed as air temperature in most cases and are typical recommendations that serve as practical guides. It is important to recognize that values describing lower critical temperature (LCT) and upper critical temperature (UCT) reflect points in a range of temperatures that relate to the thermal demands placed on the animal. That range is typically referred to as the thermal neutral zone (TNZ), or the zone of thermal neutrality. Thermal conditions above or below these points may be stressful and typically require related physiological and anatomical responses that will likely reduce the efficiency with which the animal is accomplishing its intended function. General comfort of the individual is also reduced. The terms used to describe certain characteristics of the thermal environment and the animal's relationship to it are defined in the following sections (NRC, 1981a). An expanded discussion of the basic principles involved in temperature regulation, or thermal homeostasis, appears in part 2.

LCT is the point in effective ambient temperature below which an animal must increase its rate of metabolic heat production to maintain homeothermy. Processes related to conservation of heat, including vasoconstriction in the periphery, piloerection, and behavioral adjustments to reduce heat loss from body surfaces, are at a maximum at this point.

UCT is the point above which an animal must engage physiological mechanisms to resist body temperatures rising above normal. These processes are related to cooling effected by evaporation through increased perspiration and respiration and vasodilation in the periphery to enhance heat loss from body surfaces through convection, radiation, and conduction.

The thermal neutral zone (TNZ) is the range of effective ambient temperature in which the heat from normal maintenance and productive functions of the animal in nonstressful situations offsets the heat loss to the environment without requiring an increase in rate of metabolic heat production. This range in temperature represents that between LCT and UCT. This zone may also be referred to as the thermal comfort zone or zone of thermal neutrality.

Designing the thermal environment recognizes factors influencing the effective temperature (ambient temperature, air movement, humidity, and temperature of surfaces to which the animals are exposed). Planning should also include a capacity for short-term modification in the event of emergencies that may be related to weather or facility failure. Examples are water sprinklers to enhance evaporation and alternative ventilation such as being able to open buildings for natural ventilation or utilize backup power sources. Feed facilities should also accommodate dietary changes that may be related to alleviation of thermal influences. The amount of feed provided may be greater for animals threatened by cold stress. In other cases the character of the feed may be changed to alter energy density. Another management consideration is allowance for possible altered animal population density if deficiencies in temperature and air movement occur.

The scientific literature generally refers to environmental temperature in degrees Celsius, or centigrade. However, many applied publications serving the livestock industry in the United States express temperature in degrees Fahrenheit. Reference materials quoted have, in some cases, been adapted by converting degrees Fahrenheit to degrees Celsius. Generally, however, such materials reflected in this document are cited in the form of the original source. Thus, conversion formulas are presented and may be useful as one reviews environmental considerations in animal well-being.

Temperature conversion formulas:

$$°\text{Centigrade} = [(5/9)(°F)] - 17.8$$

or

$$°\text{Centigrade} = [(0.56)(°F)] - 17.8$$

$$°\text{Fahrenheit} = [(9/5)(°C)] + 32$$

or

$$°\text{Fahrenheit} = [(1.8)(°C)] + 32$$

Temperature comparison scale:

°C	−30	−20	−10	0	10	20	30	40	100
°F	−22	−4	14	32	50	68	86	104	212

In the discussions that follow, the values cited as recommended temperatures are those that would typically be considered in the TNZ. It should be recognized also that these values may be adjusted over time based on research and experience that aid in the refinement of environmental requirements. The stages of the life cycle used as categories of animals for which the recommendations apply are those used commonly to relate stage of development and stage of production at a given time in the animal's life or production cycle. Recommended temperature ranges for swine at various stages of the life or production cycle were summarized by Harmon and Xin (1995) and are listed in table 6.1.

Thermal Considerations and Requirements of Swine in Cold Environments

Newborn Pigs to Weaning at about 3 Weeks of Age (The Neonatal Period)

The newborn pig, even though a member of a litter and nursing a dam with adequate milk flow, has very limited ability to conserve heat because of low levels of subcutaneous fat and limited hair coat to provide insulation. Because chilling of the newborn can be the major cause of pig loss, the thermal environment is an important management consideration. Pigs of this age exhibit the typical anatomical responses to cold such as vasoconstriction at the periphery of the body and piloerection as a means of heat conservation; however, they do not have adequate

TABLE 6.1   Optimum Temperature Ranges for Swine in Confinement[a,b]

Age of Animal (wk)	Weight of Animal (kg)	Low Temperature[a] (°C)	High Temperature[a] (°C)
At birth		33	35
3	5	30	32
4	7	29	31
5	9	29	30
6	11	27	30
7	14	26	29
8	17	25	29
9	21	23	28
10	25	21	28
11	31	20	27
12	36	19	27
13	42	18	27
14	47	17	27
15	53	16	27
16	58	15	27
17–19	64–78	14	27
20–22	85–98	12	27
24–26	109–118	11	27
Sow, lactating		16	24
Sow, gestating		13	27
Boar		13	24

[a]*Source:* Harmon, J., and H. Xin, *Environmental Guidelines for Confinement Swine Housing*, Ext. PM-1586a Iowa State University Cooperative Extension Service (Ames, Ia.: Iowa State University, 1995), p. 1, 2
[b]For pigs maintained in groups of ten to thirty animals on slotted floors in insulated buildings with air movement less than 50 ft/sec. If animals are on wet floors, in buildings that are poorly insulated, or in drafts, these ambient temperature values should be reduced approximately 2 to 3°C, 2 to 3°C, and 3 to 5°C, respectively. Sows and boars are assumed to be housed individually on partially slatted floors, in insulated buildings, and in minimum drafts. If on wet floors, in poorly insulated buildings, or in the presence of drafts, the temperature must be increased approximately 3°C, 3°C, and 5°C, respectively.

physical characteristics and metabolic resources at this time to resist cold stress (Mount, 1963, 1964). Pigs maintained in groups demonstrate huddling behavior that greatly reduces heat loss (Mount, 1960). An example to illustrate the influence of huddling behavior is that the thermal recommendation for the resting area for a litter of pigs is on the order of 5°C lower than that for an isolated piglet at this stage. This difference in recommendation points to the severe negative effects that can result from a cold environment if the pig is isolated.

Metabolic activity and related heat production increase rapidly during the first day of the pig's life, but the level is not adequate to achieve thermoneutral metabolic rate at this stage. Curtis (1985a) concluded from a review that stores of carbohydrates are the primary body source of energy to support this increasing metabolic activity because fat reserves are very low at birth. Shunting of blood flow to muscles rather than fat depots forms the basis for the conclusion that the response was related to glycolysis rather than gluconeogenesis and lipolysis, although the latter does occur in the newborn if regulatory systems require it. Pigs contain no brown fat (a rapidly available energy source from lipid reserves), as do lambs. These tissue and metabolic characteristics of newborn pigs normally result in about 7 days being required for pigs to achieve a thermoneutral metabolic rate. This level of metabolic activity is typically about three times the metabolic rate observed at birth. A typical pattern for attaining this level of metabolic activity involves an increasing level during the first day. This level doubles at about 3 days and at 6 to 7 days attains the necessary level of three times that existing at birth. The level of three times the rate observed at birth appears to be typical for other species in reaching a thermoneutral level. However, lambs and calves normally attain this level of metabolic activity and heat production within 1 day of birth.

Early survival of pigs is closely related to the immunity developed as a result of consumption of immunoglobulins in colostrum during the first 24 to 48 hours following birth. If pigs experience cold stress, they are likely to have lower blood levels of immunoglobulins. This effect appears to be due to a reduction in nursing activity of the cold-stressed pigs rather than lack of ability to absorb immunoprotein from the digestive tract (LeDividich and Noblet, 1981). Pigs suffering from cold stress typically are weakened physically and, as a result, are less likely to exhibit normal competitive behavior in efforts to nurse and huddle. Such behavior increases the severity of an inadequate thermal environment. Curtis (1985a) concluded that predisposition to diarrhea may be related to cold stress. This view supports the conclusion that cold stress can be related to a broad array of pathological problems. In this case the loss of liquid and related dehydration associated with diarrhea increases the severity of cold stress impacts by influencing a number of normal metabolic functions. Early consumption of milk is critical in providing energy and other nutrients; thus, the stress-related reduction in intake has negative impacts beyond the influence on immune function.

Generally, concerns relative to thermal stress are those of losses of animals and performance and efficiency. Morphological effects of early cold stress on pigs were demonstrated by LeDividich et al. (1992), who showed that pigs reared at 12°C were strikingly different in appearance at 30 kg live weight compared with pigs reared at a temperature of 25°C. The marked difference in appearance of littermates is illustrated in figure 6.1. In these studies, littermates reared at 12°C, 20°C (controls), and 28°C reflected differences in fat distribution. The cold-exposed pigs had 9 percent greater back fat and 17 percent less leaf fat than the controls reared at 20°C. Those reared at 28°C had 5 percent less back fat and 20 percent more leaf fat than the controls. T. L. Mader et al. (1997) observed a similar effect in cattle exposed to moderate cold stress.

The mentioned observations point to the importance of proper management related to the thermal environment. The environmental temperature requirement for newborn pigs, if maintained alone, is 33 to 35°C (Mount, 1959, 1963). For pigs of this age that are maintained in groups, the suggested environmental temperature for the creep-nest area is normally 30 to 33°C but can be reduced as the pigs develop.

Factors other than air temperature in the facility for newborn pigs are also important considerations to limit heat loss. Walls and ceilings of the farrowing

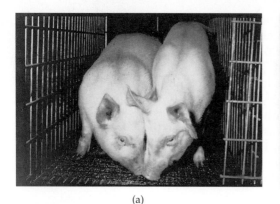

(a)   (b)

FIGURE 6.1 Physical appearance of pigs reared to 30 kg live weight. The pigs viewed from both front and rear were from litters reared at 12°C (pig on the left in each photograph) and 25°C (pig on the right in each photograph). (From LeDividich et al., 1992) (Photographs courtesy of Jean LeDividich; Jacky Chevalier, photographer, Institute National de la Recherche, Agronomique Center de Recherches de Rennes)

facility should have surface temperatures of no less than 2°C below the air temperature (Mount, 1964a) to minimize heat loss by radiation.

Air speed in farrowing units should be kept below 0.15 m/sec (Mount, 1966) to minimize potential excessive heat loss by convection.

Adams et al. (1980) concluded that supplemental heat lamps (250 W) were beneficial at a room temperature of 21°C for newborn pigs up to 3 days of age. However, subsequent research has shown a trend to increased survival rates related to the use of lamps for 21 days. The ability to maintain adequate room air temperature obviously influences the need for heat lamps or heat mats and the length of time these items are effective.

Dietary approaches to enhance pig survival may also be considered. For example, Pettigrew (1981) concluded that supplemental dietary fat for the sow in late gestation may represent an effective approach to enhanced pig survival rates by providing a higher-energy intake through the milk. This effect is more likely to provide significant enhancement in piglet survivability when typical survival rates for a given enterprise are below about 80 percent. Environments that typically result in rates higher than 80 percent may not be improved appreciably by this approach. This example of the complexity of management to prevent cold stress illustrates that approaches to animal well-being can be somewhat flexible in some instances.

Weanling Pigs

Weanling pigs are assumed to be about 3 weeks of age and at a weight of approximately 5 kg. Pigs at this stage of development have temperature requirements that are especially critical immediately after weaning, when the pigs must adapt to a postweaning diet. This adjustment is in itself a stressor and compounds problems that may be associated with an inadequate thermal environment.

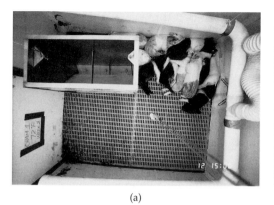

(a)

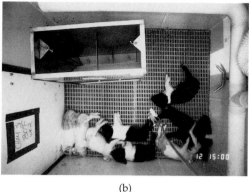

(b)

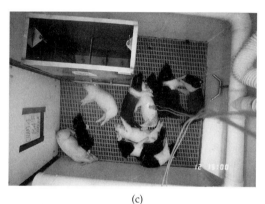

(c)

FIGURE 6.2 Effect of environmental temperature on behavior of nursery pigs. The pigs reflect level of comfort by resting distribution in the nursery pen. The groups are experiencing pen temperatures of (*a*) 72°F, (*b*) 80°F, and (*c*) 84°F. (Photographs courtesy of Hongwei Xin and Jay Harmon, Iowa State University, Ames)

The pigs shown in figure 6.2 illustrate the effect of environmental temperature on huddling and resting behavior of nursery pigs. Observing animals for behavioral responses to thermal conditions is an important management tool.

The suggested environmental temperature range is 28 to 30°C at weaning and for the 2 weeks following weaning (McCracken and Caldwell, 1980). After this period, thermal recommendations are reduced to 20°C when pigs are in the 15- to 20-kg weight range (Curtis, 1985a; Holmes and Close, 1977). The recommendation assumes adequate nutrition.

Research on temperature fluctuation within a reasonable range has produced mixed results. Some studies indicate that pigs will tolerate a range from 26°C in the afternoon to 15°C as a nighttime low at one such cycle per day. This finding was based on the work of Curtis and Morris (1982), in which pigs were allowed to regulate the temperature themselves. Brumm and Shelton (1993) conducted research to determine acceptable within-day flexibility in nursery temperature and developed recommendations for daily variations and gradual reductions in level in the interest of fuel efficiency. Temperatures during a typical 5-week nursery period may be regulated to those indicated in the following chart, which reflect acceptable within-day variation (6°C reduction for the 16-hour evening period) and temperature reductions of 2°F per week.

1st week	86.0°F	Constant temperature
2nd week	84.0°F	8:00 A.M. to 4:00 P.M.
	78.0°F	4:00 P.M. to 8:00 A.M.
3rd week	82.0°F	8:00 A.M. to 4:00 P.M.
	74.0°F	4:00 P.M. to 8:00 A.M.
4th week	80.0°F	8:00 A.M. to 4:00 P.M.
	76.0°F	4:00 P.M. to 8:00 A.M.
5th week	78.0°F	8:00 A.M. to 4:00 P.M.
	72.0°F	4:00 P.M. to 8:00 A.M.

The adjustments outlined in the preceding scenario suggest management flexibility for energy cost reductions without compromising animal well-being.

Growing and Finishing Pigs

Air temperature recommendations for growing pigs in the weight range of 35 to 70 kg are in the range of 16 to 24°C (NPPC, 1992). At weights of 70 kg to market weight (100 kg), the recommended temperature is 10 to 24°C (NPPC, 1992).

The given values are for pigs fed on well-insulated, solid floors. If the animals are maintained on slatted floors, the required temperature may be 3 to 4°C higher, depending on the type of floor material. For pigs maintained on partly wet slotted floors, temperature requirements will be even higher. Temperatures below the recommended range are likely to be associated with cold temperature stress responses and the associated influences on performance and efficiency.

Breeding Herd

Restricted feeding is commonly practiced as a means of controlling condition in sows and boars. Thus, attention to plans for adjusting the level of energy allowed during periods of severe cold is a critical management issue. Lower critical temperatures provide an indication of environmental conditions below which stressful conditions may exist for such animals. The LCT for sows fed at low levels (about one-half of the maintenance level) is on the order of 20°C; for fleshy sows fed at a maintenance level and others fed at levels above maintenance, the LCT is lower. In general, air temperature recommendations for sows and boars are in the range of 16 to 24°C (NPPC, 1992; Harmon and Xin, 1995).

Potential damage to testicular tissue and function is a concern in regions of severe cold. Cold apparently has little influence, however, if it is not sufficiently severe to cause frostbite of the scrotum. Frostbite of the scrotum resulting in swelling and elevated testicular temperature may cause sufficient tissue damage to have a negative influence on reproductive capability.

Recommended environmental temperatures for swine during different phases of production and the life cycle are summarized in table 6.2. Except for newborn pigs, the ranges in temperatures reflect the flexibility allowed by the animal's ability to regulate body temperature without the negative influences of thermal stress. A wider range is possible for short periods. However, if the temperature is below the LCT or above the UCT, management procedures should be instituted to modify the environment to prevent both acute and chronic stress effects from extreme

TABLE 6.2   Recommended Effective Thermal Environments for Swine At Various Stages in the Production System Along with Lower and Upper Critical Temperatures[a]

Stage of Life Cycle and Weight Range	Recommended Temperature (°C)	Lower Intervention Temperature[b] (°C)	Upper Intervention Temperature[c] (°C)
Lactating sow	16–28	10	33
Piglets (creep area)	33	—	—
Prenursery, 5–5 kg	27–33	16	35
Nursery, 15–35 kg	18–27	5	35
Growing, 35–70 kg	16–24	−4	35
Finishing, 70–100 kg	10–24	−15	35
Sows and boars	16–24	−15	33

[a]Values reported in NPPC (1992) are cited as adapted from NRC (1981a), DeShazer and Overhults (1982), and Hahn (1985)
[b]Measures to aid in reduced heat loss should be applied when the temperature approaches these levels; examples are bedding, supplemental heat, and other environmental modifications
[c]At these levels, cooling should be considered by improved ventilation, insulation, air movement, sprinkling, or other measures

temperatures. At lower levels, supplemental heat, bedding, and grouping to allow huddling may be helpful measures. At higher levels, supplemental cooling, enhanced ventilation, and reducing or eliminating bedding and its insulation effects are approaches to assist in the prevention of thermal stress.

Thermal Considerations and Requirements of Swine in Hot Environments

Heat stress in swine is most common in animals that have a high level of body fat, such as animals being finished and nearing market weight, sows in the breeding herd, especially those on high levels of feed intake during the lactation period, and boars. Thermal recommendations follow (NPPC, 1992).

Newborn Pigs to Weaning Stage
Pigs are unable to meet the demands of heat stress imposed by an environment warmer than about 39°C. Recommended farrowing house temperature is 27°C for gestating sows and 24°C for lactating sows. The pig's creep-nest area should be maintained at 33- 35°C (Harmon and Xin, 1995).

Nursery
Thermal recommendations initially are similar to those for the nesting area in the farrowing environment. Recommended temperature ranges are 29 to 30°C for pigs at about 9 kg and 26 to 28°C for pigs at about 15 to 20 kg. Note Table 6.1 for factors that require

adjustments to the typically suggested environmental temperatures. Supplemental cooling should be available to prevent room temperatures warmer than 35°C.

Growing and Finishing

In general, the recommended temperature range of 16 to 27°C is optimal for pigs to 50 kg and 11 to 27°C from 50 to 120 kg during the growing and finishing period. Temperatures above 30 to 32°C may result in some level of heat stress and those 35°C or higher may produce severe heat stress. Marketing finished swine during hot weather can be very damaging and result in significant death loss caused by heat stress. Supplemental cooling such as sprinklers in trucks and pens and increased ventilation can be critical. Minimizing the amount of straw used as bedding or the use of wet sand bedding instead of straw should be considered.

Breeding and Gestation

The recommended temperature range for sows in the breeding herd is 16 to 24°C (Harmon and Xin, 1995) and should not exceed 32°C. Heat stress in sows, occurring at 32 to 37°C, may result in reductions in conception rate and embryo survival. High humidity increases the hazard. Increased incidence of silent estrus may occur as a result of heat stress. The most sensitive periods in which heat stress may influence reproductive efficiency of sows appear to be 2 to 3 weeks before mating (the effect may be increased silent estrus and reduced conception rate), the first 2 or 3 weeks of pregnancy (reduced embryonic survival may result), and the last 2 weeks of pregnancy (increased stillbirths and sow death rate during farrowing may occur). Females appear to be much more resistant to heat stress immediately before and during the time of mating and during midpregnancy.

High scrotal temperature has been shown to lower semen quality by damaging primary spermatocytes and spermatids. Exposing boars to a temperature of 33°C and high levels of humidity for 3 days may result in reduced semen quality. Recommended air temperature for boars is 13 to 24°C. Temperatures above the 33 to 35°C range should be avoided.

Ventilation

The ventilation requirements shown in table 6.3 are those reported by Murphy et al., 1991. The dramatic differences in temperatures for varying thermal conditions provide an example of the interaction between temperature and ventilation as they influence the microenvironment of the animal. The increases in ventilation volume associated with increasing weight and condition provide an example of the interaction between the animal's anatomical and body composition characteristics and the environment as they influence temperature regulation.

Humidity

The general recommendation for relative humidity levels in swine facilities is from 50 to 80 percent (Harmon and Xin, 1985). Levels in or above the upper end of this

TABLE 6.3 Recommended Fan Capacities (at 1/8-Inch Static Pressure) Per Animal or Per Sow and Litter[a]

Stage of the Animal's Life Cycle	Unit Weight (kg)	Ventilation Rates (cfm[b]/head or cfm/sow and litter)[c]		
		Cold Weather	Mild Weather	Hot Weather[d]
Sow and litter	180	20	80	500[e]
Prenursery pig	5–15	2	10	25
Nursery pig	15–35	3	15	35
Growing pig	35–70	7	24	75
Finishing pig	70–100	10	35	120
Gestating sow	150	12	40	150
Boar or breeding sow	180	14	50	300

[a]*Source:* Murphy, J. P., D. D. Jones, and L. L. Christian, *Pork Industry Handbook (PIH-60)*, Ext. AS-496 Iowa State University Cooperative Extension Service (Ames, Ia.: Iowa State University, 1991), p. 2
[b]cfm, cubic feet per minute
[c]The rate for each season is the total capacity needed. For sow and litter: 20 cfm/unit (cold weather) + 60 cfm/unit = 80 cfm/unit (mild) add 420 cfm/unit for a total hot weather rate of 500 cfm/unit.
Cold weather rate: In some cases this airflow needs to be adjustable due to a change in the number of animals in the room or due to their growth. Ideally, at least one fan should operate at all times when the inside temperature is above 35°F. Set a thermostat to shut the fan off when the inside temperature drops below 35°F and activate an alarm to notify the operator. This fan should supply the cubic feet per minute rate listed under "Cold Weather." The fan should exhaust the air from above any stored liquid manure.
Mild weather rate: Provide additional airflow, thermostatically set to start in 3 to 5° steps from the lowest desired temperature to prevent sudden drops in temperature. These fans, together with the cold weather fans, provide the capacity for outdoor temperatures up to about 55°F.
Hot weather rate: Provide additional fans to supply the cubic feet per minute rates listed under "Hot Weather." Some or all of these fans should be operated when the inside building temperature is above 75°F. Hot weather rate airflow capacity of sows and litters and breeding animals can be reduced somewhat by utilizing drip cooling or zone cooling (water evaporation or mechanical air-conditioning) of sows and boars.
[d]These rates may be reduced when supplemental cooling is available in hot weather and may be increased when air velocities on pigs are low in summer.
[e]500 cfm is the generally recommended hot weather rate in farrowing; however, local recommendations range from 250 cfm in northern areas of the United States to 1,000 cfm or more in the southeast and southwest.

range may result in condensation in the housing facility. Levels in the lower end and below this range may be associated with increased dust problems and reduced air quality. Harmon and Xin (1995) suggest that relative humidity levels in confinement units be kept within the range of 50 to 60 percent. Bacterial growth and viral growth are limited at these levels, as are respiratory infections.

Air Quality

Air quality must be considered from the standpoint of both animals and humans present in the facility. Consideration is given to gases, dust, and microorganisms. The following levels of toxic gases should not be exceeded in swine buildings (based on 8 hours of daily exposure):

TABLE 6.4   Effects of Hydrogen Sulfide Exposure on Humans and Swine

Exposure Level	Effect or Symptom
	On humans
10 ppm[a]	Eye irritation
20 ppm for more than 20 minutes	Irritation to the eyes, nose, and throat
50 to 100 ppm	Vomiting, nausea, and diarrhea
200 ppm for 1 hour	Dizziness, nervous system depression, and increased susceptibility to pneumonia
500 ppm for 30 minutes	Nausea, excitement, and unconsciousness
600 ppm and above	Rapid death
	On swine
20 ppm, exposed continually	Fear of light, loss of appetite, and nervousness
200 ppm	Possible pulmonary edema (water in the lungs) with breathing difficulties and possible loss of consciousness and death

[a]ppm, parts per million
Source: Barker, J., S. Curtis, O. Hogsett, and F. Humenik, Safety in swine production systems. In *Pork Industry Handbook,* Ext. AS-572 Iowa State University Extension Service (Ames, Ia.: Iowa State University, 1986), p. 2. See also 1989 *Guide to Occupational Exposure Values,* American Conference of Governmental Industrial Hygienisis, Cincinatti, Ohio

Compound	Internal air should not exceed (ppm)[1,2]
Ammonia	25
Hydrogen sulfide	10
Carbon monoxide	50
Carbon dioxide	5,000
Methane	1,000

Special precautions are necessary when under-floor manure storage is disturbed by pumping for removal because of the possibility of high levels of hydrogen sulfide and methane. Recommended levels of these compounds in the air of swine buildings for well-being of the pigs are lower than the OSHA guidelines. The recommendations reported in Meyer, et al. (1991) are as follows:

Compound	Internal air should not exceed (ppm)[1]
Ammonia	10
Carbon dioxide	3,000
Hydrogen sulfide	5

Because of the importance of exercising proper safety precautions relative to air quality in swine facilities, the tables 6.4, 6.5, 6.6, 6.7, and 6.8 are reproduced from the Pork Industry Handbook (Barker et al., 1986).

Dust control may be important for both animals and workers. The OSHA allowable dust levels for workers exposed for 8 hours daily (with and without face

[1] ppm, parts per million
[2] OSHA (1989)

TABLE 6.5 Effects of Ammonia Gas Exposure on Humans and Swine

Exposure Level	Effect or Symptom
	On humans
6–20 ppm[a] and above	Eye irritant and respiratory problems
100 ppm for 1 hour	Irritation to mucous surfaces
400 ppm for 1 hour	Irritation to eyes, nose, and throat
700 ppm	Immediate irritation to eyes, nose, and throat
5,000 ppm	Respiratory spasms and rapid suffocation
10,000 ppm and above	Death
	On swine
50 ppm	Reductions in performance and health; long-term exposure increases the possibility of pneumonia and other respiratory diseases
100 ppm	Sneezing, salivation, and loss of appetite, thereby reducing animal performance
300 ppm and above	Immediate irritation of nose and mouth
	Prolonged exposure causes extremely shallow and irregular breathing followed by convulsions

[a]ppm, parts per million
Source: See *1989 Guide to Occupational Exposure Values* and Barker et al., 1986, p. 2

TABLE 6.6 Effects of Excessive Carbon Dioxide Exposure on Humans and Swine

Exposure Level	Effect or Symptom
	On humans
60,000 ppm[a] for 30 minutes	Heavy breathing, drowsiness, and headaches
100,000 ppm (10 percent) and above	Narcotic effect, dizziness, and unconsciousness
225,000 ppm (25 percent) and above	Death
	On swine
40,000 ppm	Increased rate of breathing
90,000 ppm	Discomfort
200,000 ppm (20 percent)	Cannot be tolerated by market hogs for more than 1 hour

[a]ppm, parts per million
Source: See *1989 Guide to Occupational Exposure Values* and Barker et al., 1986, p. 2

masks) are 5 mg/m^3 for respirable dust (5 μm or less particle size) and 15 mg/m^3 for total dust (Barker et al., 1986). NPPC (1992) concludes that pigs are tolerant of higher levels of dust; however, the lower of the two requirements for animals and humans is the determining design and management factor in dust control.

The movement of airborne microorganisms should be considered in facility design as a component of overall biological security. For this reason, engineers should consider relative air pressure among different components of buildings to minimize the transport of contagious diseases.

Factors influencing dust include physical activity, humidity, air movement, feed ingredients, and feed delivery systems. Control measures include adding fat or moisture to feeds, adjusting feeders, using closed feed conveyances, and cleaning the interior surfaces (pens and equipment).

Noise

NPPC (1992) suggests that normal noise levels in production units do not appear to have negative effects. Noise levels in pig working areas may exceed levels acceptable for humans, however. Thus, noise protection devices such as ear covers may be appropriate for those working in the facility.

Lighting

Recommended levels of lighting are 20 foot-candles for special inspection areas; 15 foot-candles for breeding, gestation, and farrowing areas; 10 foot-candles for nurseries; and 5 foot-candles for growing-finishing areas. Photoperiod management does not appear to be critical for swine (NPPC, 1992); however, some research suggests that lactating sows and their litters may respond positively to 16-hour photoperiods. Trends toward better pig performance and return to estrus have been reported (Mabry et al.,

TABLE 6.7   Effects of Methane Exposure on Humans and Swine

Exposure Level	Effect or Symptom
50,000–150,000 ppm[a]	Potentially explosive
500,000 ppm	Asphyxiation

[a]ppm, parts per million
Source: See *1989 Guide to Occupational Exposure Values* and Barker et al., 1986, p. 3

TABLE 6.8   Effects of Carbon Monoxide Exposure on Humans and Swine

Exposure Level	Effect or Symptom
	On humans
50 ppm[a] for 8 hours	Fatigue and headaches
500 ppm for 3 hours	Chronic headaches, nausea, and impaired mental ability
1,000 ppm for 1 hour	Convulsions and coma after prolonged exposure
4,000 ppm and over	Death
	On swine
200–250 ppm	Baby pigs are less vigorous
150 ppm and over	Causes abortions in late-term sows and an increased incidence of stillborn pigs
	Reduces growth rate of young pigs

[a]ppm, parts per million
Source: See *1989 Guide to Occupational Exposure Values* and Barker et al., 1986, p. 3

1983; Stevenson et al., 1983). Zimmerman et al. (1980) reported inconclusive results using 16-hour light and 8-hour dark daily periods for developing boars.

Water Availability

Typical water requirements are listed in table 6.9 (MWPS-8, 1983) but may vary greatly depending on environmental influences.

Location of waterers must be such that all pigs have good access. Crowding may restrict access, especially for less aggressive animals. Since nipple waterers are commonly used today in confined swine housing, the recommendations in table 6.10 are critical (MWPS-8, 1983).

Feed Availability and Equipment Design

Major considerations in the design of feeding equipment are size and convenience of the eating space, adjustments for control of flow rate and wastage, cleaning ease, and safety characteristics (e.g., possible injury by catching foot and possible cutting by sharp metal edges). General recommendations for feeder space are summarized in table 6.11 (MWPS-8, 1983).

TABLE 6.9   Water Requirements of Swine

Animal Type	Water Requirements (gallons/day)
Sow and litter	8
Nursery pig	1
Growing pig	3
Finishing pig	4
Gestating sow	6
Boar	8

Source: MWPS-8, *Swine Housing and Equipment Handbook,* 4th ed. (Ames, Ia.: Midwest Plan Service, 1983), p. 3, by permission

TABLE 6.10   Recommendations for Nipple Waterers[a]

	Pigs up to 12 lb	Pigs 12–30 lb	Pigs 30–75 lb	Pigs 75–100 lb	Pigs 100–240 lb	Sows and Boars
Height (inches)	4.6	6–12	12–18	18–24	24–30	30–36
Pigs/nipple	litter	10	10	12–15	12–15	12–15
Minimum flow rate (qt/min)	0.2	0.2	0.4	0.5	0.67	1.0

[a]Minimum of 2 waterers per pen located 14 inches apart for nursing pigs and 24 inches apart for larger animals
Source: MWPS-8, *Swine Housing and Equipment Handbook,* 4th ed. (Ames, Ia.: Midwest Plan Service, 1983), p. 3, by permission

TABLE 6.11   Feeder Space Allowances for Swine

Sows in the breeding herd: 1 foot of self-feeder space for an individual sow or 2 feet/sow for group-fed animals	
Pigs, 12–30 lb	2 pigs/self-feeder space
Pigs, 30–50 lb	3 pigs/self-feeder space
Pigs, 50–75 lb	4 pigs/self-feeder space
Pigs, 75–220 lb	4–5 pigs/self-feeder space

Source: MWPS-8, *Swine Housing and Equipment Handbook,* 4th ed. (Ames, Ia.: Midwest Plan Service, 1983), p. 3, by permission

Housing Systems for Sows

Housing for sows has been the subject of extensive investigation for several years. In general, facilities studied and compared include group housing, gestation stalls, and tethering (neck and girth) as systems during mating and gestation. Design variations on pens and stalls are being actively evaluated at present, but additional time is needed to reach concrete recommendations. Such design characteristics include provision of space and configuration for more freedom of movement.

Group housing of sows is attractive from the standpoint of providing an enriched environment. The system is difficult to manage to ensure appropriate individual attention and to minimize fighting, which can be severe and result in injury to the animals. Problems associated with aggressive competition are increased because of the common practice of limiting feed intake to control sow weight (condition) during the gestation period. Providing individual feeding stalls for the sows may be helpful in this regard.

Gestation stalls provide excellent control of management and feeding practices that must be administered on an individual basis. The level of confinement and character of the space provided are associated with various stereotypic behaviors. Comparisons between group housing and individual gestation stalls suggest that, with proper management of each system, reproductive performance and efficiency can be similar.

Neck tethering of sows, when compared with group housing (Janssens et al., 1994; Barnett and Hemsworth, 1991), resulted in greater stress responses and lower overall pregnancy rates, respectively. McGlone et al. (1994) concluded that girth tethering systems were undesirable from a welfare standpoint and that use should be discouraged. Sows in crates showed more oral and nasal stereotypies, drinking, and sitting than tethered sows but had larger litters. Comparisons between group housing and gestation stalls is an ongoing research activity at several locations. To date, however, results are still equivocal (Barnett and Hemsworth, 1991).

Stereotypic behavior in confined facilities may be reduced by providing even limited amounts of fibrous materials such as straw or beet pulp. It is likely also that greater attention should be given to genetic selection of animals that show better adaptation to the facility of choice.

Concerns are commonly expressed about the effect of stall confinement on foot and leg pathology and maintenance of muscle mass in the animal. Limited work to date suggests that both may be affected. Thus, efforts to design facilities to allow greater freedom of movement are important.

Housing the individual sow and litter in properly designed farrowing stalls is widely accepted as the superior system in terms of management and pig survival. Pig survival is greater in farrowing stalls because fewer pigs are crushed in these systems. Efforts are under way, however, to design such specialized facilities to provide greater freedom of movement for the sow.

Space Requirements

Space requirements for swine maintained in confinement facilities are shown in table 6.12 (MWPS-8, 1983) . Allowances for swine maintained in housing units with outside lots are shown in table 6.13 (MWPS-8, 1983). Space allowances are instructive regarding factors that influence environmental needs. Body size and expected function of the animal are obviously important. Type of flooring and thermal conditions, for example, are more subtle influences but critically important. Recognizing the potential influences of inadequate space on several aberrant behavioral responses, including increased aggression and stereotypic behaviors, is also critical.

Floors and Floor Surfaces

Slatted floors and partially slatted floor configurations have become routine floor designs in an increasing proportion of swine housing. Newborn pigs with sows on slatted floors need floor covering as temperature protection from below. Floor covering may be provided by laying mats over the pigs' resting and creep areas. For other swine, partially slatted floors are more commonly recommended so that the animal has some nonslatted resting areas. The proportion of slatted floor recommended is normally from one-third to one-half of the floor area.

Recommendations on slat width are based on the widest slat width possible that also cleans well. Common width recommendations for concrete slats are 4 to 5 inches for sow stalls and 5 to 8 inches for growing or finishing pigs. If slatted floors are used in a nursery, a slat width of 4 to 6 inches is suggested. Fewer and wider slats can result in fewer foot injuries (Fritschen and Muehling, 1989). More recently, some metal and fiberglass slats have been designed with widths as narrow as 1 inch. Edges of slats should have a pencil-round edge to minimize foot injuries.

Spacing between slats (slots) is a consideration, because it influences adequacy of self-cleaning of the floor, as well as possible foot and leg injuries. Recommended slot widths are typically 3/8 inch for sows and litters and for prenursery pigs. Slot widths of 1 inch are recommended for nursery pigs, growing or finishing pigs, and mature swine (MWPS-8, 1983).

The use of slats for nursery floors is declining in favor of elevated expanded metal, woven wire, plastic, and so forth. If slats are used for nursery floors, they should be either narrow to prevent legs from going through or wide enough not to entrap the animal by holding the leg.

Production systems involving elevated floors for farrowing stalls and nurseries appear to be superior for a number of reasons, including rapid separation of excreta from the pigs, drier pigs and floor surface area, warmer air and less draft, and the opportunity to use materials less abrasive than concrete (Fritschen and Muehling, 1989). If expanded metal is used, care should be taken to avoid a design

TABLE 6.12   Space Allowances for Swine in Confined Facilities

Type of Animal	Enclosed Housing Weight (lb)	Area (ft^2)
Prenursery	12–30	2–2.5
Nursery	30–75	3–4
Growing	75–150	6
Finishing	150–220	8

	Solid Floor (ft^2)	Totally or Partly Slotted Floor ^{a} (ft^2)	Animals per Pen	Individual Crate or Stall Size
Breeding swine (lb)				
Gilts, 250–300	40	24	up to 6	
Sows, 300–500	48	30	up to 6	
Boars, 300–500	60	40	1	2'4" × 7'
Gestating swine (lb)				
Gilts, 250–300	20	14	6–12	1'10" × 6'
Sows, 300–500	24	14	6–12	2'0" × 7'

^{a}Or flushed open gutter; open gutter not recommended in breeding because of slick floors
Source: MWPS-8, *Swine Housing and Equipment Handbook*, 4th ed. (Ames, Ia.: Midwest Plan Service, 1983), p. 3, by permission

TABLE 6.13   Space Allowances for Swine Maintained in Open Housing With Outside Lot Facilities

	Weight (lb)	Inside Space (ft^2/head)	Outside Space^a (ft^2/head)^b
Nursery pig	30–75	3–4	6–8
Growing or finishing pig	75–220	5–6	12–15
Gestating sow	325	8	14
Boar	400	40	40

^{a}Considerable space is required for manure storage and drying to prevent wet conditions
^{b}When shaded space is provided, allowances should be 15 to 20 ft^2 per sow, 20 to 30 ft^2 per sow and litter, 4 ft^2 for pigs to 100 lb, and 6 ft^2 for pigs above 100 lb; the allowance refers to space in the shaded area
Source: MWPS-8, *Swine Housing and Equipment Handbook*, 4th ed. (Ames, Ia.: Midwest Plan Service, 1983), p. 3, by permission

with sharp metal edges. Woven round wire, plastic-coated metals, fiberglass, and plastic screens are in common use and appear to offer comfortable footing. It is not clear, however, whether coated materials are superior to uncoated metal with no sharp edges. Fritschen and Muehling also comment on smoothness of concrete floors and concrete slats. Floor smoothness is a design consideration to prevent excessive abrasiveness and slipping. A finish described as a machine-trowel finish is recommended for concrete slats.

In areas where hogs are driven and in working areas, concrete floors should be scored to minimize slippage. Grandin (1992a) suggests the use of 1.5 inch raised expanded metal mesh to imprint the concrete surface for such areas. In addition, floors with a slope exceeding 1/2 to 3/4 inch per foot should be scored.

Regardless of design, concrete flooring material, as a result of surface porosity, can retain microorganisms. Thus, careful cleaning and disinfecting procedures with residual effects are recommended.

Trucking Considerations

The following considerations and recommendations are from Grandin (1992b). Hogs should be transported during the coolest part of the day in hot weather. If the temperature is above 60°F, use wet sand or wet shavings for bedding. Truck vents should be open, and any temporary slats on the truck racks should be removed for better ventilation. If the temperature is above 80°F, the animals can be wet thoroughly before loading; use of a truck sprinkler is also an effective approach to reducing heat stress. Avoid use of straw for bedding in hot weather because it reduces heat loss by the animal's normal cooling mechanisms.

In cold weather, use straw for bedding. Close truck vents and add grain bed slats if possible. Do not allow rain to fall on the animals. Grandin reports losses of 50 percent resulting from wet hogs and freezing temperature during transport. The space recommendations listed in table 6.14 are for hogs of varying weights when temperatures are below 70°F. When the Livestock Safety Index is in the alert zone, these values must be reduced by 10 to 20 percent.

Other suggestions for safe transport are the use of partitions to prevent hogs from piling up and for separation of pigs not previously housed or penned together to minimize fighting. If hogs are to be slaughtered the same day, withholding feed 6 to 8 hours before loading is a common recommendation, although DeSmet et al. (1996) found no advantage to withholding feed overnight before

TABLE 6.14   Minimum Truck Space Requirements for Hogs

Average Weight (lb)	Number of Hogs per Running Foot of Truck Floor (92-Inch Internal Truck Width[a])
100	3.3
150	2.6
200	2.2
250	1.8
300	1.6
350	1.4
400	1.2

[a]Each foot of truck length in this case represents (92/12) 7.666 ft^2. Thus the square foot allowance per animal may be calculated by dividing the number in the column per foot of truck length into this value. Example: 7.666/3.3 = 3.48 ft^2 per animal for pigs at 200 lb.
Source: Grandin, T., *Livestock Trucking Guide* (Bowling Green, Ky: Livestock Conservation Institute, 1992), p. 6, by permission

shipment. If the animals are to be slaughtered the next day, a light feeding before loading may be suggested. Free access to water should be provided at all times before and after transport.

DeSmet et al. (1996) found that a rest period of 2 to 3 or 3 to 4 hours for slaughter hogs after arrival at the abattoir and before slaughter improved pork quality traits (loin temperature and pH after 40 minutes, lean meat color, and drip loss). The effect was more pronounced in pigs that were homozygous or heterozygous for the porcine stress gene. The meat quality of homozygous stress-resistant pigs was improved by these rest periods but less so.

DESIGNING THE PHYSICAL ENVIRONMENT FOR SHEEP

The Thermal Environment

Sheep are maintained in an extremely wide range of climatic environments. Their highly developed ability to cope with extremes of cold and their ability to survive in hot, arid regions makes this wide range possible. On the other hand, sheep production occurs only to a limited degree in hot, humid regions. Thus, certain types of sheep are more suitable to the various geographic regions, except those described by Thwaites (1985) as hot-wet climates.

Thermal Considerations and Requirements
for Sheep in Cold Environments

It is important to give careful consideration to the effective thermal adaptation of various types and breeds of sheep in relation to environmental criteria. LCTs for sheep and lambs are described in a later section. These ranges represent the low end of the thermal neutral zone; however, the values are for sheep in good condition in terms of health and nutrition. Animals in weakened condition have much higher LCTs. LCTs reported in the literature vary for adult and growing sheep due presumably to great variation in type of sheep, wool length, and condition.

Newborn lambs reflect a much more rapid development to adapt to cold than pigs and chicks. Lambs must increase metabolic rate by about three times that at birth to reach full homeothermic capability. This required increase in metabolic rate level appears to be comparable in other species. In lambs, this adaptation occurs within the first few hours after birth (as compared with up to a week in pigs), and rates up to five times that at birth have been observed. For newborn lambs, LCTs are up to 18–20° C and a minimum temperature of 22–30° C is suggested for the first two weeks (Hahn, 1985).

Feeder lambs on full feed show good performance and efficiency at temperatures in the range from 0 to 20°C. For adult sheep, LCTs for newly shorn animals is on the order of 15-20°C. For full-fleece adult sheep, LCTs may range as low as −10°C or below.

Based on the provided values, it is clear that newborn lambs should be dried and provided protection from wind and wetting. For the first 24 hours, supplemental heat is often critical in cold climates. Heat lamps are routinely used for this purpose. In providing lamps for newborn lambs, precautions must be taken for fire prevention and possible lamp contact injury to the animals. Such precautions include providing appropriate electrical equipment for moist conditions, location of lamps at least 18 inches above the lamb's reach, shielding the lamps from contact by the dam, and careful attention to possible power overload. MWPS-3 (1994) suggests a maximum of twelve 150-W lamps or seven 250-W lamps on a 20-amp, 12-gauge circuit. On 14-gauge circuits, maximums are nine 150-W bulbs or five 250-W bulbs.

Adult sheep require only wind protection and shelter from rain and snow in severely cold weather. Special attention should be given to protection for newly shorn ewes during winter storm conditions because shearing reduces the ewe's LCT dramatically. In general, windbreaks and open-shed-type structures afford adequate environmental protection for sheep with the exception of the newborn.

Thermal Considerations and Requirements for Sheep in Hot Environments

Planning for hot seasons should be based on the UCTs of sheep, which are on the order of 25°C for newborn lambs, 30 to 32°C for adult sheep, and 25°C for full-feed growing sheep (Hahn, 1985). Some lambs and adult sheep can adapt to temperatures up to about 40°C. Heat stress can reduce reproductive efficiency of rams; thus, facility planning should recognize this potential damage to reproductive performance. In rams, semen damage is closely correlated with scrotal and testicular temperature. The higher the scrotal temperature, the earlier semen damage is initiated and the longer it persists (Moule and Waites, 1963). Environmental temperatures of 27 to 30°C may be sufficient to lower semen quality. Heat stress in ewes may reduce the number of multiple ovulations. In general, however, ewes appear to be more resistant to heat stress than are cows and sows.

Ventilation

Since open sheds or other types of buildings provide adequate protection for sheep, the design should provide for adequate natural ventilation. The main objective in sheep housing design is to provide a dry, draft-free environment rather than closing building ventilation openings to increase temperature during cold weather. Closing such systems increases relative humidity, condensation of moisture, and related problems. In buildings that are to be closed to provide a so-called warm building, power ventilation is essential and requires engineering design specifications.

Humidity

As indicated earlier, sheep are not well adapted to hot, wet climates. In other climates, depending on the type of sheep, adaptability to levels of heat and humidity is wide. Adequate ventilation is the primary consideration in designing building environments for sheep to prevent buildup of moisture in the air and condensation.

Air Quality

Sheep maintained in well-ventilated open or closed buildings should normally have acceptable air quality. Safety precautions should be followed in production systems involving slatted floors with anaerobic manure pits within buildings. These precautions are discussed under design of environments for swine.

Noise

Limited noise is associated with sheep enterprises; therefore, no specific design for sound management need be considered. Sheep appear to be adaptable to normal external noise and probably habituate to typical conveyance noises such as planes, highways, and the like, as well as equipment noises.

Lighting

Artificial lighting requirements (MWPS-3, 1994) are determined by that needed for management functions and other operations (table 6.15). Sheep are seasonal breeders and normally breed in the fall. Management to alter the photoperiod to result in changing the breeding season to produce fall lambs is a highly specialized system and requires maintaining animals in complete darkness for extended periods. Success is variable, and the process is economically prohibitive in most production situations. The goal of fall lamb production is to have lambs ready for market during the time of the year when prices are normally higher.

Water Availability

Daily water requirements for sheep are shown in table 6.16 (MWPS-3, 1994). These are mean values and may vary depending on climatic conditions. Waterer space allowances are outlined in table 6.17 (MWPS-3, 1994). They are typically based on production systems that provide frequent and free access to water, eliminating problems related to crowding.

Feed Availability

Feeder space requirements are shown in table 6.18 (MWPS-3, 1994). When animals are limit-fed, they must have sufficient space to eat simultaneously in order to allow equal access and minimize the negative effects of dominant animals. If self-

TABLE 6.15   Light Levels for Sheep Housing[a]

Operation or Activity	Illumination (Foot-candles)	White Fluorescent (40 W)	Standard Incandescent (100 W)	(150 W)
			Building Area (W/ft2)	
Lambing	15	0.42	1.72	1.50
Lamb feeding and gestation	5	0.14	0.58	0.50
Feed storage and processing	10	0.28	1.15	1.00
Record keeping and office	50	1.38	5.72	5.00
Inspection and handling	20	0.55	2.29	2.00

[a]Values assume lights located for uniform lighting at about 5 foot-candles/watt incandescent and 20 foot-candles/watt fluorescent; some areas may require additional task lighting; values shown are for 8-foot-high ceilings
Source: MWPS-3, *Sheep Housing and Equipment Handbook,* 4th ed. (Ames, Ia.: Midwest Plan Service, 1994), p. 76, by permission

TABLE 6.16   Daily Water Requirements of Sheep

	Gallons/Day
Rams	2–3
Dry ewes	2
Ewes with lambs	3
5- to 20-lb lambs	0.1–0.3
Feeder lambs	1.5

Source: MWPS-3, *Sheep Housing and Equipment Handbook,* 4th ed. (Ames, Ia.: Midwest Plan Service, 1994), p. v, by permission

TABLE 6.17   Waterer Space Allowances for Sheep[a]

Type of Sheep	Average Waterer Allowance		Gallons/Day[a]
	Head/Bowl or Nipple	Head/Ft Tank	
Rams	10	2	2–3
Dry ewes	40–50	15–25	2
Ewes with lambs	40–50	15–25	3
Feeder lambs	50–75	25–40	1.5

[a]5- to 30-lb lambs will consume 0.1 to 0.3 gallons of water daily on the average
Source: MWPS-3, *Sheep Housing and Equipment Handbook,* 4th ed. (Ames, Ia.: Midwest Plan Service, 1994), p. v, by permission

TABLE 6.18   Feeder Space Recommendations for Sheep

	Limit-Fed (inches/Animal)	Self-Fed (inches/Animal)
Rams	12	6
Dry ewes	16–20	4–6
Ewes with lambs	16–20	6–8
Feeder lambs	9–12	1–2

Source: MWPS-3, *Sheep Housing and Equipment Handbook,* 4th ed. (Ames, Ia.: Midwest Plan Service, 1994), p. v, by permission

fed, all animals do not normally attempt to eat at one time; therefore, the amount of feeder space required is markedly reduced.

Space Requirements

Building and lot space allowances for sheep are summarized in table 6.19 (MWPS-3, 1994). Sheep are typically produced, at least to weaning, under extensive production systems. The species is adaptable to more intensive systems, making space considerations as important as with the other species. Allowances are expressed for

TABLE 6.19 Space Allowances for Sheep

	Building Space (ft²/head)		Lot Space (ft²/head)	
	Solid Floors	Slatted Floors	Dirt	Paved
Rams	20–30	14–20	25–40	16
Dry ewes	12–16	8–10	25–40	16
Ewes with lambs	15–20[a]	10–12[a]	30–50	20
Feeder lambs	8–10	4–5	20–30	10

[a]For ewes with lambing rates above 170 percent, increase space by 5 ft²/head; creep space for lambs should be 1.5 to 2 ft²/lamb
Source: MWPS-3, *Sheep Housing and Equipment Handbook,* 4th ed. (Ames, Ia.: Midwest Plan Service, 1994), p. v, by permission

animals in closed buildings, as well as those maintained in outside pens. The differential between inside and outside space allowances and also that between solid and slatted floors in buildings relate largely to the need for space in manure management, sanitation, and maintenance of good footing conditions for the animals.

Group Sizes

Maintaining sheep in groups of limited size facilitates the prevention of young lambs from being separated from the dam and management observation including detection of health problems. The recommendations for group sizes in table 6.20 (MWPS-3, 1994) serve as general guides based on producer experience under midwestern U.S. conditions. The recommendations may not be typical of those practiced under more extensive conditions. The concepts, however, may aid in achieving higher productivity in the sheep enterprise in both intensive and extensive types of operations. Characteristics related to the animal's innate flocking behavior are important in handling and are considered under handling procedures.

Floors and Floor Surfaces

Concrete paved lots should have a broom finish and inside floors should be less rough (machine-trowel finish) to increase ease of cleaning. If slatted floors are used, the width of slats should be in the 3- to 4-inch range with 1/2-inch spacing. The slat surface should be finished reasonably smooth for self-cleaning. Expanded metal also works well for sheep floors. MWPS-3 (1994) recommends an unflattened 3/4-inch number 9 expanded metal for floors for sheep. This surface provides more secure footing than flattened metal mesh material.

Trucking Considerations

Transport sheep at night or in the early morning during hot weather. Avoid exposure that will wet the animals and allow dipped animals to dry after dipping and

TABLE 6.20   Recommended Group Sizes for Sheep[a]

	Number of Ewes with Lambs before Weaning				
	Birth to 1 day	2–4 days	5–7 days	8–14 days	14 days to weaning
	Maximum Group Size				
Single lambs per ewe	lambing jugs	10 ewes + lambs	20 ewes + lambs	40 ewes + lambs	50–100 + lambs
Twins lambs per ewe	lambing jugs	5 ewes + lambs	10 ewes + lambs	20 ewes + lambs	50–100 + lambs

[a]Pregnant ewes, 200; ewes about to lamb, 50; early weaned lambs, 50 animals per group
Source: MWPS-3, *Sheep Housing and Equipment Handbook,* 4th ed. (Ames, Ia.: Midwest Plan Service, 1994), p. 3, by permission

TABLE 6.21   Truck Space Requirements for Sheep

Weight (lb)	Number of Sheep per Running Foot of Truck Floor[a,b] (92-Inch Internal Truck Width)
60	3.6
80	3.0
100	2.7
120	2.4

[a]Reduce value by 5 percent if animals have heavy or wet fleeces
[b]To convert to area per animal, divide 7.666 by the appropriate value in the column since this represents the area of truck space per running foot
Source: Grandin, T., *Livestock Trucking Guide,* 1992b, (Bowling Green, Ky.: Livestock Conservation Institute, 1992b), p. 12, by permission

before loading. A rest stop should be arranged if transport will exceed 48 hours. For animals traveling less than 8 hours, feed and water should be withheld 15 to 18 hours before loading. For trips longer than 8 hours, sheep should be fed lightly 2 to 3 hours before loading, and water should be withheld for several hours before loading (Grandin, 1992b). Truck space requirements are shown in table 6.21.

DESIGNING THE PHYSICAL ENVIRONMENT FOR CATTLE

The Thermal Environment

Beef cattle are produced in an extremely wide range of thermal environments. They vary in their adaptability, as do sheep, based on breed characteristics, available environmental protection, and feed supply. Animals with Zebu breeding are more resistant to heat and to insect populations associated with hot climates and those of a tropical nature. The range of environments for commercial dairy production is normally less variable because of the necessity to provide more favorable support for the demanding requirements associated with lactation. However, animals for milk production are maintained throughout the world and often in environmental conditions that simply will not support the high levels of production characteristic of commercial dairies. However, such animals provide an important function in supplying food to many.

The ability of cattle to adapt to a wide variety of environmental conditions and our understanding of these capabilities and the demands of such environments are demonstrated by the sophisticated procedures to determine nutritive requirements for variable environments. NRC (1981a, 1996) provides excellent examples of technology to meet the needs of cattle and in turn their general well-being. Some of these procedures are illustrated in the section covering design of the dietary environment.

Thermal Considerations and Requirements for Cattle in Cold Environments

Newborn Calves If calves are dried at birth, because they have a rapidly increasing ability to attain thermoneutral metabolic rate within the first day, they are reasonably resistant to cold. This response capability is similar to that of lambs. However, LCT of the newborn is around 8° to 10°C (Hahn, 1985; Gonzalez and Blaxter, 1962). Thus, in many parts of the world, calves born in early spring are introduced into an environment that presents conditions leading to cold stress. Protection against wind and wet conditions is important. Predisposition to diseases such as diarrhea and varied respiratory problems is greatly increased by stress. As with all newborns, early nursing is critical to attaining and sustaining homeothermy.

Feedlot Cattle Under applied conditions, Milligan and Christison (1974) reported that cattle in a Canadian feedlot environment performed well in a range of mean monthly temperatures between −8 and 18°C. Weight gain was highest at a range of 11 to 18°C.

Cattle under *ad libitum* feeding conditions have very high rates of heat production and thus low critical temperatures (−30°C or below, Hahn, 1985; Webster, 1970). In severe weather, however, feed consumption may be reduced (animals refuse to leave shelter to eat or feed consumption may be restricted by wet, freezing feed), which compounds cold stress effects.

In extreme cold (−8°C and below) the maintenance energy requirement increases and may reduce efficiency of feed conversion dramatically, even though the animals may consume enough additional feed to maintain normal rate of gain.

An example of difficulty in adapting to changing demands related to cold is the high death loss resulting from winter storms when cattle of tropical and southern climates are relocated to a cold environment. Such cattle perform satisfactorily in high plains and midwestern U.S. feedlots if properly managed to allow acclimatization and if provided normal feedlot protection in the form of windbreaks and shelter in wet conditions.

Breeding Herd LCT for pregnant cows is around −11 to 23°C (Webster, 1970). Cattle exhibit wide adaptation to thermal environments, and performance is acceptable under a wide range of environmental conditions. Prolonged cold stress, however, may result in anovulation and acyclicity. As with sheep, environmental protection requirements are minimal except for newborn calves. Adult cattle and those being fed for high growth rates and finishing require limited protection during severe

TABLE 6.22 Estimates of Lower and Upper Critical Temperatures for Cattle

	LCT[a] (°C)	UCT[a] (°C)
Beef cow (pregnant)	-11 to -23[b]	Range for all classes is from 25, if animals are on full feed, to 28 to 32 for calves and cows.
Newborn calf (3 days)	12.8[c]	
Calf (20 days)	8.2[c]	
Growing calf (0.8 kg gain)	−16 to −21[b]	
Feeder calf (1.5 kg gain)	−31 to −38[b]	

[a]LCT, lower critical temperature; UCT, upper critical temperature
[b]*Source:* Webster, A.J.F. 1970. Direct effects of cold weather on the efficiency of beef production in different regions of Canada. *Can. J. Anim. Sci.* 50:563
[c]*Source:* Gonzalez-Jiminez, E., and K.L. Blaxter. 1962. The metabolism and thermoregulation of calves in the first month of life. *Can. J. Anim. Sci.* 50:563

periods. Such protection is commonly afforded by windbreaks or open-front buildings. In many areas minimal but effective protection is provided by land terrain, forests, and other environmental characteristics

Thermal Considerations and Requirements for Cattle in Hot Environments

Calves Calves can readily adapt to temperatures of 30°C or more; therefore, the possibility of heat stress in young calves is limited in most production systems. Penning calves without shade in extreme heat should be avoided.

Feeder Cattle Feedlot performance normally remains at acceptable levels at average daily temperatures up to 28 to 30°C, depending on relative humidity. Above this level some heat stress may be apparent, which in turn results in reduced performance and efficiency.

Mature Breeding Cattle Adult cattle are typically adaptable to temperatures of 30°C or more. High scrotal temperature may damage spermatogonia, primary spermatocytes, and spermatids. Bulls exposed to 37°C and 80 percent relative humidity for 17 days (12 hr/day) showed reduced sperm concentration and output (Casaday et al., 1953).

Heat stress in cows tends to increase the length of the estrous cycle and the duration of estrus. Conception rates appear to be inversely related to average daily temperature-humidity index (THI) when sufficiently high to cause stress effects. Properly designed shade can be helpful and may be critical during periods of high temperature.

Lactating Dairy Cattle Heat prostation results when constant environmental temperature is above body temperature (CDGCAA, 1988). Johnson (1985) reported that the critical THI for milk production is 72.

The decline in milk production appears to relate to a reduction in feed intake. Some dairy managers develop feeding schedules to take advantage of the cooler nighttime temperatures in hot regions. Tropical breeds have evolved as somewhat more resistant to the impacts of higher temperatures, although U.S. and to a great extent European dairies are based on the Holstein or Friesian breeds.

Properly designed shades may assist cattle in coping with extremes in heat. These features are commonly used in more intensive systems such as pasture grazing and intensive systems in midwestern, southwestern, and southern areas of the United States.

Ventilation

Since most cattle production systems involve open buildings, planning of facilities should ensure adequate ventilation in both hot and cold conditions. The goal in facility design must be to provide a draft-free, dry internal environment to prevent the buildup of moisture in the air and condensation. In buildings without adequate natural ventilation, forced-air ventilation is essential. Recommended ventilation rates for cattle are shown in table 6.23 (MWPS-6, 1987; MWPS-7, 1995).

Humidity

If proper ventilation is provided in cattle environments, humidity is a problem only during severe periods. At such times shade and open facilities to allow air movement are critical.

Air Quality

Since most beef production occurs in open buildings or lots, adequate natural ventilation and air movement eliminate most air quality problems. Confinement

TABLE 6.23   Recommended Ventilation Rates for Beef and Dairy Cattle Facilities[a]

	Ventilating Rates cfm[b]/Animal		
	Cold Weather	**Mild Weather**	**Hot Weather**
Calves, 0–2 mo	15	50	100
Feeder calves and dairy replacements, 2–12 mo	20	60	130
Yearlings and dairy replacements, 12–24 mo	30	80	180
Beef cow or dairy cow, 1,400 lb	50	170	470
Milk room			600
Milking parlor		100 cfm/stall	400 cfm/stall

[a]An alternative cold weather rate can be calculated by dividing the room or building volume (in cubic feet) by 15; the hot weather rate can be calculated by dividing the volume by 1.5
[b]cfm, cubic feet per minute
Sources: MWPS-6, *Beef Housing and Equipment Handbook*, 4th ed. (Ames, Ia.: Midwest Plan Service, 1987), p. 1.2, by permission; MWPS-7, *Dairy Freestall Housing and Equipment Handbook*, 5th ed. (Ames, Iowa: Midwest Plan Service, 1995), p. 4.4, by permission

buildings over anaerobic manure pits provide potential dangers to both animals and workers from toxic gases. Some dairy facilities utilize closed buildings or a combination of closed and open housing. Dairy facility ventilation rates are shown in table 6.23. Refer to the air quality section for swine for more detailed information on toxic gases associated with animal facilities and related precautions.

Noise

Noise in beef and dairy production enterprises is not a common problem. Cattle appear to be able to adapt and habituate to routine noise levels associated with planes, trains, highways, and equipment operation.

Lighting

Artificial lighting is not required in all cases for satisfactory animal performance. Proper lighting, however, is important to the operation for several reasons. Benefits include reduced predator threat, a quieting effect on the animals, better feed consumption because of more nighttime eating during hot weather, and reduced feed bunk space required because of 24-hour feeding (this may also allow better access for less aggressive animals and may be a stress-reducing effect for newly added cattle arriving in unfamiliar surroundings) (MWPS-6, 1987).

Levels of recommended lighting are shown in table 6.24 (MWPS-6, 1987; MWPS-7, 1995). Cattle management does not involve management of photoperiod or length of light during the 24-hour day. Thus, such lighting is not a consideration in beef and dairy production. However, some photoperiod influence occurs in cattle in that there is a greater tendency for cows to calve in the spring than at other times of the year. The effect is not of sufficient magnitude to be considered a management factor.

Water Availability

In extensive production systems, the distances animals must travel to water must be considered; since the daily migration to the water source is usually accomplished by the entire herd, the available water supply at any given time is also important. In intensive operations, animals typically have free access to water at all times. The recommendations in table 6.25 relate to more intensive situations, although those for pasture conditions are applicable to extensive systems as well. Tables 6.26 and 6.27 summarize the estimated daily water allowances (MWPS-6, 1987; MWPS-7, 1995) required for beef and dairy cattle. The demand for water associated with milk production should be noted.

Feed Availability

Feeder space allowances are shown in table 6.28 for beef cattle (MWPS-6, 1987) and table 6.29 for dairy cattle (MWPS-7, 1995). The feeding system utilized in beef production is variable and adds flexibility to management opportunities. Such choices require careful evaluation of feeder space to minimize the effects of competition

TABLE 6.24   Recommended Light Levels for Beef and Dairy Cattle Buildings

	Light Level	Standard Cool White Fluorescent (40 W)	Standard Incandescent	
			(100 W)	(150 W)
	Foot-Candles	Watt/ft^2 of Building Area		
Beef cattle facilities				
Housing	5	0.14	0.58	0.50
Calving barn	20	0.55	2.29	2.00
Loading platform	15	0.42	1.72	1.50
Animal inspection and handling	20	0.55	2.29	2.00
Along feed bunk	10	0.28	1.15	1.00
Feed storage and processing	10	0.28	1.15	1.00
Haymow	3	0.09	0.35	0.30
Office	50	1.38	5.72	5.00
Toilet room	30	0.83	3.43	3.00
Dairy cattle facilities[a]				
Dairy cow housing	7	0.20	0.80	0.70
Dairy calf housing	10	0.28	1.15	1.00
Milking area	20	0.55	2.29	2.00
Equipment washing area	100	2.75	11.44	10.00
Milk handling	20	0.55	2.29	2.00

[a]Lighting specifications from other areas in dairy facilities are the same as for beef cattle
Sources: MWPS-6, *Beef Housing and Equipment Handbook,* 4th ed. (Ames, Ia.: Midwest Plan Service, 1987), p. 10.3, by permission; MWPS-7, *Dairy Freestall Housing and Equipment Handbook,* 5th ed. (Ames, Ia.: Midwest Plan Service, 1995), p. 11.3, by permission

TABLE 6.25   Waterer Space Recommendations for Cattle

	Animals per Cup or Bowl		Animals per Foot of Accessible Tank Perimeter	
	Lot	Pasture	Lot	Pasture
Feeder cattle				
Calves, 400–800 lb	25	18	16	10
Finishing, 800–1,200 lb	20	—	16	—
Bred heifers, 800 lb	20	15	16	10
Cows, 1,000 lb	20	15	16	10
Cows, 1,300 lb	18	14	16	10
Bulls, 1,500 lb	16	10	9	7

Sources: MWPS-6, *Beef Housing and Equipment Handbook,* 4th ed. (Ames, Ia.: Midwest Plan Service, 1987), p. 1.1, by permission; MWPS-7, *Dairy Freestall Housing and Equipment Handbook,* 5th ed. (Ames, Iowa: Midwest Plan Service, 1995), p. 1.2, by permission

TABLE 6.26   Estimated Daily Water Requirements for Beef Cattle

	Cold Weather (gallons/head/day)	Hot Weather (gallons/head/day)
Feeder cattle		
Calves, 400–800 lb	4–7	8–15
Finishing, 800–1,200 lb	8–11	15–22
Bred heifers, 800 lb	7	15
Cows, 1,000 lb	9	18
Cows, 1,300 lb	13	25
Bulls, 1,500 lb	14	27

Source: MWPS-6, *Beef Housing and Equipment Handbook,* 4th ed. (Ames, Ia.: Midwest Plan Service, 1987), p. 1.1, by permission

TABLE 6.27   Estimated Daily Water Requirements for Dairy Cattle

	Gallons/Head/Day
Calves, 1–1.5 gallons/100 lb body weight	6–10
Heifers	10–15
Dry cows	20–30
Milking cows	35–45

Source: MWPS-7, *Dairy Freestall Housing and Equipment Handbook,* 5th ed. (Ames, Ia.: Midwest Plan Service, 1995), p. 1.2, by permission

TABLE 6.28   Recommended Feeder Space Allowances for Beef Cattle

	1× Day Feeding	2× Day Feeding	Self-Fed Grain	Self-Fed Roughage
	Inches/Animal			
Feeder cattle				
Calves, 400–800 lb	18–22	9–11	3–4	9–10
Finishing, 800–1,200 lb	22–26	11–13	4–6	10–11
Bred heifers, 800 lb	22–26	11–13	4–6	11–12
Cows, 1,000 lb	24–30	12–15	5–6	12–13
Cows, 1,300 lb	26–30	12–15	5–6	13–14
Bulls, 1,500 lb	30–36	—	—	—

Source: MWPS-6, *Beef Housing and Equipment Handbook,* 4th ed. (Ames, Ia.: Midwest Plan Service, 1987), p. 1.1, by permission

TABLE 6.29 Recommended Feeder Space Allowances For Dairy Cattle

	Age (months)					Mature Cow
	3–4	5–8	9–12	13–15	16–24	
	Inches/Animal					
Self-feeder						
Hay or silage	4	4	5	6	6	6
Mixed ration or grain	12	12	15	18	18	18
Once-daily feeding						
Hay, silage, or ration	12	18	22	26	26	26–30

Source: MWPS-7, *Dairy Freestall Housing and Equipment Handbook,* 5th ed. (Ames, Ia.: Midwest Plan Service, 1995), p. 1.1, by permission

TABLE 6.30 Recommended Space Allowances for Beef Cattle

	Open Lot Space				Barn Space		
	Paved	Unpaved with Mound	Mound	Unpaved without Mound	Barn with Lot	Barn without Lot	Enclosed Barn with Slotted Floor
	ft²/Head						ft²/1,000 lb
Feeder cattle							
Calves, 400–800 lb	40–50	150–300	20–25	300–600	15–20	20–25	17–20
Finishing,							
800–1,200 lb	50–60	250–500	30–35	400–800	20–25	30–35	17–20
Bred heifers, 800 lb	50–60	250–500	30–35	400–800	20–25	30–35	—
Cows, 1,000 lb	60–75	300–500	40–45	500–800	20–25	35–40	—
Cows, 1,300 lb	60–75	300–500	40–45	500–800	25–30	40–50	—
Bulls, 1,500 lb	—	1,200	50–60	1,500	40	45–50	—

Source: MWPS-6, *Beef Housing and Equipment Handbook,* 4th ed. (Ames, Ia.: Midwest Plan Service, 1987), p. 1.1, by permission

and to provide ample opportunity for the animals to consume an adequate amount of feed. Nutrient density of feeds for beef cattle is probably more variable than that commonly used for the other species. Thus, eating time is correspondingly variable. This factor is an important feeding management issue.

Space Requirements

Space requirements for beef cattle (MWPS-6, 1987) are shown in table 6.30, and those for dairy cattle (MWPS-7, 1995) are shown in table 6.31. Space limitations can represent both physical and psychological stressors, as discussed in part 2. One effect is simply limited access to utilities, such as feed and water. Another is the impact of crowding on the behavioral characteristics of the animal.

TABLE 6.31   Recommended Space Allowances for Dairy Calves and Replacement Heifers

Individual Calf Housing	Pen Size (ft)	Outdoor Run, Pen Size (ft)
Up to 2 mo of age		
Calf hutch	4 × 8	4 × 6
Bedded pen	4 × 7	—
Tie stall	2 × 4	—
3–5 mo of age (group housing)		
Group calf hutch		25–30
Bedded pen		25–30

	Age in months			
Heifer Housing	5–8	9–12	13–15	16–24
	ft²/animal			
Housed resting area and outside lot system				
Resting area	25	28	32	40
Paved lot	35	40	45	50
Total confinement system				
Bedded area	25	28	32	40
Slotted floor area	12	13	17	25

Source: MWPS-7, *Dairy Freestall Housing and Equipment Handbook,* 5th ed. (Ames, Ia.: Midwest Plan Service, 1995), p. 1.1, by permission

Dairy production facilities commonly involve the use of stalls that may have either stanchions or ties for restraining the cow. Another commonly used system is referred to as the freestall system. In the latter system, except for the milking time, cows can move in and out of a stall area at will but will spend much of the day in the stall provided. Table 6.32 lists size recommendations for freestall facilities on an individual cow basis (MWPS-7, 1995). Table 6.33 provides recommended alley widths for freestall facilities.

Veal Calf Facility and Systems Considerations

Veal calf production is a highly specialized sector in the meat production industry. Typical production systems involve feeding calves liquid milk replacer from soon after birth to approximately 16 weeks of age, at which time body weight is in the range of 350 to 400 lb. Meat from calves produced in this system is referred to as milk veal. Meat from calves slaughtered at 3 to 4 weeks of age may be referred to as *bob veal.* *Red veal* is a term describing meat from grain-fed calves marketed at weights ranging from 150 to 450 lb. Calves maintained for veal production are usually housed in individual stalls or in some cases in group pens. In other instances calves may be housed in individual stalls for the first few weeks and then in group pens until marketed. Those maintained in stalls are usually tethered with a chain or strap of approximately 2 feet to allow calves to lie down and rise to their feet normally. The diet normally consists of milk replacer fed as a liquid. Some production systems involve limited iron lev-

TABLE 6.32   Cow Freestall Dimensions[a]

				Neck Rail and Brisket	
Cow weight (lb)	Freestall Width (inches)	Freestall Length		Neck Rail Height Above Stall Bed (inches)	Board Distance from Alley Side of Curb (inches)
		Side Lunge	Forward Lunge[a]		
800–1,200	42	6'6"	7'6" to 8'	37	62
1,200–1,500	45	7'	8' to 8'6"	40	66
Over 1,500	48	7'6"	8'6" to 9'	42	71

[a]Stall width measured from center to center of 2-inch pipe dividers. For wider dimensions, increase width accordingly. An additional 12 to 18 inches in stall length is required to allow the cow to thrust her head forward during the lunge process.
Source: MWPS-7, *Dairy Freestall Housing and Equipment Handbook,* 5th ed. (Ames, Ia.: Midwest Plan Service, 1995), p. 1.1, by permission

TABLE 6.33   Typical Freestall Alley Widths

Type	Width (feet)
Feed and stall access alley	10–12
Feeding alley	9–10
Access alley between two stall rows	
Solid floor	8–10
Slotted floor	6–9

Source: MWPS-7, *Dairy Freestall Housing and Equipment Handbook,* 5th ed. (Ames, Ia.: Midwest Plan Service, 1995), p. 1.1, by permission

els in the diet to produce a lighter-colored veal; however, this practice is controversial.

Stull and McMartin (1992) conducted extensive evaluations of ten commercial veal production units and concluded that the facility characteristics that follow allow efficient production and a high level of well-being among the calves housed. Individual calf stalls ranged from 19 to 22 inches in width, and dividing partitions were 30 to 36 inches high. This height, with board spacing of about 2 inches, allows calves to have contact (head to head and head to body) with other calves. Partition length of individual pens ranged from 28 to 43 inches. Floor length of individual pens ranged from 58 to 70 inches, with most in this study being 60 to 66 inches in length. These measurements of floor length include 18 to 24 inches of expanded metal or steel bar grates at the rear of the stall floor. The balance of the stall floor was made up of wooden slats with 1-inch spacing.

Tethering of calves with light-weight chain is typical for the entire feeding period; in some cases tethering is utilized after 45 days of age. When calves are not tethered, wire or wooden end gates are used at the rear of the stall. In these evaluations Stull and McMartin concluded that the performance was equal between calves with and without tethers the first 45 days. Following this period, all calves were tethered.

Measures of levels of cortisol in the blood were not significantly different between calves in the tethered and nontethered groups. Interestingly, calves maintained in groups of thirty in pens showed higher cortisol levels than those maintained in individual stalls (the group-fed calves in this evaluation were placed in the production system at about 8 weeks of age and continued to about 20 weeks of age). This finding suggests that group-housed calves may have a higher stress level than individually housed calves.

Analysis of neutrophil to lymphocyte ratio in the blood as a measure of stress at 2 weeks showed higher levels in calves that were tethered. This difference did not exist at 4 weeks or during the balance of the production period. Stull and McMartin suggest that this difference may be due to a level of stress associated with tethering initially that disappears as calves condition to this practice.

These evaluations also included one production unit utilizing group housing. In this unit satisfactory performance was observed when calves were housed in groups (after 8 weeks of age) of thirty calves in pens measuring 19 by 28 feet, or about 17 square feet per calf. Pen floors were plastic-coated expanded metal. This space allowance is less than the 25 to 30 square feet recommended by MWPS-7 (1995) for bedded pens for calves up to 5 months of age. None of the facilities in these studies attempted to limit natural or artificial light as a part of the production system. The authors did not report quantitative observations relative to stereotypic behavior in these studies.

Individual stalls are commonly used for a part or all of the veal calf production period. Advantages relate to improved individual care and observation and may contribute to improved health as well because of reduced contact with other calves and individual observation of performance and condition. Calves in individual stalls are prevented from engaging in behaviors such as sucking ears, sheaths, and other parts of conspecifics and urine drinking in the process of cross-sucking. Disadvantages often associated with individual stalls are limited or no social contact with conspecifics, lack of freedom of movement, and greater likelihood of stereotyped behaviors.

A compromise to individual stall and group pen production is a combination of the two described by Stull and McMartin (1992). Stull (1992), in observing calves kept in group pens versus stalls beyond 8 weeks of age, reported approximately equal amounts of time spent in recumbency or for shifts from recumbency to standing for each group. SVCAWS (1995) recommended that space allowances for calves maintained in individual pens be determined by the following: Pen or stall width should be equal to the calf's height at withers, and stall length should be determined by multiplying the length of the calf from the tip of the nose to the caudal edge of the pin bone by 1.1. This figure allows adequate space for forward and backward movements in normal rising and lying. Individual pens or stalls should be constructed with spaces in the sides to allow visual and tactile contact. This same committee (SVCAWS, 1995) suggested that individual space in group pens be calculated by the following: Height at withers multiplied by length of the calf from nose to the caudal edge of the pin bone multiplied by 1.1. Space allowances for typical Holstein calves calculated by this means are as follows: $1.4 \, m^2$ at 8 weeks of age, $1.8 \, m^2$ at 16 weeks of age, and $2.1 \, m^2$ at 22 weeks of age.

Design criteria for veal calves must recognize that the LCT for newborn calves until 2 weeks of age is about 10°C, with optimal temperature on the order of 18 to 21°C.

Floor and Floor Surfaces

Good footing is a critical safety factor with large animals. Concrete floors and lot paving should be finished rough, with grooving on rises and around feeders and waterers. Grooves placed diagonal to cleaning and animal traffic patterns are preferable to those set parallel. Suggested groove depth is 3/8 to 1/2 inch, and groove width should be about 1 to 1 1/2 inches. Spacing between grooves should be 4 to 8 inches. Existing floors that are slick should be grooved. Grandin (1988) indicates that the industry standard for dairy facilities is to score concrete 3/8 inch deep, 1 1/2 inches wide, and 3 to 4 inches apart. Grooves should run in the same direction as primary cow movement. High-traffic areas for dairy cows should be scored in a diamond pattern. Working areas for feedlot cattle should be grooved in a diamond or square pattern of about 8 inches with grooves about 1 inch deep (Grandin, 1988a).

Trucking Considerations

Transport should be accomplished at night or in the early morning during hot weather. Avoid shipment of wet animals or exposure to precipitation during cold weather. Allow dipped animals to dry before loading. A rest stop should be planned if transport is longer than 48 hours. Feeder calves that will be traveling for more than 18 hours should be fed a grain-based diet for at least 24 hours before shipment. Feeder calves should be fed on arrival, but water must be withheld until some feed has been consumed to prevent engorging. Feedlot cattle being readied for market should have feed withheld (with access to water) for 12 hours before shipment. No water is available during shipment, so the animals should be watered upon arrival at the processing plant. No more than 48 hours should elapse between the last feeding and slaughter (Grandin, 1992b). Space allowances for cattle in trucks are shown in table 6.34.

DESIGNING THE PHYSICAL ENVIRONMENT FOR HORSES

The Thermal Environment

Horses, like cattle and sheep, if adequate feed is available, can withstand extremes in temperature without negative stress effects. Maton et al. (1985) suggest a lower temperature comfort zone for horses of 10 to 15°C. The goal in designing the internal environment of buildings is to provide a dry, draft-free space. Thus, open-front buildings with good natural ventilation are adequate in most climates. In the case of new foals, a stall with straw bedding for the dam and foal is adequate. Closing of buildings to elevate temperature should be avoided unless adequate ventilation exists.

In hot environments horses have the ability to dissipate large amounts of heat through sweating and evaporation. Thus, shade with good air movement should be adequate protection from heat stress.

Humidity

The suggested relative humidity maximum level in horse facilities is 60 to 80 percent.

TABLE 6.34   Truck Space Requirements for Cattle

Average Weight (lb)	Number of Cattle Per Running Foot of Truck Floor[a] (92-Inch Internal Truck Width)
Cows and feedlot cattle	
600	0.9
800	0.7
1,000	0.6
1,200	0.5
1,400	0.4
Calves	
200	2.2
250	1.8
300	1.6
350	1.4
400	1.2
450	1.1

[a]To convert to area per animal, divide 7.666 by the appropriate value in the column since this represents the area of truck space per running foot
Source: Grandin, T., *Livestock Trucking Guide* (Bowling Green, Ky.: Livestock Conservation Institute), p. 11, by permission

Ventilation

Natural ventilation is adequate in most cases for horse facilities. If a building is closed to increase internal temperature, forced-air ventilation will be required to prevent the buildup of moisture in the air and condensation on internal surfaces. Similarly, if openings are not sufficient to allow free airflow through a building during summer, power ventilation is required. In power-ventilated buildings, airflow should normally be from 0.7 to 2.8 m^3/min/450 kg body weight. The 0.7 level is appropriate for temperatures of -18 to $-7°C$ (0 to 20°F), and the 2.8 level is appropriate for temperatures -1 to 10°C (30 to 50°F) (MWPS, 1987).

Air Quality

Building design should normally provide good ventilation, eliminating the problem of air quality if proper building sanitation is maintained. Since horses are not normally maintained on slatted floors with manure pits below, this source of possibly toxic gases is not present.

Lighting

Lighting requirements are based on that required for movement of horses and management operations. Because horses are seasonal breeders, managers accomplish day-length-control procedures by lengthening the light period in order to move the breeding season forward or earlier in the spring. The management scenario is referred to as light priming.

Light Priming in Horses

The goal of light priming is to cause mares to cycle and ovulate at an earlier date than normal (normal breeding season is spring and summer) by increasing day length during a critical period. An added goal may be to cause stallions to demonstrate higher levels of libido and to be more successful in settling mares earlier in the year.

Extended light periods in winter have been investigated with both mares and stallions. The general pattern used is to provide artificial light (in individual stalls, in group housing, or even outdoors in groups with large overhead lights) starting typically between November 15 and December 15 in an effort to have mares ovulating and stallions fertile, with more libido than normal for earlier breeding (in February, for example). The length of time for artificial lighting should be planned to supplement natural day length in winter during the priming period to approximately 16 hours.

Results suggest that mares may begin cycling 60 to 80 days sooner and may have a greater tendency to actually ovulate during the first estrous period. It should be noted that results may be variable in terms of success in achieving earlier conception.

Stallions may respond with higher libido but may not characteristically show higher than normal seasonal spermatogenesis. It is common for stallions to breed and settle females throughout the year. Enhancement of libido, however, may be sufficient reason to try light priming for some stallions, even if the technique may not increase semen quality at the same time.

Water Availability

In group housing management, one water cup should be provided for five horses. Water should be available at all times during warm weather and throughout the year, if possible, except immediately after strenuous exercise.

If horses have free access to water cups in tie stalls, these should have valves to permit turnoff so that sweating horses can be prevented from drinking before cooldown. In other situations exercised animals should not be allowed access to water troughs before cooldown. If water is allowed during the cooldown period, it should be in small amounts every 10 to 15 minutes. This schedule should prevent related digestive problems and the possibility of founder.

Feed Availability

Feeding systems for horses typically consist of a separate allowance of hay and grain or concentrate mix. Hay feeders in stalls should be designed to hold a daily hay allowance of 10 to 20 lb, and feeder space should be a minimum of 36 inches wide for light horses and a minimum 40 inches for larger animals.

Ad libitum feeding of grain or concentrates is not practiced in horse management. Thus, space requirements for self-feeders are not useful design criteria for horse facilities, except in the case of hay feeders.

Space Requirements

Space requirements for housing horses (table 6.35) are adapted from MWPS (1987) and CDGCAA (1988).

Fencing

Smaller areas where horses may be crowded or trapped and approaches to gates should be fenced with smooth wire. The height of the lowest wire or board should be high enough to prevent animals from getting their legs caught under the fence. Any single-strand wire, electric, or other fencing of low visibility should be marked

TABLE 6.35   Space Requirements for Horses

	Box Stalls[a] (ft)	Tie Stalls[b] (ft)
Mature animals		
Small	10 × 10	4.5 × 9
Medium	10 × 12	5 × 9
Large	12 × 12	5 × 12
Pony	9 × 9	3 × 6
Brood mare	12 × 12	
Foal to 2 yr old		
Small	10 × 10	4.5 × 9
Large	12 × 12	5 × 9
Stallion	14 × 14	

Ceiling height: In stable area, 8 to 10 ft, but the guide should be to allow at least 1 foot above the erect ears of the animal

In riding areas allow 14 to 16 ft for the safety of horse and rider

Alley widths (ft)	
Between rows of stalls	8–10
Behind rows of stalls	6
Feed alley in front of stalls	4–5
Door height (ft)	7–8 minimum
Door width (ft)	4–4.5 minimum

Trailers

Ceiling height (ft)	
Below 15 hands	5.6–5.9
15–16 hands	6.6–6.9
Width (ft)	
Single or tandem	4.0
Two abreast	5.6–6.6 × 5.9–10.2

[a]Wall height should be 6 to 7 ft; upper 2 to 2.5 ft of the stall wall may be of an open design that allows the animal to see through. Regardless of dimensions, the minimum width of the stall should allow the animal to turn around, which normally requires 8 ft for a mature horse. Dutch doors are commonly used to allow the animal to have the perception of more space and to observe activities in the barn area that provides some environmental enrichment. Foaling stalls should have solid walls to prevent aggression toward other horses that may be viewed or redirected aggression toward the foal.
[b]Length includes space for manger
Sources: MWPS, *Structures and Environment Handbook,* 11th ed. (Ames, Iowa: Midwest Plan Service, 1987), p. 514.1, by permission; CDGCAA, *Guide for the Care and Use of Agricultural Animals in Agricultural Research and Teaching,* 2d ed. (Champaign, Ill.: FASFAS, 1988), p. 7, by permission

with bright tape or flags to help prevent injury. Fence height recommendations are generally 4.5 to 6 ft for horses and 3.5 to 5 ft for ponies.

Floor and Floor Surfaces

Concrete passageways should be rough or grooved to provide for good footing. If horses are maintained in tie stalls, hard-surfaced floors should be covered with shavings or other material to absorb moisture and soften the surface. Maton et al. (1985) concluded that foot lesions are more common in horses maintained in tie stalls than among those in boxed stalls.

Bedding

In general, straw, wood shavings, and sand are excellent floor coverings for horses. If the animals are consuming low-roughage or pelleted feeds, sand should not be used as bedding since the horses may develop digestive problems from consuming excessive amounts of the material from the floor.

Transportation Considerations

Recommended space in a transport vehicle is approximately 90 cm in width and 2.4 m in both length and height for average-size horses (Houpt and Lieb, 1993). The space should allow the horse to balance with head and weight shift and to brace with the legs. Wooden floors are commonly used in trailers; however, other materials may be used if slipping can be avoided by the use of bedding or floor pads. Effective environmental temperature may be a critical consideration in preventing heat and cold stress. Leadon et al. (1989) suggest use of caution when temperatures are above 25°C or below 10°C.

DESIGNING THE PHYSICAL ENVIRONMENT FOR POULTRY

The Thermal Environment

Recommended effective environmental temperatures for poultry of different ages and some production systems are shown in table 6.36. The recommendations are a guideline for certain situations, but more detailed specifications are required for specialized design purposes and deviations from these ranges are acceptable under some conditions. It should be recognized also that specific genetic lines may have different requirements. Thus, specific recommendations for a given line should be obtained from the supplier of that strain. Carr and Carter (1985) suggest that observing the behavior of birds in the brooding environment is an effective approach to evaluating whether the animals are uncomfortable. Loud chirping and crowding are observed when they are cold. Chicks are spread out and calm under optimum conditions. Panting and distress are observed when the chicks are too warm.

Environmental temperatures below the LCT reduce feed efficiency in growth or egg production because of increased maintenance requirements. Excessive heat

TABLE 6.36   Recommended Thermal Environments for Poultry

Brooding chicks:	
Room temperature	20–25°C
Under hover temperature	32–35°C
Reduce the temperature 2.5°C weekly to 20°C	
Brooding turkey poults:	
Room temperature	20–25°C
Under hover temperature	35–38°C
Reduce the temperature 3°C weekly to 24°C	
Brooding ducklings:	
Room temperature	20–25°C
Under hover temperature	26.5–29.5°C
Reduce the temperature 3.3°C to 13°C	
Growing broilers	20°C
Growing turkeys	24°C
Growing ducklings	18–20°C

Source: CDGCAA, *Guide for the Care and Use of Agricultural Animals in Agricultural Research and Teaching,* 2d ed. (Champaign, Ill.: FASFAS, 1988), p. 44, by permission

results in stress, which is evidenced by panting and difficulty in accomplishing heat loss. This stress results in reduced feed consumption and thus performance. Payne (1966) reported a 1.5 percent decrease in feed consumption for each degree increase from 21 to 30°C and a 4.6 percent decrease for each degree increase from 32 to 38°C. Carter (1981) devised a guide to illustrate expected effects of different environmental temperature ranges on layers. The following list is adapted from that report.

Below 10°C	Reduction in egg production, weight gains, and feed efficiency
10 to 21°C	Feed efficiency is reduced
21 to 26°C	Ideal temperature range
26 to 29°C	Slight reduction in feed intake; limited influence on production and efficiency
29 to 32°C	Feed consumption is reduced; egg production drops; egg size and shell quality deteriorate; cooling procedures should precede the attainment of this range to maintain adequate performance
32 to 35°C	Feed consumption and performance are reduced drastically and heat prostration may occur; cooling is essential.

When cooling is required in addition to normal ventilation rates, evaporative cooling systems are used. Such systems include fogging of birds or movement of air through evaporative pads, or fog.

In addition to appropriate temperature, it is important to recognize that diurnal variation in temperature on the order of 10 to 14°C is important to ensure eggshell strength and laying performance. If such variation does not occur normally, management measures are required to achieve it. One practice is increased night ventilation of the poultry house to equalize internal and external temperatures.

Ventilation

Design of adequate mechanical ventilation for poultry facilities, as is true for other species, requires appropriate engineering and technology. Because poultry production facilities frequently require sophisticated light management, natural ventilation options may be limited. This places added emphasis on the critical nature of ventilation design considerations. Thermal requirements and related ventilation needs may also vary among genetic lines of birds. This adds to the site specific character of design elements. Thus, the reader is referred to Carr and Carter (1985) for an extensive review of ventilation planning methods and recommendations. In addition, materials relating to specific requirements for genetic lines should be obtained from the supplier of the birds to be housed. Because many poultry facilities do not provide the option for achieving adequate natural ventilation, advanced preparations to meet power needs in emergencies are critical. Heat prostration is likely to occur at temperatures above 30 to 35° C (Carter, 1981).

Humidity

The general recommendation for relative humidity levels in poultry facilities is 60 percent minimum during the first 2 to 3 weeks of age (levels below 40 percent will likely increase respiratory problems); after this period a level ranging from 30 to 80 percent appears to be acceptable for broilers and turkeys. A range from 50 to 80 percent for laying hens is a common recommendation. It is important to recognize the dramatic change in moisture production by poultry as temperature changes in a house. One thousand laying hens, for example, produce about 11 kg and 14 kg of water (respiration, defecation, and wasted drinking water), respectively, when room temperature is 7.2 or 26.7°C (MWPS, 1987). This level approaches a 30 percent increase due to change in temperature and has implications related to excess moisture content of the litter and associated ammonia production.

Air Quality

Ammonia release from litter is a common influence on air quality in poultry houses. Factors contributing to the level of ammonia in a facility are bird density, manure buildup, temperature, moisture content of the litter, and ventilation. Atmospheric ammonia levels of 70 to 100 μL/L have been shown to have negative effects on performance and may result in pathological respiratory problems such as bronchitis as well as eye irritation. Removal of toxicants by ventilation is important; however, regulating the pH of the litter may also be considered as a management tool. Products containing ferrous sulfate or superphosphate have been utilized to reduce pH from 8.0 or above to 7.0 or below, which reduces ammonia production. Control of ammonia by ventilation only can be extremely expensive in colder climates.

Barker et al. (1986) specifies the following maximum levels of various gases in animal buildings for worker safety. (A more complete review of safe levels of gases appears under air quality considerations for swine in a previous section.)

	Internal air
Gas	*must not exceed (ppm[3])*
Ammonia	25
Hydrogen sulfide	10
Carbon monoxide	25
Carbon dioxide	5,000
Methane	1,000

Noise

Normal and routine noise levels in production facilities appear to be acceptable when birds are conditioned to these stimuli. The major noise-related problem in poultry production is panic reaction (hysteria) resulting from unexpected noise, even at low levels. Poultry house workers commonly make a noise to which the birds are conditioned, such as knocking, before entering a house. A routine noise level provided by a radio is sometimes used.

Lighting

Variations in lighting (day length alteration by artificial light) are commonly used for specialized effects in the development of layers and breeders and the production aspects of poultry to a greater extent than the other farm animal enterprises.

Many factors contribute to the complexity of decision making in the management of lighting as a component of environmental design and management. Included in these factors are geographical location as related to season of the year and periods of increasing or decreasing day length, age of sexual maturity desired in a particular production system, breed of bird involved, type of housing facility, and cage arrangement in laying houses. Because of this extreme range of variables, a specific management system for lighting must be determined for a given production unit. Some examples of lighting systems will be presented in the following sections to illustrate industry practices rather than to suggest specific management recommendations.

The goal of light management in developing layers and breeders and during egg production is related to the effect of light on the reproductive process, although a certain amount of light is required for birds to feed normally. Light level may also be altered in an effort to minimize certain detrimental social behaviors such as feather and body pecking. Thus, in layer management light levels are utilized during development of the female to regulate age of sexual maturity and during the laying period to maintain optimum levels of egg production. The primary effect is related to length of day rather than light intensity.

A light management system for the laying chicken is described to provide a practical example. Lighting management is also important in the production of broilers and market turkeys as well as breeder birds, both female and male. In this

[3] ppm, parts per million

example for the pullet development period, the light intensity is specified routinely at 5 lux, except for the first 2 to 3 days, which require a level of 10 lux. The higher light level is important in assisting the chicks to exercise and learn eating and drinking behavior. At the time of sexual maturity, the level is raised to 10 to 20 lux and the day length is extended to stimulate lay. The example management plan[4] that follows is for a specific genetic strain (Hy-Line W-36) of layers.

First 2 days	Continuous (24 hours) 10 lux at bird level
2 days to 3 weeks	Length of light period is reduced to 15 hours and intensity is reduced to 5 lux at bird level
3 to 18 weeks	Maintain a constant day length of 8 to 12 hours for the entire period
At 18 weeks or expected sexual maturity (expected sexual maturity is approximately 18 weeks of age in this example; this age may vary depending on genetic strain of the birds; breeders' or suppliers' recommendations should be used for specific lines)	Increase day length to 13 hours with an intensity of 10 to 20 lux, at which point the pullets are ready for egg production. It is essential to increase the length of light to a threshold level of 13 hours at 10 to 20 lux. Light stimulation too early results in smaller egg size and reduced peak production. After this initial level of light stimulation, raise the day length by 15 to 30 minutes per week or biweekly until a level of 16 hours of light is attained. Sudden large increases in length of day should be avoided during this period. Length of day and light intensity must not be decreased during the laying period.

Guidelines for different light management systems vary somewhat with respect to genetic line, age of birds, and number of hours of light; however, three basic rules appear to be constant. These rules are as follows: (1) increasing day length is avoided during the development of the pullet to sexual maturity, (2) it is essential to increase day length above a threshold at sexual maturity, and (3) decreasing day length during the laying period should be avoided.

Intermittent lighting systems may be used as an alternative to continuous lighting during the growing period for pullets. An example program is 15 minutes of light and 45 minutes of darkness throughout the day for a total of 6 hours of light. This schedule appears to accomplish the same result as eight hours of continuous light with some energy savings. Specific recommendations for a given genetic line should be used, however.

Banks and Koen (1989) compared a continuous lighting system of 17 hours of continuous light followed by 7 hours of dark with an intermittent program of 2 hours light, 4 hours dark, 8 hours light, 10 hours dark from 21 weeks up to 52 weeks of age. No production disadvantages (body weight, egg weight, or egg size) were associated with intermittent lighting. Lighting costs were reduced by 41 percent in the case of the intermittent system.

[4] From Hy-Line, *Hy-Line Variety W-36 Management Guide: Chick, Pullet, Layer* (West Des Moines, Ia.: Hy-Line International, 1996), p. 7, by permission

Photoperiod management also influences the laying performance of turkeys. Light intensities of 16 to 32 lux applied starting at 32 to 36 weeks of age for a daily period of 15 hours appear to be adequate light stimulation following a development period with photoperiods of 6 to 8 hours daily (McGartney, 1971).

Management of lighting is also used to influence rate of sexual maturity of cocks and toms. Photoperiods of 14 to 15 hours at 30 to 40 lux are appropriate for development and reproductive performance after the first 5 days of age. From birth to five days of age, 23- to 24-hour lighting at 40 to 50 lux is recommended.

Water Availability

The amount of water required depends on environmental temperature, humidity, dietary composition, and performance levels. Generally, water is provided *ad libitum*. Total water consumption is related to total dry matter feed consumption, usually at a ratio of about 2:1 (water to feed). Adequate water is essential to body temperature regulation, and the requirement increases with increasing environmental temperature. For example, Farrell and Swain (1977) reported a consumption rate of 204 mL/day and 118 mL/day per laying hen when maintained, respectively, at 35 and 20°C. Respective values for percentage of total heat loss due to evaporation were 19.5 and 82.7. Example weekly water consumption rates in poultry maintained at environmental temperatures of 20 to 25°C reported by NRC (1994) are as follows: Leghorn hens at 20 weeks of age, 1,600 mL; broilers at 2 weeks of age, 480 mL; broilers at 6 weeks of age, 1,500 mL; turkeys at 2 weeks of age, 750 mL; and turkeys at 20 weeks of age, 7,040 mL.

Excess water consumption may create problems of excess water content of poultry house waste and related difficulties with the disposal system. Restriction of water to limit excretion or to limit feed intake can be hazardous, resulting in dehydration.

High mineral content of feed or water will increase the total water requirement to ensure adequate excretion of the unneeded nutrients.

Specific waterer space recommendations vary with different types of equipment and facility design. Equipment manufacturers provide such guidelines.

Feed Availability

Important considerations in feeding facilities are ease of cleaning, feed wastage, and accessibility to the birds to minimize undesirable behaviors related to highly competitive environments. Feeder space allowances are published by manufacturers to apply to specific types of feeders and facility designs.

Space Requirements

Space requirements for poultry of different types vary dramatically due to wide differences in size characteristics of species and genetic lines within species. Age, stage of development, purpose or function of the bird, and production system are added factors that must be considered in planning individual and group space. In general, the space requirement for Leghorn-type laying hens (22 weeks of age and older) in multiple-bird wire cages is 464 cm^2 per hen (Craig et al., 1986a, 1986b and Craig and Milliken, 1989). Smaller birds such as miniature leghorns require less space and larger genetic lines require more space as would be expected. This variability points to the extreme importance of seeking recommendations for specific genetic lines from the supplier of the genetic stock. Laying hens are normally maintained in multiple bird cages accommodating 3 to 6 birds. Individual space recommendations must be multiplied accordingly. Thus, a three bird cage should provide a minimum of 2395 cm^2.

Space recommendations for broilers (meat type) in the typical floor-litter pen system are based largely on industry experience; thus, the supplier of genetic stocks will most likely provide guidelines. In general, space guides suggest about 929 cm^2 per bird in floor litter systems for mixed sex groups of birds up to 8 weeks of age (CDGCAA, 1988).

Space recommendations for turkeys in floor-litter pen systems are 929, 1857, 3250, and 3715 cm^2, respectively, for birds below 3 kg, 3 to 7 kg, 7 to 12 kg, and above 12 kg (CDGCAA, 1988). Recommendations by Borg and Halverson (1985) include the following space guides for hens and toms, respectively, at ages of up to 4 weeks, 4 to 8 weeks, 8 to 12 weeks, and 12 to 16 weeks: hens, 0.5, 1.0, 2.1, and 2.4 sq ft; toms, 0.5, 1.2, 2.4, and 3.0 sq ft. Turkeys maintained beyond 16 weeks should be allowed about 4.0 sq ft per bird. Typical recommendations in the European communities is 450–500 cm^2 per hen and 600–700 cm^2 in Scandinavia (SVCAWS, 1996).

Additional space considerations for all birds, regardless of system, should be related to providing adequate cage dimensions to accommodate standing erect normally, turning around, resting, and ease of access to feed and water.

Floors and Floor Surfaces

A variety of floor types appear to give satisfactory results in poultry houses. The types utilized are most commonly referred to as all-litter, slat and litter, wire and litter, all-slat, and all-wire. Cage flooring is typically made up of wire fabric or plastic.

Handling Considerations in Harvesting

Gregory et al. (1992), Scott and Moran (1993), Kannan and Mench (1996), and Nicol and Saville-Weeks (1993) provide research results suggesting harvesting-handling methods that minimize distress in the harvesting process for chickens. Generally this work suggests catching by both legs and holding birds in upright position and emphasizes the potential of minimizing stress by use of mechanical harvesting equipment.

CHAPTER 7

Design Characteristics of the Dietary Environment

Nutritional needs are influenced by a variety of factors relating to the species, size of the animal, particular purpose of the animal, stage of maturity or stage of the life cycle, and physical environment in which the animal exists. The latter may be referred to as the microclimate, which is defined as the immediate area surrounding the animal. As this "envelope" changes, various stimuli change and associated responses occur. Nutrient requirements also change as alterations impact the animal.

In most cases, animals can do little to alter their own dietary resources, aside from moving to better pasture conditions in livestock systems relying on extensive grazing resources. Animals in confinement have no chance of altering their nutrient resource base. The manager of the livestock enterprise must ensure adequate nutrition in the interest of both animal well-being and economy as it relates to goals of the enterprise.

Management is also critical to ensure that the behavioral characteristics of animals do not result in inadequate nutrition. In some cases the physical character of facilities may be limiting, and weather conditions may also reduce access to feed. Agonistic behavior and the related dominance order form the basis for some diet-related management considerations to ensure equal dietary access.

Classic examples of environmental effects on nutrient and feed requirements are the depression in total feed intake that results from heat stress, the need for dietary iron in baby pigs not having access to soil containing this element, the increased energy requirement associated with increasing cold, and the change in the nutrient composition of feedstuffs with the stage of growth in plants.

Among the most important factors impacting dietary adequacy is the thermal environment. Numerous mathematical procedures are available to aid in the design of diets allowing for changing thermal demands. These formulas, which will be illustrated later, involve concepts that define the level of demand associated with changes in the thermal environment. The following terms represent some important concepts in defining certain dietary needs. Values for these parameters are required in designing feeding programs to meet specific demands. These issues, along with other related terms, are discussed in more detail relative to body temperature regulation in part 2, which considers thermal stress physiology and body temperature homeostasis. Note also the discussion of thermal demands in the previous section dealing with design of the physical environment. The levels of thermal challenge represented by temperature zones described by Curtis (1983) and NRC (1981a) aid in understanding the levels of demand associated with variable thermal environments. Some important terms and concepts are the following:

> *Degrees of Coldness.* A term used to describe the number of degrees in temperature below the thermal neutral zone. The lowest point in the thermal neutral zone is referred to as the lower critical temperature (LCT). This value is a reflection of the severity of potential thermal stress.
>
> *Lower Critical Temperature (LCT).* The lowest temperature point in the thermal neutral zone. This value represents the point below which an animal must increase its rate of metabolic heat production to maintain body temperature within the normal range for that particular animal. These processes are related to metabolic heat production and heat conservation and include vasoconstriction in the periphery to reduce heat loss from body surfaces.
>
> *Upper Critical Temperature (UCT).* The highest temperature point in the thermal neutral zone. This value represents the point above which an animal must engage physiological mechanisms to resist body temperatures rising above normal. These processes are related to cooling effected by evaporation through increased perspiration, respiration, and vasodilation in the periphery to enhance heat loss from body surfaces.
>
> *Thermal Neutral Zone (TNZ).* The range in temperature between LCT and UCT. This zone may also be referred to as the comfort zone or zone of thermal neutrality.

The information provided in this section is intended to establish key areas that require consideration in ensuring that nutrient requirements are met. Specific formulation procedures and alternative approaches to meet dietary needs by selection of certain feed ingredients are appropriately left for courses and literature in animal nutrition and feeding. Some examples, however, require limited use of such procedures to illustrate the principles involved. Consideration is also given to factors influencing composition, supply, and availability of feedstuffs and pasture as these relate to the management of animals and the resource base supporting animal enterprises.

Dietary formulation technology has advanced rapidly within the last decade. Of particular value are the approaches that include important environmental considerations in assessing nutrient needs. For example, Holden et al. (1996) designed a method for adjusting nutrient requirements of swine based on genetic and environmental deviations from those characteristic of typical standard production conditions. In this procedure levels of the required nutrients in the diet for standard conditions are adjusted for genetic capacity for lean growth, previous environmental conditions that influence the animal's level of antigen exposure, thermal environment in relation to the TNZ, physical form of the diet, and use of growth-enhancing compounds.

DESIGNING THE DIETARY ENVIRONMENT FOR SWINE

Influence of Cold on the Energy Requirements of Swine

Because very young swine are commonly maintained in more highly controlled environments, the discussion here includes examples for growing and finishing swine and those of the breeding herd.

Growing and Finishing Pigs

In swine, the recommended environmental temperatures are 16 to 24°C for growing pigs and 10 to 24°C for finishing pigs. The ranges reflect that critical temperatures are influenced by characteristics of the animal that include heat production and insulation, as well as those of the environment that may have variable influences on heat loss from the animal's body. Evaluating the cold stress level requires that an LCT be estimated for a specific animal or group, and the deviation of actual temperature from this value provides a measure expressed as degrees of coldness. Once the value is estimated for degrees of coldness, calculations can be made to assess the extent to which compensating management alternatives such as added protection and dietary changes may be required. Figure 7.1 depicts the thermoneutral range for growing and finishing pigs by charting LCT and UCT for varying weights during the normal production period (Holden et al., 1996). This information can be used to establish baseline values for determining degrees of coldness or degrees of heat stress.

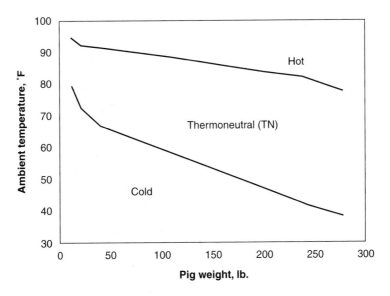

FIGURE 7.1 Thermal climate for growing and finishing pigs up to 280 lb live weight (*Source:* Holden, P., R. Ewan, M. Jurgens, T. Stahly, and D. Zimmerman, *Life Cycle Swine Nutrition*, Ext. PM-489 [Ames, Ia.: Iowa State University, 1996], p.10)

In terms of the influence of cold stress on performance, Verstegen et al. (1979) suggest that growing pigs will gain about 10 g less per day, and finishing pigs about 25 g less per day, for each degree Celsius of coldness. Feed consumption was shown to increase at about 19.5 g per degree of coldness; however, pigs typically may increase intake during periods of cold stress up to a maximum of about 15 percent because of limitations in digestive tract capacity. To compensate for the increased heat production required and to maintain rate of gain, for example, a total increased feed intake of 25 to 39 g per degree of coldness is commonly required by growing and finishing pigs (pigs 25 to 60 kg, 25 g and pigs 60 to 100 kg, 39 g of added feed daily per degree of Celsius coldness). For pigs from 20 to 100 kg in body weight in a typical growing and finishing production system, NRC (1981a) suggests that a reasonable range relative to the additional feed required to compensate for cold stress and maintain level of production is 30 to 40 g of feed per degree of coldness, which is similar to the previously mentioned value.

For example, we can estimate the level of cold stress in a group of pigs in the 60- to 100-kg range being fed a typical finishing diet and determine the effect on feed requirement and efficiency. The LCT for this group of pigs is estimated to be 16°C, and the animals experience a reduction in environmental temperature from 16 to 6°C. This shift represents 10°C of coldness (the expression of stress level) and could require an increase of 390 g of feed daily (39 g of feed per degree of coldness). The animals might typically be gaining 1.8 lb a day at a conversion rate of 2.77 lb of feed per pound of gain. Thus, 390 g, or 0.86 lb, of added feed is required daily. The level of feed being consumed before the stress period was 4.99 lb daily. The stressor raises the total feed requirement to 5.85 lb daily, or an increase of 17.2 percent to maintain a constant level of production. The amount of feed required per pound of gain would be increased from 2.77 up to 3.25 lb. This example presents a practical problem in terms of the animals' ability to totally compensate for higher levels of cold stress and maintain productivity. If digestive tract capacity limits the level to which the animal can compensate, then the only possible result is a reduction in rate of gain. In general, finishing pigs might be expected to increase consumption up to 15 percent, but in the example the required 17 percent increase may not be possible. In this instance other actions (e.g., increased energy concentration of the feed and additional bedding supplied in some systems) must be taken if the performance level is to be maintained. Openings normally providing ventilation and cooling might be reduced in size by curtains or other means; however, the required level of ventilation may not allow this option.

Recognizing that many finishing pigs are maintained in insulated buildings, cold stress is likely only in very severe weather. In addition to the building retaining heat from the pigs, the fact that they may huddle to reduce heat loss suggests that the provided example is somewhat extreme. Thus, a more typical example might be a reduction from 16 to 10°C, which would entail a compensatory feed consumption increase from 4.99 to 5.50 lb daily (39 g increase in feed allowance per degree of Celsius coldness × 6 degrees of coldness = 234 g, or 0.51 lb of additional feed) to maintain performance at a 1.8-lb daily gain. This value represents an increased daily feed allowance of 10 percent. Feed required per pound of gain would increase correspondingly from 2.77 up to 3.05 lb, however.

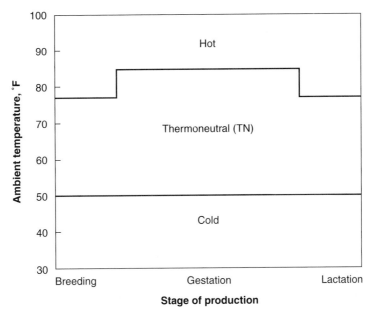

FIGURE 7.2 Thermal climate for sows and boars (*Source:* Holden, P., R. Ewan, M. Jurgens, T. Stahly, and D. Zimmerman, *Life Cycle Swine Nutrition*, Ext. PM-489 [Ames, Ia.: Iowa State University, 1996], p.17)

If a compensatory level of feed is to be maintained over an extended period, the nutrient density of protein and other ingredients may be reduced. For example, if finishing pigs are consuming a diet at the rate of 6 lb daily and the protein content is 12 percent and compensation for cold stress raises consumption to 6.6 lb, then the protein level could be reduced to about 11 percent based on a daily protein requirement of 0.72 lb (6 × 0.12).

Feed ingredients other than protein sources may be adjusted; however, each must be considered carefully in relation to a possible relationship in aiding the animal to cope with a stressful condition.

Breeding Animals

Breeding animals are commonly fed at limited levels during much of the year as a means of weight control. Cold obviously increases the maintenance requirement, and the feed level should be adjusted accordingly. Figure 7.2 shows recommended limits of the thermal environment for swine in the breeding herd (Holden et al., 1996). The upper limit reflects the UCT, while the lower limit reflects the LCT; the area between depicts the TNZ.

Influence of Heat on the Energy Requirements of Swine

Growing and Finishing Pigs

It is well established that feed consumption declines when environmental temperature for growing and finishing pigs exceeds about 30°C and is greatly influenced at levels of 35 to 38°C.

A major portion of swine production is accomplished in the more humid regions of the world. Thus, the impact of humidity along with heat is a prominent consideration. Holmes and Close (1977) concluded that at 30°C an increase of 18 percent in relative humidity is equivalent to an increase of 1°C in air temperature.

A basic guide is to assume that a stressful heat load may exist when the mean daily temperature is 5°C above the optimum range. It is also established that finishing hogs may reduce consumption of feed on the average of 30 to 60 g daily per degree Celsius of heat stress. This figure may be associated typically with a decrease of 10 to 30 g daily in weight gain, although higher values have been observed. An example could be a situation in which the environmental temperature increases from 30 to 35°C or 5°C of heat stress level. Assume that the animals are consuming 6 lb of feed daily at a 30°C environmental temperature. Consumption could drop by 40 g per degree of heat stress, or 200 g (0.44 lb). Thus, consumption would be 5.56 lb daily, and rate of gain will be reduced by an average of 30 g per degree of heat stress per day, or 150 g (0.33 lb). If the pig was gaining 1.90 lb daily in the 30°C environment, this rate will be reduced to 1.57 lb, and feed conversion will change from 3.15 lb per pound of gain up to 3.54 lb. This 12 percent reduction in efficiency is significant.

The reduced level of feed intake in the preceding example results in reduced consumption of all nutrients. Thus, the levels of protein, minerals, and other nutrients should be increased if the stress is to continue beyond a few days.

An important consideration in assessing the level of heat stress is to recognize that mean daily temperature is important. Animals will tolerate rather wide variations in temperature during the 24-hour day without important negative impacts on dietary requirements or performance. Thus, Curtis (1983) suggests that in practical production situations the mean daily microenvironmental temperature be used to assess the potential stress level. A 35°C environment for finishing pigs, if held constant, may be stressful. However, if 35°C is a peak afternoon temperature only and a daily mean does not exceed 25°C, it is likely that pig performance will be at or near optimum.

A variety of means are used to mitigate potential heat stress (e.g., insulated buildings, sprays of water to enhance cooling by evaporation, and air movement). In terms of assessing the impact of heat on dietary requirements, it must be kept in mind that the microenvironment of the animal is the key. Air temperature in a building is only one factor in determining stress level. If the effects of air temperature, even though high, are mitigated in the animals' microenvironment, then heat stress may be reduced. Some work suggests that using fat as a source of a portion of dietary energy may assist in minimizing heat stress (Stahly et al., 1979a), and the appropriate use of amino acids as a part of the protein requirement may also be helpful (Stahly et al., 1979b).

Breeding Animals

Heat stress is most likely to influence sows by depressing feed intake below optimum during lactation. Alleviation of the stress is the most important management consideration; however, increased nutrient density should be considered.

Energy Requirements

Energy Requirements for Maintenance

The maintenance requirement is the energy needed for all body functions. It is influenced by factors such as environmental temperature, activity level, stress, body weight (metabolic size), and body composition. In general, the base level for the maintenance requirement is established by a regression factor, determined through research, multiplied by metabolic size, which is expressed as body weight $^{0.75}$ or $W^{0.75}$. In some cases the formula is $EBW^{0.75}$.

Expression of energy units are digestible energy (DE), metabolizable energy (ME), and net energy (NE). In expressing the energy required for maintenance, the forms are, respectively, DE_m, ME_m, and NE_m in calories (c), kilocalories (kcal), or megacalories (Mcal).

Factors influencing the maintenance requirement have been evaluated in terms of degree of effect. Such factors are used in adjusting the basal maintenance requirement to meet the needs for different conditions and added functions.

Thermogenesis elevates the energy requirement when the ambient temperature is below the critical temperature. The critical temperature (T_c) is the point below which an animal must increase heat production in the effort to maintain normal body temperature. The energy cost of cold thermogenesis can be described by the following equation (NRC, 1988):

$$DEH_c(\text{kcal DE/day}) = 0.326W + 23.65(T_c - T)$$

W = animal weight (kg)
T_c = critical temperature (°C)
T = ambient temperature (°C)
DEH_c = digestible energy to meet body heat requirements

Verstegen et al. (1982) estimated that growing pigs (25 to 60 kg) need an additional 80 kcal of metabolizable energy to compensate for each 1°C below T_c and 125 kcal of additional metabolizable energy daily during the finishing period (60 to 100 kg) for this compensation. In terms of added feed allowance (for typical formulations used during this period), these values reflect a respective increased need of 25 g and 39 g of feed per 1°C below T_c.

For pregnant sows the metabolizable energy requirement for maintenance increases by about 4 percent for each degree below the sow's T_c (Verstegen et al., 1987).

Energy Requirements for Sows

Pregnancy   Guidelines proposed for weight gain (NRC, 1988) during gestation (for the first three or four pregnancies) suggest a weight increase of 45 kg (25 kg net for the sow and 20 kg for products of conception). The daily base energy requirement for a pregnant sow weighing 140 kg at the time of mating (an allowance to provide the target weight increase of 45 kg during gestation) is expressed in the following example (NRC, 1988):

Weight at mating (kg)	140
Mean gestation weight (kg)	162.5
Energy requirement (Mcal DE/day)	
For maintenance	5
For gestation weight gain	1.29
Total (DE Mcal)	6.29
Estimated daily feed allowance (corn–soybean meal diet containing 3.34 Mcal/kg) (kg)	1.9

Thus, if the environmental temperature drops 5°C below T_c (e.g., 10°C as compared with 15°C, which is the T_c estimated for gestating sows in this example) for the 140-kg sow and the increase in energy required for maintenance to compensate is 4 percent for each degree below T_c, or a total increase of 20 percent, the adjusted feed allowance is calculated as follows:

$$\frac{5 \times 1.20}{3.34} + \frac{1.29}{3.34} = 2.17 \text{ kg/day}$$

Lactation   The necessary energy allowance during lactation is based on DE_m, which is about 110 kcal/$W^{0.75}$/day plus DE for milk, which is 2 kcal of DE per kilogram of milk produced.

Estimating the impact of environmental temperature on the energy requirements of a 165-kg lactating sow is accomplished as follows:

Weight of lactating sow (kg)	165
Daily milk yield (kg)	6.25
Energy requirement (Mcal DE/day)	
For maintenance	5
For milk production	12.5
Total	17.5
Estimated daily feed allowance (corn–soybean meal diet containing 3.34 Mcal/kg) (kg)	5.3

In this example the influence of a reduction in environmental temperature of 8°C below T_c is on the maintenance energy component at a rate of 4 percent increase per degree below T_c, or a total increase in the maintenance of 32 percent.

The adjusted feed allowance to compensate for the temperature reduction is calculated as follows:

$$\frac{5 \times 1.32}{3.34} + \frac{12.5}{3.34} = 5.7 \text{ kg daily}$$

Lactating sows may reduce feed intake during periods of heat stress. Some evidence suggests that the addition of fat to the diet may assist in keeping total energy intake up. This finding is thought to be related to the lower heat increment associated with fat consumption as compared with typical diets without added fat. Fat additions to the diet for sows during late gestation and lactation have resulted in increased milk yields, milk fat levels, and pig survival. Other effects of fat additions have been a reduction in the weight loss of sows during lactation and some reduction in interval from weaning to mating (NRC, 1988).

Energy Requirements for Boars
The principles outlined for sows apply to the maintenance component of nutrient requirements of males. Since limit feeding is practiced during much of the year for boars, the influence of severe cold should be recognized and feed allowances adjusted accordingly.

Energy Requirements for Nursing Pigs
Since nursing pigs are allowed creep feeds along with the milk consumed, and good environmental control is normally provided, it is assumed that nutrient requirements are met. Because the manager exerts little control over nutrient intake during the neonatal period, the practical approach to managing cold stress during this period is through modification of the environment rather than through dietary manipulation.

Energy Requirements for Growing and Finishing Pigs
Pigs in the range from 15 to 100 kg (from weaning to market weight) are normally fed *ad libitum.* Thus, compensation for increases in energy demand is accomplished by the pig. Modest increases in demand for energy associated with cold cause an increase in feed intake. Consideration may be given to adjusting nutrient density for protein and possibly other nutrients, as discussed previously. Severe cold or heat stress may reduce total feed intake, which again suggests the need to alter the energy density of the total feed. Adding fat to the diet during warm periods may enhance energy consumption and gain because of the lower heat increment associated with fat as an ingredient.

It is important to recognize that for each degree of coldness ($^{\circ}$C below T_c), growing and finishing pigs 15 to 110 kg in weight will require about 30 to 40 g of additional feed to maintain rate of gain. Associated effects on feed efficiency must be considered along with impacts on length of the feeding period to market weight and facility turnover rates.

Protein and Amino Acid Requirements

Protein is an important component of virtually all systems relating to an animal's functional processes. Hormone synthesis and cell formation are examples that emphasize the importance of adequate protein nutrition in the animal's ability to respond to stress.

Swine are monogastrics and therefore cannot synthesize adequate amounts of the essential amino acids that make up protein; ruminants, on the other hand, have a large microbial population to accomplish this production for the host animal. As a result, the formulation of diets for monogastrics must recognize both kinds and amounts of the essential amino acids. Some amino acids such as lysine and methionine, the commonly limiting amino acids in swine diets, are available in synthetic form for use in feeds. Because of this species characteristic, proper dietary composition in this regard is essential for a high level of animal well-being.

The major consideration in terms of altering the nutritional character of feeds in response to a stressful environment is to recognize that such influences may alter feed intake. In the case of stress, nutrient density may be adjusted in relation to altered intakes; in cases where cold stress, for example, increases intake, a reduction in density of protein in the feed may be appropriate from an economic standpoint.

Boars, because of leaner composition of weight gain and restricted feed intake, require a higher percentage of protein than barrows. Specific requirements are 18 and 16 percent crude protein, respectively, for 20- to 50-kg and 50- to 110-kg boars.

Gilts, although requiring slightly less protein as a percentage of the diet than boars, have a higher requirement than barrows. This difference is demonstrated by a crude protein requirement of 16 percent for gilts 20 to 50 kg in weight and 15 percent for those 50 to 100 kg in weight. These values are based on NRC (1988) requirements for developing replacement gilts.

Technology for formulating diets to meet the nutrient requirements of swine has advanced significantly in the last few years. Determining the protein requirement of a specific group of swine illustrates the elements of a current procedure (Holden et al., 1996) for adjusting the requirement considering genetic and environmental influences on nutrient requirements. Procedures require (1) estimating the pig's capacity for lean growth per day, (2) defining the influence of previous environmental factors on antigen exposure (level of immune system activation), (3) determining the deviation from the animal's TNZ in the current thermal environment, (4) determining the physical character of the feed, (5) selecting the appropriate values related to influence of the animal's sex, (6) adjusting the nutrient requirements for environmental conditions under which the pigs are maintained and how these deviate from standard production conditions, and (7) adjusting standard requirements for a given stage of development up or down based on the animal and environmental characteristics.

Protein requirements increase with increasing rate of protein deposition in the animal. Thus, pigs that have higher genetic potential for protein growth have higher requirements. Boars have higher requirements than barrows and gilts because testosterone stimulates a higher level of protein deposition.

As growing pigs advance in weight and maturity, the protein requirement as a percentage of the diet declines. For example, pigs with a moderate level of lean growth capacity at 5, 10, 20, 50, and 100 kg in weight have respective protein requirements of approximately 27, 24, 20, 16, and 13 percent. Pigs characterized by a low protein growth capacity at the same weights have respective protein requirements of approximately 26, 22, 19, 14, and 11 percent of the diet.

Mineral Requirements

Both major and trace minerals are critical in a variety of metabolic processes, including acid-base balance, catalysts for metabolic reaction, osmotic processes, oxygen transport, and hormone systems.

Specific examples are the role of iodine in thyroxine synthesis and this hormone's function in the regulation of metabolic rate and the role of iron in hemoglobin synthesis and oxygen transport. Both of these systems are vital to stress responses.

Aside from the essential nature of minerals in proper nutrition and metabolism, certain minerals (e.g., selenium) that have toxic properties must be considered in providing the proper dietary environment.

The net effectiveness of some minerals depends on their ratios with other minerals. The most prominent consideration in this regard is the ratio of calcium (Ca) to phosphorus (P). The suggested ratio of Ca to P for swine is typically around 1.25:1. High Ca to P ratios interfere with phosphorus absorption, which may be reflected in lower rates of growth and bone calcification. Sodium and chlorine are critical in acid-base balance in the body and thus essential to the stress response. Recommended levels of salt in swine diets are 0.4 percent for gestation, 0.5 percent for lactation and 0.20 to 0.25 percent for growing and finishing pigs.

High levels of salt can be tolerated by swine; however, if water is limited simultaneously, toxicity may result. Pigs maintained in areas with high-saline water supplies are more subject to salt toxicity if excessive levels exist in the feeds. Proper levels of salt, (sodium chloride [NaCl]), in iodized form normally satisfy the iodine requirement of swine.

Copper is important in the synthesis of hemoglobin and in several enzyme systems. Diets are typically fortified with copper in sulfate, carbonate, or chloride forms. Toxicity may result when the diet includes levels of copper above 250 mg/kg.

Cobalt is important in the synthesis of vitamin B_{12}. There is no well-established dietary requirement for cobalt. Levels above 400 parts per million (ppm) in the diet may be toxic, however.

Iron (Fe) is essential for hemoglobin formation and thus for functional red cells in the blood. Newborn pigs are deficient in iron because placental transfer from the dam is poor; mammary transfer of iron is also poor. Thus, supplemental iron is a critical factor in pig survival. Iron injection (100 to 200 mg Fe) soon after birth is the common approach to correcting this problem. Oral administration of chelated iron in the first few hours after birth is also effective. The iron requirement in diets for creep-fed pigs before weaning is 100 mg/kg of feed and for postweaning up to 10 kg in body weight. Following this period, levels of 80, 60, and 40 mg/kg of feed are appropriate for growing and finishing pigs in respective weight categories of 10 to 20 kg, 20 to 50 kg, and 50 to 110 kg. Iron levels for gestating sows, gilts, and boars should be on the order of 150 mg/day, with lactating sows requiring 425 mg/day.

Selenium deficiencies may occur in certain geographical regions because of low content in the soil. In these regions supplemental dietary selenium is required. Environmental stress may also increase the incidence of selenium deficiency. Since selenium may cause toxicity at high levels, strict limits are imposed. Acceptable dietary levels are 0.3 mg/kg in diets for pigs up to 20 kg and 0.1 ppm for other swine. The 0.3 mg/kg level is being considered by the U.S. Food and Drug Administration (FDA) for all swine diets.

Zinc is an important component of insulin and several enzymes. The dietary requirement is about 50 mg/kg for growing pigs. Gestating sows, lactating sows, and boars should have zinc allowances of 50 mg/kg of feed. Parakeratosis (thickening of skin and irritation shown on the abdominal area and rear legs) is a classical sign of zinc deficiency in swine. Zinc oxide at a level of 3,000 parts per billion (ppb) in nursery pig diets may improve performance, possibly because of reduced *Escherichia coli* infection.

Current recommendations may include the addition of chromium at 200 ppb for sows for enhanced litter size and farrowing rate and for growing pigs for possible improvement in growth rate and carcass leanness (Holden et al., 1996).

Trace minerals other than those discussed are not usually required as specific dietary additions. In practice, however, trace mineral mixes are routinely added to feeds as insurance against deficiencies. Although this practice is common, there is some disagreement about its effectiveness.

Vitamin Requirements

Fat-Soluble Vitamins

The fat-soluble vitamins (A, D, E, and K) are all critically involved directly in metabolic or tissue resistance to stress. For example, vitamin A is involved in the maintenance of epithelial tissue, which resists entry of harmful agents into the body. Vitamin D helps to maintain calcium and phosphorus homeostasis. Vitamin E is involved in the maintenance of muscle tissue and thus normal functioning of both cardiac and skeletal muscle. Vitamin K is involved in blood clotting through the synthesis of prothrombin and other factors involved in blood clotting. It is also involved in calcium and vitamin D metabolism.

An additional consideration is because vitamins A and E are not synthesized in the body of animals, the requirement must be met through dietary sources. Vitamin D is converted from precursors (7-dehydrocholesterol) in the skin by the action of sunlight to cholecalciferol, or vitamin D_3. Vitamin D_2 is found in plants as a result of ultraviolet light converting ergosterol. For animals sheltered from sunlight (or ultraviolet light), dietary vitamin D is essential.

Because of this dietary dependence, vitamins A and E are critical considerations in planning for animals' nutritional needs. Supplemental vitamin D is less critical, depending on access to sunlight. Vitamin K is usually adequate because of naturally occurring amounts in feeds or microbial synthesis in the intestinal tract. Another factor that warrants special dietary consideration is the extreme variability of fat-soluble vitamins in feeds. This variability is due to the instability of these compounds in the feedstuffs and losses during storage. Storage time and conditions such as temperature and humidity are important considerations in the loss of these vitamins. Commercial vitamin products are almost always chemically or physically stabilized to ensure the effectiveness of such supplemental sources. Manufacturers provide information about the stability of various products.

Water-Soluble Vitamins

The water-soluble vitamins include those commonly referred to as the B vitamin complex and vitamin C (ascorbic acid).

The B vitamins are biotin, folicin (folic acid group), niacin (nicotinic acid), pantothenic acid, riboflavin (B_2), thiamin (B_1), pyridoxine (B_6), and cyanocobalamin (B_{12}). Choline, while not truly a B vitamin, is usually considered along with this vitamin group in formulating diets.

The B vitamins have a critical role in virtually all metabolic transformations in the body. A major role is that of a cofactor relationship in enzyme systems. Their importance in stress responses relates to their involvement in the mobilization of nutrient reserves through gluconeogenesis, glycolysis, lipolysis, and neural transmission.

The B vitamins normally considered in terms of dietary additions are choline (baby pigs and females), niacin, riboflavin, thiamin, and cyanocobalamin. Other B vitamins, due to content of feeds or synthesis within the animal's body, are not usually considered required dietary supplementations. It is important, however, to evaluate feeding programs for special conditions that may uniquely influence the demand for supplementation.

Vitamin C is involved in the synthesis of norepinephrine, which is a critical neurotransmitter in part of the autonomic nervous system stress-response system. Other important functions include involvement in iron metabolism and collagen formation, which is necessary for the normal development of cartilage and bone growth. Dietary supplementation of vitamin C is not required for swine.

Water Requirements

Water is obviously vital for life. Examples of its importance in metabolism related to acid-base balance and heat regulation, together with the relationship of these to stress responses, underscore the importance of this nutrient in maintaining the fitness of animals to cope with the environment.

Water accounts for 50 to 80 percent of the animal body, depending on the stage of growth. Body water represents a balance among water consumption, metabolic water, water in feedstuffs, and losses in the form of milk production, respiration, skin evaporation, and excretion. Because all of these factors may be related to environmental conditions, water balance is a critical consideration in stress physiology.

Since many factors influence both water consumption and loss, no specific requirements are quoted here. A liberal supply of quality water should be provided to all animals. In most swine production systems, provision of free access to water at all times is advisable.

In considering water supplies, toxic elements may be cause for concern. High saline level or nitrite content may affect water quality negatively. Microbial contamination is also a common problem in maintaining water quality.

DESIGNING THE DIETARY ENVIRONMENT FOR SHEEP

Influence of Cold Stress on the Energy Requirements of Sheep

Feeder Lambs

Practical guidelines suggest that lambs perform at optimum rates in the temperature range of 5 to 25°C. Some evidence suggests that it may be helpful to increase the level of roughage in the diet for finishing feedlot lambs when the temperature drops below 5°C. The LCT for lambs on full feed is in the area of 20°C for newly shorn animals and −15°C or below for those in full fleece and on a full-feed diet. Should feed consumption drop because of severe cold or other conditions, adjustments in the energy density of the diet should be considered. Normally, however, if lambs have access to finishing rations or good-quality hay and protection from wet conditions, extremely low temperatures have minimal effect on performance because of the typical LCT of this species.

Ewes

The LCT of full-fleece adult sheep with an adequate quantity of quality roughage or roughage and supplement is sufficiently low (well below 0°C) to minimize the need for altering the feeding program to meet requirements related to changes in the thermal environment. A danger of thermal stress exists, however, in the case of ewes that are newly shorn. In this case the LCT may be 20°C or higher if wet conditions prevail. Then larger allowances of good-quality hay with adequate protein are essential. An alternative is additional higher-energy supplemental feed such as grain and protein supplements fortified with adequate vitamins and minerals as required.

Recognizing that the newly shorn ewe and newborn lamb are probably the two most vulnerable animals in sheep production systems requires that high priority be placed on the animals' nutritional condition and their protection from wind and wetting. Providing an abundant amount of straw bedding for newly shorn ewes or lambs along with shelter in wet climates permits them to protect themselves from threatening cold in emergencies.

Influence of Heat Stress on the Energy Requirements of Sheep

Feeder Lambs

In cases where heat stress results in reduced feed intake, the nutrient density of the feed can be adjusted to ensure that requirements are met. If the animals have shade and are not subjected to a stressful level of activity, heat stress is not likely to occur in lambs.

Ewes

Although ewes may show some evidence of heat stress during peak daily temperature periods, no nutritional adjustments are necessary. Heat stress may occur in ewes if in full fleece when housed before and during parturition. However, the normal practice is to shear ewes before this time.

Energy Requirements

Energy Requirements for Maintenance

The energy value of feeds and the energy requirement for sheep are expressed as ME or NE. The energy requirement for maintenance is reflected as NE_m and that for gain as NE_g.

The various expressions of energy (e.g., gross energy, digestible energy, metabolizable energy, and net energy) are discussed in detail in National Research Council publications relating to nutrient requirements for the various species. In the case of sheep, refer to NRC (1985).

The basic formulas utilized to estimate the maintenance energy requirement for sheep, as for all other farm animals, involve the use of $W^{0.75}$ as an expression of metabolic body size. The maintenance energy requirement for a 50-kg ewe is 2 Mcal/day of metabolizable energy as compared with 2.4 and 2.8 Mcal/day for 70- and 90-kg ewes, respectively.

Considering a diet of midbloom alfalfa hay containing 1.82 Mcal/kg, a 70-kg ewe could be maintained on 1.32 kg of hay daily in the absence of cold stress. In cold, the length of fleece is a major factor in determining the LCT for ewes. For example, NRC (1985) indicates respective LCTs of 25 to 31°C for shorn sheep and −3°C for those with full fleece. Thus, shorn sheep should have shelter and essentially *ad libitum* roughage of good quality during severe cold.

Energy Requirements for Ewes

The reproductive stages normally considered in ensuring adequate nutrition are prebreeding (flushing), gestation, and lactation. In addition, special consideration must be given to the proper development of replacement females.

Flushing   Flushing is a term used to describe a dietary management practice. The general procedure is to increase the feeding level for a period of about 2 weeks before breeding. The relative level of nutrient intake is comparable to that recommended for late gestation. This dietary enrichment is usually accomplished by providing pasture of superior quality, higher levels of quality hay, and supplemental grain.

In terms of energy level, requirements reflect a daily allowance of metabolizable energy to be 3.8 Mcal/day for a 70-kg ewe 2 weeks before breeding as compared with an early gestation level of 2.8 Mcal/day and a maintenance level of 2.4 Mcal/day. This dietary energy allowance during flushing is 35 percent above that recommended for the first 15 weeks of gestation. This level of energy should not be continued for an extended time during gestation in order to prevent excess body condition.

Pregnancy   Energy requirements for gestating ewes are normally quoted for the first 15 weeks and the final 4 weeks before parturition. The higher level during late pregnancy is necessary to prepare the animal for the demand and potential stress associated with lactation.

The level of energy in the diet for ewes can be related to the development of a toxic syndrome commonly termed *pregnancy disease* or *pregnancy toxemia*. The metabolic disorder is ketosis, or acetonemia (high level of ketones in the blood). This metabolic disorder can be associated with either underfeeding or more likely with overfeeding of the pregnant female. It is more commonly found in ewes carrying multiple fetuses because of their higher nutrient demand. The basic cause of the disorder is related to rapid demand for energy and the mobilization of fat during late pregnancy. Animals are predisposed to it by stressors such as severe weather, shearing, and predator attacks. Prevention involves maintaining proper condition of the ewe and an adequate level of energy as the demands of late pregnancy increase. A practical approach is providing more feed to ewes carrying multiple offspring.

Another factor in establishing a level of energy for pregnant ewes is related to the expected percentage of the lamb crop involving multiple births. NRC (1985), for example, expresses requirements for ewes in the last 4 weeks of gestation in two categories: lamb crop expectations of 130 to 150 percent and 180 to 225 percent. The relative difference of 15 percent in daily energy allowance is clearly an important consideration in providing an appropriate nutritional environment for the ewe and is directly related to the prevention of adverse effects of stress associated with the increased demands of providing for multiple offspring.

Lactation   Diets for lactating ewes must consider both stage of lactation and the number of lambs nursed. Twinning is common in this species; the impact of the relatively higher demand of nursing twins is illustrated by the daily requirement of 6.6 Mcal metabolizable energy for a 70-kg ewe with twins as compared with 5.9 Mcal for a ewe of the same weight nursing a single lamb, a 12 percent increase in demand. In general, ewes nursing twins produce 20 to 40 percent more milk than contemporaries nursing singles. This increased milk production explains the higher energy recommendations during late gestation for ewe flocks producing relatively more multiple births. Nutrient requirement reference sources provide for adjustments during late pregnancy according to the expected percentage of lamb crop.

Lower levels of dietary energy during the last third of lactation are related to the normal downward slope of the lactation curve. This period is characterized by 30 to 40 percent less milk production than the first half of lactation.

Development of Replacement Females

Proper development of replacement females, particularly in systems of production that may involve limited pasture resources, is a critical management issue to ensure that young females enter the enterprise at a stage of fitness to cope with the demands of reproduction and lactation. The concept of biologically determined priorities is one that managers must understand in order to provide an appropriate nutritional environment to accommodate the continuing growth of the female, conception for the first pregnancy, and preparation for lactation and conception for the next pregnancy. A young female entering her first breeding season may fail reproductively if the available level of energy is low and the level of stored fat is extremely low. However, good conception rates are possible if the growth rate to first breeding has been moderate. The extreme demand that may result in reproductive failure is more likely to occur at the time of breeding for the second pregnancy if energy supplies are below requirements; such failure may be due to the heavy demand for lactation. Thus, the most critical time in terms of energy supplies and reserves is during the periods of late pregnancy, lactation, and before the breeding season for the second pregnancy. Special attention to the energy requirement is warranted for all species during this time period. The biological concept involved is that the animals have adaptive mechanisms to first ensure the ability to reproduce later by protecting the life of the dam. The second biological priority is to protect and provide for the offspring at side rather than conceive for the next one. Thus, the period between that of high demand for lactation and the next breeding season provides a time for recovery of reproductive fitness if adequate dietary energy supplies and amounts of other nutrients are available. Sheep's typically longer postlactation period before the next possible conception may be a factor in their greater adaptability to harsh and stressful environments characterized by sparse feed (range) supplies.

Generally recommended developmental guidelines for replacement females are as follows (NRC, 1985):

> Ewe lambs (except for Finn crosses) should be developed to weigh about 50 kg at the time of the first breeding season. During gestation a level of energy is provided to support a daily gain of 0.12 to 0.16 kg. During lactation the ewe should be fed at a level to continue growth as well as support the milk yield, which may require approximately 1 kg of grain along with *ad libitum* forage of good quality. The lamb then should be weaned at 6 weeks of age to permit the ewe to make appropriate recovery before the subsequent breeding period.
>
> Replacement ewes should reach about 65 percent of expected mature weight by the beginning of the breeding season and 75 percent of that weight by lambing time.

The example provided illustrates a carefully orchestrated management scenario to prevent the damaging impacts of stress on reproductive performance.

Development of Ram Lambs

Young rams should be developed without excessive condition. Requirements for energy must recognize the greater capacity for growth in males than in females and thus be higher than for females of equal weight. Daily gain recommendations for young rams range from about 0.75 kg for 40-kg lambs to 0.55 kg daily for 100-kg rams. These values demonstrate the concept of reducing the energy allowance as the rams mature to prevent excess fatness.

Growing and Finishing Lambs

Lambs being fed for market are usually in the weight range of 15 to 50 kg. The energy requirement is established by a combination of NE_m (for maintenance) and NE_g (for gain). The determination of these values is based on the following formula (NRC, 1985):

$$NE_m = 56 \text{ kcal} \cdot W^{0.75} \cdot d^{-1}$$
$$NE_g \text{ for small mature-weight lambs} = 317 \text{ kcal} \cdot W^{0.75} \cdot LWG \cdot d^{-1}$$
$$NE_g \text{ for medium mature-weight lambs} = 276 \text{ kcal} \cdot W^{0.75} \cdot LWG \cdot d^{-1}$$
$$NE_g \text{ for large mature-weight lambs} = 234 \text{ kcal} \cdot W^{0.75} \cdot LWG \cdot d^{-1}$$

In the given formulas for estimating energy requirements, NE_m is the daily energy required for maintenance and NE_g is the daily energy required for gain. W is body weight, and d^{-1} represents 1 day (i.e., the formula is for determining the daily caloric requirement).

Because feedlot lambs are usually afforded some protection and have a very low LCT due to their wool coats and high levels of feed consumption, adjusting the feed level for variation in degree of coldness is not usually required.

Enterotoxemia (overeating disease) is a special consideration for lambs fed high-grain diets. It may also affect suckling lambs and creep-fed lambs. Certain stressors (e.g., heavy parasite load) may increase predisposition to this pathological condition. The disease is caused by a toxin produced by *Clostridium perfringens* type D. Protection is afforded by vaccination of the dam 2 to 4 weeks before parturition for passive immunity in the offspring, which can then be vaccinated before 6 weeks of age to protect them until they reach market weight. In cases where lambs have not been provided passive immunity through the dam, they are vaccinated with *C. perfringens* type D toxoid. Care should be given to the time required for development of immunity, which is 2 to 4 weeks.

Protein Requirements

Two basic principles of feeding ruminants in terms of protein requirements deserve special note:

> Because the newly born ruminant does not yet have a functioning rumen, the enormous capacity of the contents of this organ to compensate for dietary shortages (B vitamins and amino acids) later in life is not developed. This function is partially developed by 2 weeks of age and is fully developed normally after 6 to 8 weeks of age.
>
> A fully developed ruminant digestive system typically meets needs for B vitamins and eliminates the necessity of considering protein quality (kind and amount of amino acids) to the extent required in designing feeding programs for monogastrics. However, the need for attention to amino acid balance in ruminant diets is currently being reevaluated.

Other concepts of importance in protein nutrition relate to the growth curve, to the modest demands of pregnancy, and to the great demands of lactation. These are illustrated by the fact that a 10-kg lamb gaining 250 g daily requires 157 g of crude protein. This lamb at 50 kg in weight gains 425 g daily and requires 240 g of crude protein. The younger lamb requires 0.63 g of dietary protein per gram of gain, whereas the older lamb requires only 0.56 g of protein per gram of gain. As the animal matures, protein deposition relative to that for fat is reduced. In terms of lactation demand, a 60-kg nonlactating ewe requires 161 g of crude protein daily as compared with a requirement of 295 g daily for the same ewe at the onset of lactation if nursing a single lamb or 336 g if nursing twins.

Mineral Requirements

Sodium, chlorine, calcium, phosphorus, magnesium, potassium, and sulfur are the major minerals normally necessary for a balanced sheep diet. Trace minerals normally considered of dietary importance are iodine, iron, molybdenum, copper, cobalt, manganese, zinc, and selenium.

Some general guidelines to consider in evaluating the mineral adequacy of the diet are as follows:

Typical requirements for sodium chloride (NaCl) are met by levels of 0.5 percent of mixed feeds or 1.0 percent of a concentrate mix. If salt is not restricted for animals in drylot, lambs will consume 5 to 10 g daily and ewes 15 to 30 g daily. Amounts from 220 to 340 g per ewe per month should be allowed under range conditions.

The calcium (Ca) to phosphorus (P) ratio should be around 1.5 to 2.0:1. Castrated and intact males should have a ratio of 2:1 to minimize predisposition for urinary calculi.

Supplemental sulfur may be required if sheep are fed high-urea or high-concentrate diets.

If sheep are maintained in an area of deficient or excess selenium, local dietary recommendations should be followed.

Vitamin Requirements

Fat-Soluble Vitamins

Vitamin A or its precursor carotene is required in the diet of sheep. However, supplementation is not normally required if animals are grazing green pasture because green growing plants have ample supplies of carotene. Sheep consuming green, leafy hays should ingest ample quantities of the vitamin unless the material has been stored for a year or more. Animals consuming dried grass on the range or brown hay should be given supplemental vitamin A as part of the protein supplement; however, animals consuming green pasture or other forms of feed high in carotene or vitamin A have the ability to store the vitamin in sufficient quantities to last 4 to 6 months (NRC, 1985).

Vitamin D is normally supplied in ample quantities through sun-cured hays or exposure of the animal to sunlight. Young lambs are normally supplied adequate amounts through milk and exposure to sunlight. In production systems where animals are not exposed to sunlight or fed sun-cured hays, vitamin D supplementation can be important in preventing rickets in lambs and osteomalacia in mature sheep.

Vitamin E is not stored in the body in significant quantities and therefore may be of greater concern in typical production systems than are vitamins A and D. Nutritional inadequacy is associated with white muscle disease, or nutritional muscular dystrophy. There is a nutritional interaction between vitamin E and selenium; therefore, a sparing effect exists between these nutrients. Sheep consuming normal quantities of pasture, dehydrated legumes, or grain at levels to support good performance are not likely to develop vitamin E deficiencies. Vitamin E injection of lambs corrects symptoms of white muscle disease in a matter of hours.

Vitamin K, while essential, is normally not of concern in considering dietary additions for sheep because of its presence in the usual feedstuffs used and synthesis by rumen microorganisms.

Water-Soluble Vitamins

B vitamins and vitamin C are not normally required as dietary additions for animals with functioning rumina. Early weaned lambs may benefit from B vitamin fortification of creep feeds for a few weeks, however.

Water Requirements

As with other species, a significant relationship exists between dry matter feed intake and water intake. Higher levels of feed intake increase the requirement for water. If sufficient water is not available, feed intake will decline. Late stages of pregnancy and lactation increase water demands by two to three times that for other stages of the reproductive cycle. Butcher (1970) reported that sheep may consume twelve times as much water in summer as in winter but will subsist on once-a-day watering if temperatures do not exceed 40°C. Butcher also showed that ewes could subsist on snow as the water source in a temperature range of 0 to 21°C. NRC (1985) suggests that lactating ewes would likely be stressed in this situation.

DESIGNING THE DIETARY ENVIRONMENT FOR CATTLE

Effect of Cold on the Dietary Requirements of Cattle

The National Research Council (NRC, 1981a) has summarized the impact of temperature on the voluntary feed intake of cattle (table 7.1). In general the thermal zone associated with what is termed feed intake is 15 to 25°C.

The basal net energy requirement for maintenance in Mcal/day for cattle is calculated by using W (weight of the animal) (NRC, 1984; Lofgreen and Garrett, 1968) or EBW (empty body weight of the animal) (NRC, 1996) as follows:

$$NE_m = 0.077\ W^{0.75}$$

or

$$NE_m = 0.077\ EBW^{0.75}$$

Calculating $W^{0.75}$ may be done as follows to arrive at metabolic size, which can then be multiplied by the appropriate regression value to determine energy demand:

Find the log of the weight (W) and multiply by 0.75. Find the antilog of the result, which is $W^{0.75}$. For example, to convert a weight of 600 kg to $W^{0.75}$,

$$\text{Metabolic body size} = \text{antilog of } (\log 600 \text{ kg} \times 0.75)$$
$$= 2.778 \times 0.75 = 2.0835$$
$$\text{antilog of } 2.0855 = 121.23 \text{ kg}$$

Using an electronic calculator with exponential key Y^x, using the same inputs enter 600 kg, press Y^x key, enter 0.75, and press = key.

$$\text{Metabolic body size} = 600 \text{ kg}^{0.75} = 121.23 \text{ kg}$$

TABLE 7.1 Effect of Cold and Heat Stress on Feed Intake

Temperature	Percentage Change in Feed Intake Daily	Water Intake per Kilogram Dry Matter of Feed Consumed
>35°C	Reduction of 10 to 35 percent on full feed	8–15
	Reduction of 5 to 20 percent on maintenance diet	
25 to 35°C	Reduced intake of 3 to 10 percent	4–10
15 to 25°C	Normal intake as indicated by live weight and type of diet	3–5
5 to 15°C	Increased by 2 to 5 percent	3–4
−5 to 5°C	Increased by 3 to 8 percent	2–4
−15° to −5°C	Increased by 5 to 10 percent	2–3
<−15°	Increased by 8 to 25 percent	2–3

Source: NRC, *Effect of Environment on Nutrient Requirements of Domestic Animals* (Washington, D.C.: National Academy Press, 1981), p. 60, by permission

The NE_m value then is adjusted to determine the impact of thermal stress. Such adjustments relate to the LCT, which, except for young calves, is low in cattle; as a result, cold stress occurs normally only in winter temperature extremes and through wet conditions conducive to chilling.

Young calves are commonly exposed to cold stress during the winter. For example, table 7.2 lists the LCT for a week-old calf to be about 7.7°C; the maintenance energy requirement increases by 2.83 percent per degree Celcius below this LCT. Specifically for this animal, the maintenance energy requirement is 2.93 Mcal/day, thus 0.0283×2.93 reflects the increase of 2.83 percent °C below LCT. If the temperature drops to freezing (0°C), the daily energy requirement increases by 0.64 Mcal to 3.57 Mcal, an increase of 21.8 percent. Wet, cold conditions could easily double this increased demand. It is virtually impossible for the calf to compensate for this increase, and stress effects that ultimately result in hypothermia will occur.

For comparison and to illustrate the cold resistance of cattle, a yearling steer (300 kg) on full feed gaining 1.1 kg daily with an LCT of −9.5°C could cope with declining temperature by increasing feed intake to compensate for the 1.3 percent increase in energy demand per degree Celsius below LCT (table 7.2).

The data shown in table 7.2 provide a basis for evaluating the effect of cold on the energy requirements of cattle (NRC, 1981a).

In adjusting the diet to compensate for changes in feed intake due to cold, the following practical example may be helpful. The concept of nutrient density adjustment is the same as presented earlier for swine.

A group of steers on full feed in an environment within the TNZ might be expected to increase feed consumption by 10 percent if the temperature drops 7 or 8°C below the LCT. A feed containing 11 percent crude protein at a daily consumption of 14 lb would provide 699 g of protein daily. At the higher level of intake the protein content of this feed must continue to supply the 699 g of protein, but the percentage in the diet at a daily consumption of 15.5 lb (110 percent of 14 lb) can be adjusted to 9.9 percent.

TABLE 7.2   Effect of Cold on The Energy Requirements of Cattle

Animal	Body Weight (kg)	Heat Produced (Energy Requirements) without Cold Stress (Mcal/day)	LCT (°C)	Increased Energy Requirement per Degree Celcius below LCT	
				Mcal ME[b]/day	*Percentage*
Calf, 1 wk					
Dry climate, low wind	50	2.93	7.7	0.083	2.83
Heifer, 6 mo, gaining 0.5 kg/day					
Dry, low wind	100	5.91	−17	0.097	1.64
Wet, 10-mph[a] wind	100	5.91	9.9	0.194	3.28
Yearling steer, gaining 1.1 kg/day					
Dry, low wind					
Wet snow, mud, 10-mph wind	300	15.62	−34.1	0.202	1.30
	300	15.62	−9.5	0.310	1.99
Dry, pregnant cow					
Dry, low wind	500	16.4	−25	0.237	1.45
Wet snow, 10-mph wind	500	16.4	−7.3	0.334	2.04

[a]mph, miles per hour
[b]ME, metabolizable energy
Source: NRC, *Effect of Environment on Nutrient Requirements of Domestic Animals* (Washington, D.C.: National Academy Press, 1981), p. 68, by permission

As with swine, adjusting the percentage of other nutrient sources for cattle must be considered individually to ensure adequacy in terms of those that may assist in meeting stress-imposed demands.

Beef cows maintained outside, which is typical of the majority of beef production systems in the United States, are routinely subjected to severe cold. Consumption of large amounts of high-fiber feeds, however, results in very low LCTs. Even in cold, wet conditions, the LCT of an adequately fed pregnant beef cow is well below 0°C. The value provided by NRC (1981a) is −7.3°C. For the same animal under dry conditions, the LCT is −25°C. The critical relationship between level of nutrition and reproductive performance provides a warning to be certain that nutrient requirements are met. However, meeting the requirements is most critical for females nursing the first calf and entering the breeding season to conceive the second calf. Inadequate nutrition during and before this period triggers stress responses that prevent conception for the next pregnancy.

An example of the change in feed requirements for maintenance of a dry pregnant beef cow in two environments and associated feed and energy requirements is illustrated in table 7.3 (NRC, 1981a).

TABLE 7.3   Influence of Cold on Energy and Feed Requirements for a Dry Pregnant Beef Cow

	Thermal Neutral Zone	Cold, Dry Conditions (−25°C)	Severe Storm of Snow or Wind (−25°C)
Mcal ME[a]/day	16.4	23.1	27.5
Bromegrass hay daily (kg)	8.8	12.9	15.4[b]

[a]ME, metabolizable energy
[b]The cow may not be able to consume this amount of bromegrass hay; thus, supplemental grain or a combination of grain and higher-energy roughage is required to prevent cold stress from resulting in normal responses (e.g., weight loss and predisposition to disease)
Source: NRC, *Effect of Environment on Nutrient Requirements of Domestic Animals* (Washington, D.C.: National Academy Press, 1981), p. 74, by permission

Effect of Heat on the Dietary Requirements of Cattle

Feeder Cattle

In general, performance of cattle being finished in the feedlot is optimum if the minimum daily temperature is no less than 15°C and the average daily temperature is below 30°C. When minimum temperature reaches 27°C and daily temperatures average 35°C or more, feed intake and daily weight gain may be reduced.

Results of studies to determine the quantitative effect of heat stress on rate of gain in feedlot cattle do not provide adjustment formulas that are broadly applicable. It is clear, however, that performance is influenced negatively by high ambient temperatures. In regions where humidity is low and limited in variability, direct assessment of air temperature on performance is reasonable and some operations have accumulated such estimates. In regions where humidity is a much greater factor, assessing the possibility of heat stress requires the use of indices reflecting temperature and humidity combined. Curtis (1983) concluded that in midwestern U.S. studies, daily gain is reduced when a Fahrenheit-humidity index exceeds 69. This index is calculated as follows:

$$15 + 0.4 \text{ (dry bulb temperature} + \text{wet bulb temperature)}$$

Dry bulb temperature, commonly referred to as air temperature, is that measured by an ordinary thermometer. *Wet bulb temperature* is measured by a thermometer in which the bulb is covered with a fabric that has been wet with distilled water. Air is forced over the wet bulb at a given rate. If the air is dry (low humidity), evaporation of water occurs in the fabric covering the bulb and reduces the temperature as compared with dry bulb (air) temperature. Under conditions of high humidity, little or no evaporation may occur. Thus, wet bulb temperature corresponds more closely with dry bulb temperature. When water vapor in the air (humidity) is high, little evaporation can occur at the skin level. Thus a major avenue for heat loss and temperature regulation is greatly restricted and heat stress is more likely to occur.

Dairy Cattle

Heat stress, if sufficient to elevate body temperature, depresses milk production. This result is associated with reductions in feed consumption, a typical response of all farm animals when exposed to heat stress. Daytime temperatures below 30°C are not likely to result in heat stress if cows have access to shade unless humidity exceeds 30 to 40 percent. Even in hot periods, cooler nighttime temperatures aid in the prevention of heat stress. Applied research indicates that variation in temperature during the 24-hour day in summer in most areas alleviates much of the effect of any stress that might occur at peak temperature during midday. Hahn and Osburn (1969) developed estimates of expected production losses typical of different regions of the United States during the 122-day summer period. These losses ranged from below 50 kg per cow in the extreme northern areas to 450 kg in the extreme south for cows with normal milk production in the area of 30 kg daily. Corresponding values for lower-producing cows (22.5 kg/day) were under 22 kg and 340 kg in the respective areas.

Dairy operations commonly experience reductions in levels of milk fat during the summer months. The reasons for this are not clear. If there is a reduction in fiber intake volatile fatty acid (VFA) patterns in the rumen may be altered and in turn milk fat synthesis influenced. Since total feed consumption may be reduced in hot weather, the daily requirement for individual nutrients must be met in the lower total amount of feed consumed. Thus, nutrient density becomes a management issue. Approaches to the calculation of dietary adjustments and nutrient density when feed intake changes are reviewed in the sections relating to swine and beef cattle finishing diets.

Management approaches to reducing heat stress include feeding the animals at night, providing shade, avoiding obstructions to air movement, and misting of water in resting areas to provide cooling by evaporation.

Beef Cows

Heat stress is not normally a management issue for beef cows because most are kept on pasture during the growing season. Access to shade can provide some escape from heat stress, but in general most areas have sufficient temperature variation within the 24-hour day to minimize any effects of summer heat. However, if beef cows are subjected to prolonged heat stress, conception rates may be reduced.

Energy Requirements

Energy Requirements for Maintenance

Beef cattle, like sheep, are maintained in a great range of climatic conditions and types of facilities. Climates commonly vary from extensive and arid range conditions such as in the Great Basin of the United States to highly productive pasture conditions of the Midwest to tropical conditions characterized by high levels of heat and humidity. Unlike sheep some breeds of beef cattle have been adapted to hot, humid regions. Those breeds or breed crosses utilizing Zebu are examples.

Maintenance energy requirements reflect the need for the level of caloric intake to maintain body weight and condition. Deviations in the thermal environment outside of the thermoneutral zone alter the level of energy required for maintenance. In general, this zone, in which cattle are relatively free of thermal stress, is in the area of 15 to 25°C. Any given level of production (e.g., gain in weight, pregnancy, and lactation) requires a specific increment of energy above that required for maintenance.

The potentially wide variation in the thermal environment for cattle represents a challenge for the manager of production enterprises to see that proper dietary adjustments are made as the environment changes to ensure satisfactory performance and well-being of the animals. On the other hand, adjustments for short-term thermal challenges to the animal may not be economically advantageous because of the animal's ability to cope for a limited time through normal stress-response mechanisms.

The maintenance energy requirement for beef cattle is determined as follows (NRC, 1981a; Lofgreen and Garrett, 1968):

$$\mathrm{NE}_m \text{ (Mcal/day)} = aW^{0.75}$$

or

$$\mathrm{NE}_m \text{ (Mcal/day)} = 0.077\ W^{0.75}$$

a = regression factor reflecting the relationship between body mass and the energy demand for maintenance
W = body weight

For example, a 400-kg steer in the TNZ has a daily net energy requirement of

$$\mathrm{NE}_m \text{ Mcal/day} = 0.077 \times 400^{0.75}$$

$$\mathrm{NE}_m \text{ Mcal/day} = 0.077 \times 89.44$$

$$\mathrm{NE}_m \text{ Mcal/day} = 6.88$$

The regression factor (a) used in the preceding formula may be altered by the ability of cattle to become acclimated. For example, if a group of cattle is subjected to cold below the typical TNZ the body will alter physiological mechanisms to increase metabolic rate and the rate at which body stores may be mobilized to meet cold stress. This increases the maintenance requirement. Sufficient research has been accomplished to permit adjustments in (a) in the formula to recognize acclimatization. The adjustment factor is the addition or subtraction of 0.0007 to (a), respectively, for each degree Celcius to which the animals have been subjected to thermal environments below and above 20°C. If a group of animals is acclimatized to a 10°C environment, then 0.077 would be adjusted by adding 0.007 (0.0007 × 10). Thus the formula becomes

$$\mathrm{NE}_m \text{ Mcal} = 0.084\ W^{0.75}$$

The maintenance energy requirement for a 400-kg steer so acclimatized would be

$$\text{NE}_m \text{ Mcal/day} = 0.084 \times 400^{0.75}$$

$$\text{NE}_m \text{ Mcal/day} = 0.084 \times 89.44$$

$$\text{NE}_m \text{ Mcal/day} = 7.51$$

It is important to recognize that the adjustments are for use in evaluating energy demands in the longer term such as for seasonal changes or relocation of cattle. Adjustments in energy for acute thermal demands must be in addition to the component of the maintenance requirement associated with acclimatization.

The value of 6.88 Mcal (maintenance for a 400 kg steer) corresponds with the value found in NRC (1984) nutrient requirement data. Thus, if this caloric level is provided to the animal daily, one would expect it to maintain the 400 kg in weight for a period of time if the thermal environment remains in the TNZ. If the animal is to gain, for example, 1.3 kg daily, an additional 7.16 Mcal will be required. If the temperature drops below the TNZ, then gain will be reduced to compensate for the added demand for maintenance. An additional energy allowance during acute periods of cold normally meets this increased demand unless conditions are severe.

Adult cattle exposed to acute cold stress (environments below the LCT) may not require adjustment in dietary energy in all cases to promote well-being. However, the performance level may be reduced. Under conditions in which additional dietary energy is required to compensate for cold stress and maintain productivity, the appropriate increase in energy allowance can be determined. It is first necessary to estimate the LCT for the animal. This calculation is performed by using the following formula (NRC, 1981a):

$$\text{LCT} = T_c - I(\text{HE} - H_e)$$

LCT = lower critical temperature
T_c = core temperature (°C), usually assumed to be 39°C
I = total insulation (i.e., tissue plus external insulation [°C/Mcal/m^2/day])
H_e = heat of evaporation (Mcal/m^2/day)
HE = heat production (Mcal/m^2/day)

HE is calculated by the following equation:

$$\text{HE} = (\text{ME} - \text{NE}_p)/A$$

ME = metabolizable energy intake (Mcal/day)
NE_p = net energy for production (Mcal/day)
A = surface area (m^2)(calculated by $A = 0.09W^{0.75}$)

The increased energy required to maintain productivity in an environment colder than the animal's LCT is calculated as follows:

$$\text{ME} = A(\text{LCT} - T)/I$$

T = effective ambient temperature
ME = increase in maintenance energy requirement (Mcal/day)

Based on work by Blaxter (1977) and Webster (1974, 1976), estimated values for tissue and external insulation effects that can be used in the preceding formulas are shown in table 7.4 (NRC, 1981a).

Based on calculations utilizing the preceding procedure, LCTs were determined for various classes of cattle (NRC, 1981a). These calculations are summarized in table 7.5.

The values in the table provide a vivid illustration of environmental effects on the level of energy required to maintain homeothermy and to allow the animal to cope with thermal stress. The values also demonstrate the hardiness of beef cattle if the animals are in a good state of nutrition and have appropriate environmental protection.

TABLE 7.4   Tissue Insulation Factors

Tissue Insulation (I_t)	I_t (°C/Mcal/m²/day)
Newborn calf	2.5
Month-old calf	6.5
Yearling	5.5–8.0
Adult cattle	6.0–12.0
External Insulation (Ie)	Ie (°C/Mcal/m2/day) in Relation to Coat Depth

Wind Speed (mph)	<5mm	10 mm	20 mm	30 mm
<1	7	11.0	14	17.0
4	5	7.5	10	13.5
8	4	5.5	8	9.0
16	3	4.0	5	6.5

Source: NRC, *Effect of Environment on Nutrient Requirements of Domestic Animals* (Washington, D.C.: National Academy Press, 1981), p. 67, by permission

TABLE 7.5   Lower Critical Temperature for Cattle and the Influence on Energy Requirements

	LCT (°C)	Increased Requirement per °C below LCT	
		Mcal ME[a]/day	(Percentage)
Calf, 1 wk old	7.7	0.083	2.83
Heifer, 6 mo			
Low wind	−17.0	−0.097	1.64
Wet, 10-mph wind	9.9	0.144	3.28
Yearling steer			
Low wind	−34.0	0.202	1.30
Wet snow, mud, 10-mph wind	−9.5	0.310	1.99
Dry pregnant cow			
Low wind	−25.0	0.237	1.45
Wet snow, 10-mph wind	−7.3	0.334	2.04

[a]ME, metabolizable energy
Source: NRC, *Effect of Environment on Nutrient Requirements of Domestic Animals* (Washington, D.C.: National Academy Press, 1981), p. 68, by permission

Beef cattle being finished in the feedlot consume large quantities of feed, generally have heavy hair coats during winter, and are provided some environmental protection such as windbreaks or sheds. Under these circumstances LCTs are very low. However, acute cold stress periods that may occur require some adjustment in the feeding program. If total feed allowance is increased, nutrient density becomes a management consideration.

Examples of estimated energy requirements and performance of 300-kg steers maintained under different environmental conditions are shown in table 7.6 (NRC, 1981a). These values demonstrate the influence of thermal demand on the feed energy remaining for production after that required for maintenance is met.

An additional example illustrating the influence of a range of environmental conditions on cattle is shown in table 7.7 (NRC, 1981a). In this case the animal is a 500-kg pregnant beef cow maintained in variable thermal environments.

The adjustment for the NE_m energy value of bromegrass hay (from NRC composition table) for the example in the severely cold environment is calculated as follows:

$A = 1.87$ Mcal/kg $+ 1.87[(0.0010)(-25°C - 20°C)]$
$A = 1.87 + 1.87[(0.0010)(-45)]$
$A = 1.87 + (1.87)(-0.045)$
$A = 1.87 + (-0.084)$
$A = 1.79$ (the adjusted value of 1.79 Mcal/kg should be noted in comparison with the 1.87 Mcal/kg value for this feedstuff when it is fed in a thermal neutral environment)

TABLE 7.6   Influence of Environmental Conditions on Energy Requirements and Expected Performance of Steers

		Environmental Conditions		
	TNZ^a	Hot and Dry[b]	Cold and Wet[c]	Cold and Dry[d]
Daily feed (kg DM)	7.60	6.80	8.00	8.40
Daily NE_m required Mcal (adjusted for acclimatization temperature and for direct heat stress)	5.55	5.39	6.56	7.32
Feed need for maintenance (kg DM)	3.03	2.93	3.62	4.07
Feed left for gain (kg DM)	4.57	3.91	4.36	4.29
NE_g available for deposit as gain (Mcal)	5.26	4.54	4.97	4.85
Expected weight gain (kg/day)	1.20	1.06	1.13	1.10
Feed dry matter per unit gain	6.30	6.50	7.10	7.60

[a]Nonstressful conditions
[b]Hot, dry environment with average ambient temperatures of 30°C, but 35°C daytime high resulted in rapid, shallow breathing
[c]Cold, wet environment with average temperature of 0°C with presence of mud and wet snow
[d]Cold, dry environment with average temperature of −15°C, effective wind protection, and dry bedding
Source: NRC, *Effect of Environment on Nutrient Requirements of Domestic Animals* (Washington, D.C.: National Academy Press, 1981), p. 72, by permission

TABLE 7.7   Environmental Influences on Energy Requirements of a Dry Pregnant Beef Cow

		Environmental Conditions		
	TNZ^a	Hot and Dry[b]	Cold and Dry[c]	Cold and Wet (Winter Storm)[d]
Maintenance energy requirement, Mcal ME[e]/day				
Adjusted for acclimatization temperature	16.4	14.9	23.1	21.6
Added for direct heat or cold stress	0.0	1.0	0.0	5.9
Total	16.4	15.9	23.1	27.5
Diet on DM basis (bromegrass hay)				
Energy content adjusted for environmental conditions (Mcal/kg)[f]	1.87	1.90	1.79	1.79
Feed requirement (kg/day)	8.8	8.4	12.9	15.4

[a]Nonstressful conditions
[b]Hot, dry environment with average temperatures of 30°C, with 35°C daytime high resulting in short, shallow breathing
[c]Cold, dry environment with average temperature of −25°C with dry bedded area and effective wind protection
[d]Cold, dry environment with seasonal temperatures usually of −15°C, but requirements based on needs for a period of several days of a winter storm with air temperature of −25°C, 10-mph winds, drifting snow, and no bedding or shelter
[e]ME, metabolizable energy
[f]In the given example the feed (bromegrass hay) was adjusted for environmental conditions since the ability of ruminants to digest roughage depends to some degree on the thermal environment. Such adjustments change the energy values for feedstuffs used in formulation and estimation equations. In general, roughages are somewhat more digestible in warm environments than in cold environments. The adjustment in energy value for the thermal environment is calculated as follows:

$$A = B + B[C_f(T - 20)]$$
A = value adjusted for environment
B = diet component value from NRC feed composition table
C_f = correction factor
T = effective ambient temperature (°C)

Correction factor (C_f) for effect of temperature on diet digestibility

Diet component	
Dry matter	0.0016
Energy component (ME, NE, total digestible nutrients [TDN])	0.0010
Acid detergent fiber	0.0037
Nitrogen (crude protein)	0.0011

Source: NRC, *Effect of Environment on Nutrient Requirements of Domestic Animals* (Washington, D.C.: National Academy Press, 1981), p. 74, by permission

Energy Requirements for Females

Nutrient requirements for cows in the breeding herd are usually considered in three categories related to demand: dry pregnant cows during the middle third of gestation, dry pregnant cows during the last third of pregnancy, and cows nursing calves (average or superior levels of lactation).

Because young beef females have special demands, additional areas for consideration are those for proper development of the replacement heifer and those including the demands for pregnancy and lactation, especially through the first three calf crops for females calving first at 2 years of age. Maturity is usually considered to be reached after the third pregnancy, although some additional size of the cow may accrue until 5 or 6 years of age.

The use of visual condition scores is increasing as a technology for evaluating the adequacy of feeding programs for beef cows. Descriptions for such scores are outlined by NRC (1996) as adapted from Herd and Sprott, 1986. Scores of 1 through 3 reflect lower levels of body condition and are likely to be related to unacceptable levels of well-being, necessitating an increased energy allowance. Scores of 4 should be considered marginal in terms of well-being. Scores of 5, 6, and 7 could be judged as adequate for both well-being and performance. Scores of 8 and 9 are likely associated with excessive levels of fatness and should be avoided. The condition scores adapted from Herd and Sprott (1986) are summarized in table 7.8.

Pregnancy   Beef cows are typically bred during the first 2 to 4 months postpartum; energy demands above maintenance are for lactation and possibly some recovery following a stressful winter period in the more severe climates. Many beef cows produce such low levels of milk that lactation does not represent a large demand. More highly productive cows have greater requirements; however, if pasture conditions are reasonably good, there should be no negative impacts of lactation stress on conception rates.

If pasture supplies are low, supplemental feeding may be beneficial. Very high levels of milk production, typical of dairy breeds utilized as beef cows, may reflect a type of lactation effect on conception rates that is not well understood. There are apparently breed differences in this regard, and, when encountered, even high levels of energy do not correct the problem.

In areas of limited feed supplies (such as arid range areas or areas of low-quality dry grass during the winter), proper energy and protein levels during late pregnancy can be critical to good reproductive performance during the following breeding season. If deficiencies occur during this period, it is difficult to correct for them by heavy feeding after parturition and before the breeding season. Thus, an adequate level of feeding before parturition is critical to good conception rates during the breeding season that follows, reflecting the longer-term importance of conditioning for animals to cope with potentially stressful conditions.

In the latter stages of gestation, the recommended energy requirement increases about 16 percent above that for the middle third of pregnancy. This is a modest increase compared to the 72 percent increase above mid-gestation and 48

TABLE 7.8   Cow Condition Scores

Score	Body Fat (Percentage)	Appearance of Cow
1	3.8	Emaciated; bone structure of shoulder, ribs, back, hooks, and pins sharp to touch and easily visible; little evidence of fat deposits or muscling
2	7.5	Very thin; little evidence of fat deposits but some muscling in hindquarters; the spinous processes feel sharp to the touch and are easily seen with space between them
3	11.3	Thin; beginning of fat cover over the loin, back, and foreribs; backbone still highly visible; processes of the spine can be identified individually by touch and may still be visible; spaces between the processes are less pronounced
4	15.1	Borderline; foreribs not noticeable, twelfth and thirteenth ribs still noticeable to the eye, particularly in cattle with a big spring of rib and ribs wide apart; the transverse spinous processes can be identified only by palpation (with slight pressure) and feel rounded rather than sharp; full but straightness of muscling in the hindquarters
5	18.9	Moderate[a]; twelfth and thirteenth ribs not visible to the eye unless animal has been shrunk; the transverse spinous processes can be felt only with firm pressure (and feel rounded) and are not noticeable to the eye; spaces between processes are not visible and only distinguishable with firm pressure; areas on each side of the tail head are fairly well filled but not mounded
6	22.6	Good[b]; ribs fully covered and not noticeable to the eye; hindquarters plump and full; noticeable sponginess to covering of foreribs and each side of the tail head; firm pressure now required to feel transverse processes
7	26.4	Very good; ends of the spinous processes can be felt only with very firm pressure; spaces between processes can barely be distinguished; abundant fat cover on either side of tail head with some patchiness evident
8	30.2	Fat; animal taking on a smooth, blocky appearance; bone structure disappearing from sight; fat cover thick and spongy, with patchiness likely
9	33.9	Very fat; bone structure not seen or easily felt; tail head buried in fat; animal's mobility may actually be impaired by excess of fat

[a]Moderate is generally considered adequate for effective reproduction, lactation, and overall well-being
[b]The good score may not be a cost-effective goal in most production situations and is most likely to exceed the level required for overall well-being
Source: NRC, *Nutrient Requirements of Beef Cattle* (Washington, D.C.: National Academy Press, 1996), p.214, by permission

percent increase above late gestation for lactation in beef cows of superior milking ability.

Lactation   Unlike ewes, beef cows have few multiple births. Thus the energy demand for lactation depends on cow size, environmental effects, and level of milk production.

Development of Replacement Females

Because there is great variation in the mature size of beef animals, recommendations for heifer development may be expressed in relation to expected mature weight. The general guideline is that heifers should attain about 55-65 percent of expected mature weight at the time of breeding for the first calf. The first breeding takes place at about 15 months of age for heifers to calve at 2 years of age, which is characteristic of many beef cattle management systems.

As with female sheep during early pregnancies and lactations, this time in the animal's life cycle is most sensitive to negative stress effects on reproductive performance. Decreased reproductive performance may result because biological priorities are first for protection of the dam for future reproduction and second for care of the calf at side rather than conception. Only after the dam's condition is maintained at a reasonable level and the calf at side is provided for does the biological system allow for pregnancy. The most critical time in the young beef female's reproductive life, relative to adequate energy allowances, is the period of the first pregnancy and subsequent lactation and rebreeding. The period associated with the second calf and rebreeding for the third is also critical for the female but less so. It does, however, warrant special attention if feed supplies are limited.

Development of Young Bulls

Energy requirements for bulls are based on the concept of good growth early in life and limited levels of energy with advancing maturity. As examples, NRC (1997) recommends a rate of gain of 1 kg/day for 300-kg bulls and daily gains of 0.9, 0.7, 0.5, and 0.3 for bulls in the respective weights of 400, 500, 600, and 700 kg. For mature bulls, the energy requirement is basically determined by maintenance needs and any recovery of condition lost during the breeding season.

Growing and Finishing Cattle

Typical beef production systems involve (1) growing calves to yearlings at weights of 300 to 350 kg and then finishing them for market on high-grain diets to about 500 to 600 kg or (2) placing calves directly in the feedlot at 200 to 225 kg (immediately after weaning) and feeding high-grain diets to market weights of 500 to 600 kg.

Principles involved in meeting energy requirements for growing and finishing cattle are straightforward and utilize the $NE_m + NE_g$ system for determining energy needs for meeting environmental influences and the requirement for production at specified levels. The system is advanced and is utilized in computer applications (NRC, 1996) for feedlot management to ensure that the animal's nutritional needs are met.

In general, because maintenance requirements for cattle are high, systems of production emphasizing rate of gain and related high-energy diets are the most efficient.

Low environmental temperatures increase energy requirements, which is shown in higher feed intake up to a point. When weather conditions are very severe, feed consumption may drop. Managers of the nutrition program must recognize that these conditions may warrant adjustments in the diet's nutrient density.

Energy requirements are influenced by sex. Thus, these differences must be taken into account in establishing dietary criteria for energy. Maintenance energy requirements appear to be similar for steers and heifers. Bulls have higher basal energy requirements that range from 10 to 15 percent higher than those of heifers.

The genetic makeup of the animal may also influence the maintenance energy requirement. For example, animals of dairy or dual-purpose breeding have higher maintenance requirements on the order of 15 to 20 percent higher than beef breeds. In general, animals that have the genetic potential for high productivity have higher maintenance energy requirements (NRC, 1996). This finding provides some explanation about why highly productive animals do not successfully cope with environments characterized by limited feed supplies.

Cattle, as is characteristic of other species, must have energy allowances above maintenance for gain or combinations of gain, lactation, and any other productive functions that require added feed allowances. Thus, considerations for growing cattle include body size, previous conditioning (acclimatization) to the thermal environment, and previous level of nutrition as it relates to potential compensatory gains, sex, breed, and age. Lactating animals may still be growing or recovering from previous limitations in feed supplies. Computer programs are available (NRC, 1996) for combining all of these factors in making accurate assessments of the animal's needs for a given set of circumstances.

Because of the capacity of the digestive tract and capabilities of the ruminant digestive system, cattle can utilize a wide range of feeds. Low-quality roughages such as straw and ground corncobs can be used to a limited extent even in situations designed to give highest levels of performance. In fact, such feeds may be beneficial in providing a part of the necessary bulk in the diet for normal rumen function and health. Such low-grade feeds, however, must be limited because of their low energy content. Higher-grade roughages such as alfalfa hay and corn silage may be fed at higher levels without reducing performance. Because of these relationships, growing and finishing beef animals tolerate a wide range in ratio of concentrate to roughage without a reduction in rate of gain. Diets made up of concentrates (grains and protein supplement) and good-quality roughages (hays or silages on an air-dry basis) may vary from 70:30 up to 90:10 in concentrate to roughage ratio and basically produce the same daily gain. Sudden changes in the concentrate to roughage ratio in the diet should be avoided, however, to allow the rumen microbial population to adjust to the different feed material.

Very-low-fiber diets for cattle (90 percent concentrate and above) may predispose animals to liver abscesses and rumen parakeratosis. Dietary antibiotics reduce the incidence of liver abscesses. These common problems are discussed in part 3 in relation to environmental conditions that result in tissue and metabolic disorders. In this case the diet appears to be the stressor.

Protein Requirements

Except for calves under 6 to 8 weeks of age, all cattle have a functional rumen. The microbial population has a major role in the synthesis of amino acids and some vitamins. Thus, ruminants can utilize nonprotein nitrogen as well as protein as a source of nitrogen for amino acid synthesis. This capability makes it possible to use nonprotein nitrogen (NPN) compounds as major dietary components to meet the protein needs of cattle and sheep. Urea is such a material and can be used to make up as much as one-third of the total nitrogen in the diet of growing and finishing cattle and up to about one-fourth in diets for pregnant and lactating cows if, in the latter case, an adequate, readily digestible supply of carbohydrates is included. Cattle on finishing rations consume large quantities of readily digestible carbohydrates in the form of starch in the grain consumed. Cows often subsist on range grass in the winter and have no readily digestible carbohydrate source available. For this reason, grain may be used along with oilseed meals and urea in range supplements to provide added starch to make more efficient urea utilization possible. Improperly used, urea can be toxic.

Because of the ability of rumen microorganisms to synthesize amino acids, the protein quality issue, so critical in monogastrics, has largely been ignored in many feeding programs for ruminants; however, the use of nonprotein nitrogen supplements requires careful attention to the microorganism's requirement for amino acids. Increasing evidence will probably result in attention being given to amino acid balance in cattle diets in the future.

Characteristics of the growth curve influence the relative requirement for protein. Thus, as a percentage of the diet, the protein level is reduced as animals grow and mature. Protein requirements increase moderately with pregnancy and dramatically with lactation. This nutrient, like energy, is critically important in conditioning females for the stress of lactation and must be given special consideration in feeding young females during the first and second pregnancies and lactations.

Mineral Requirements

Cattle require sodium, calcium, phosphorus, magnesium, potassium, sulfur, chlorine, iodine, iron, copper, cobalt, manganese, zinc, and selenium.

Sodium and chlorine needs are met by adding salt (NaCl) at the rate of 0.1 percent of the dietary dry matter. Cattle also consume adequate amounts in free-choice granular or block form. In some production systems, salt may be mixed at high levels (up to 30 percent) in a protein mixture for self-feeding the necessary supplemental protein. The high salt level is used for limiting feed intake (e.g., in range supplementation). Use of this practice requires that adequate water supplies be available. Salt toxicity is likely if the water supply is interrupted. Weather that freezes water sources represents a common hazard when this management practice is used.

Sodium and potassium should receive special consideration when animals are under stress since an imbalance may result in shrinkage and loss of body tissue water. Higher-than-normal levels of these two minerals are now recommended during periods of stress, such as before and after shipment of animals and during the initial phases of finishing programs.

The calcium to phosphorus ratio for most cattle diets is 1:1. Low-phosphorus soils are a common problem in many range areas; therefore, this nutrient warrants special consideration in beef herd management. In addition, the phosphorus content of many grasses is greatly reduced after frost. The usual approach to correcting this problem is provision of free-choice mineral mixtures and fortified supplements.

The trace minerals are normally supplied by free-choice trace mineral mixtures or as part of mixed diets. Requirements are not well established.

Vitamin Requirements

Fat-Soluble Vitamins

Vitamin A is the major fat-soluble vitamin for consideration in cattle diets and is typically added to all mixed diets for growing and finishing cattle and to range supplements. Cattle and calves consuming green pasture do not need additional vitamin A because of the carotene content of grasses and legumes. Additional vitamin E is commonly recommended for cattle experiencing stressful conditions such as shipment or severe weather. Vitamin E is also added to beef cattle finishing feeds as an approach to enhance the shelf life of beef.

Water-Soluble Vitamins

B vitamins and vitamin C are not commonly added to beef cattle diets because of their synthesis by rumen microorganisms. However, addition of niacin is often recommended for cattle enduring periods of stress.

Water Requirements

The water requirement is influenced by the size of the animal, composition of gain, pregnancy, lactation activity, thermal environment, and feed intake. The thermal environment and feed intake are major factors. At a temperature of 5°C cattle consume about 3 kg of water per kilogram of dry matter ingested, and at 32°C they consume 8 kg of water per kilogram of dry matter.

DESIGNING THE DIETARY ENVIRONMENT FOR DAIRY CATTLE

Energy Requirements

Most principles of nutrition apply to both beef and dairy cattle; however, because the productivity for dairy animals is so high, some critical areas relating specifically to this specialized animal must be covered in any consideration of the interactions among nutrition, genetics, performance expectations, and environmental influences.

Terms used to express energy requirements and feed values in designing diets for developing dairy replacement heifers and for the development and maintenance of bulls are as follows:

NE_m = net energy value of feeds and requirements for maintenance in nonlactating animals

NE_g = net energy value of feeds and requirements for deposition of body tissue in nonlactating animals

NEL is the term used to express net energy requirements for lactation. Because the efficiency with which energy is utilized by lactating cows is similar for maintenance or lactation, the NEL value is used to express energy needs for the maintenance component of the diet. An additional NEL energy allowance for milk production reflects both the quantity and fat content of the milk produced and is added to the maintenance level to meet the production component of the requirement. For example, a 600-kg mature lactating cow requires 9.70 Mcal NEL for maintenance and 16 Mcal for the production of 25 kg of milk containing 3 percent fat (25 kg milk × 0.64 NEL per kilogram of 3 percent milk). Thus the cow will need a total of 25.70 Mcal NEL daily from its diet to meet the requirements for both maintenance and production.

Mature Nonlactating Pregnant Cows

Important physiological changes occur in mammary tissue during the period from termination of lactation to parturition. This time is commonly referred to as the dry period, during which an advanced stage of pregnancy exists. During this period the mammary tissue undergoes a three-phase process. These phases are termed (1) active involution of the mammary secretory tissue, (2) steady-state involution, which is a true rest period for the tissue, and (3) colostrogenesis and lactogenesis.

The energy level required for maintenance plus gestation for the nonlactating dairy cow during the last 2 months of pregnancy is increased about 30 percent over the maintenance requirement of the lactating cow of similar weight. This period is extremely important for recovery from the previous lactation period and final stages of fetal growth. Failure to provide adequate energy at this time may contribute to a reduced conception rate and lower milk production in the next period of lactation. It should be noted that adequate energy levels are also very important postpartum in the interest of good reproductive performance as well as milk production. NEL requirements and feed values found in nutrient requirement data sources such as NRC (1989a) are used to formulate diets for animals during this period. Failure to provide an adequate energy balance at this time may result in excess tissue mobilization and predispose the cow to development of a metabolic disorder called fatty liver soon after parturition.

Mature Lactating Pregnant Cows

Energy needs for mature lactating pregnant cows are met by providing for the NEL requirement for maintenance and for the level of milk produced and its fat content.

Activity level should be considered for cows obtaining a significant portion of their diet from grazing since recommended nutrient allowances do not provide for the long distances that must be covered for feed and water. The added requirement for travel is about 3 percent for each kilometer of movement. In general, to support grazing, maintenance allowances should be increased by 10 percent for good pasture areas and up to 20 percent for sparse pastures (NRC, 1989a).

Because of the large amounts of feed consumed and normally adequate shelter, dairy cows are likely to encounter cold stress only in severe conditions. A recommended increase in feed allowance is up to 8 percent in total energy allowance for winter storm conditions.

Critical temperatures for cows range from 5 to 25°C, with maximum energy efficiency assumed to be in the 13 to 18°C range.

First- and Second-Calf Lactating Heifers

To allow for the growth of young lactating cows, NRC (1989a) recommends that maintenance level allowances be increased by 20 percent during the first lactation and 10 percent during the second lactation. A growth allowance is also provided during dry periods for heifers following first and second lactations.

Development of Replacement Heifers

The first consideration in the development of calves is to ensure that the animals receive an adequate amount of colostrum the first 8 to 12 hours after birth in order to allow absorption of antibodies (colostral immunoglobulins) from this milk. Absorption of colostrum antibodies normally ceases about 24 hours after birth. Failure to take this approach to strengthening the calf's immunity greatly increases potential problems in health management. Continued feeding of colostrum after this period may be of benefit for some calves since it provides a more concentrated source of nutrients than regular milk.

Milk or milk replacer should be fed for at least 28 days following birth, and continuation to 60 days of age may be beneficial, depending on the level of dry milk replacer consumed. If calves are eating 1.5 lb of dry milk replacer at 28 days of age, weaning is appropriate at this time.

Good pasture, quality hay, and concentrates are alternative sources of feed for calves. Calves under 6 months of age have insufficient digestive tract capacity to depend solely on pasture, however. Strong growth rates should be maintained and excessive fatness should be avoided since it may impair future milk-producing ability. Recommended levels of gain are well below the potential rate of gain if higher energy levels are fed. Thus, higher-roughage diets are appropriate. After calves reach 250 kg, good-quality hay, silage, or pasture provides ample energy to sustain acceptable growth rates.

Because the dairy breeds vary in body size, nutrient requirement recommendations are made for small and large breeds. For example, the energy requirement for a Jersey heifer (small breed) expected to weigh 100 kg at 26 weeks of age might be compared with a Holstein heifer at 16 weeks of age also weighing 100 kg. In this case, the recommended rate of gain for the Jersey is 500 g daily as compared with 700 g daily for the Holstein. The daily NE_m requirement is 2.43 Mcal for each calf, but the respective recommended NE_g requirements are 1.05 and 1.47 Mcal daily.

General guidelines for rate of development in heifers (NRC, 1989a) suggest that calves attain an approximately 100 percent increase in weight from birth to the age of 8 to 9 months, 50 percent of mature size at 15 to 17 months of age, and 75 to 80 percent of mature size by the first parturition at 24 to 27 months of age.

Development of Young Bulls

As is the case for young females, nutrient requirements for bulls being developed as sires reflect differences in potential mature size. The developmental rate recommended is one of good growth without excessive fatness. Allowances normally provide for faster daily gains in developing bulls than in heifers.

Veal Calf Production

Typical systems for production of veal involve placing calves on feed for this purpose soon after birth (or up to about 8 weeks of age) and providing a liquid diet of milk replacer to about 16 weeks of age, at which time the calves will have attained body weights in the 350- to 400-lb range. To minimize health problems, the calves should receive colostrum for at least the first 24 to 48 hours. The importance of newborn calves consuming colostrum during these early hours after birth cannot be overemphasized. Without this intake of colostral immunoglobulins to build a temporary immunity, health management is likely to be much more difficult and increased death loss can be expected. This problem is more prominent among calves going into veal production systems and results often by the practice of saving the colostrum for extended feeding to replacement females. Best management of calves going into this system is to ensure early consumption of normal amounts of colostrum.

The diet should contain sufficient iron to prevent anemia. In systems designed to produce veal of lighter color, Stull and McMartin (1992) concluded from studies involving ten commercial veal calf producers that the iron content of the diet can be reduced during the period from 8 to 16 weeks of age. This practice lowers the hemoglobin content of the blood but does not result in the negative effects normally associated with anemia; however, it produces blood hemoglobin levels that are considered marginally anemic (7.0 to 7.9 g/dL). Levels below 7.0 g/dL are considered clinically anemic. In these herds Stull and McMartin reported average dietary iron content of 209 ppm Fe during the early part of the 16-week production period compared with 32 ppm Fe during the last half of the period.

The protein content of veal calf diets typically is about 22 percent at the beginning of the production period and reduced gradually to about 17 percent toward the end of the feeding period. Correspondingly, the fat content of the liquid diet is raised from 17 to 20 percent over this period.

Some veal calf producers provide water in addition to the liquid in milk replacer for calves during the entire production period. Others do so after about 45 days. Both systems appear to be satisfactory.

Protein Requirements

The significance of protein in the diet of lactating cows is illustrated by the fact that a cow producing 30 kg of milk daily is making about 1 kg of new protein daily for the milk produced (NRC, 1989a). This amount of protein is similar to that in 6 to 7 kg of body weight gain. This fact, along with recognition of the extreme energy demands for lactation, underscores the exposure to potential stress conditions with the onset of lactation. As with any ruminant, dairy cattle can synthesize amino acids from natural protein and nonprotein nitrogen. Thus, properly managed, urea may be used in diets for all classes of dairy animals as soon as a functional rumen develops completely (approximately 6 weeks of age). Dairy cattle, like all ruminants, can utilize urea for a part of the nitrogen (crude protein) requirement because of the presence of the rumen microbial population. When it is used in high-roughage diets, guidelines suggest that urea should make up no more than 7 percent of the total dietary crude protein. When included in high-energy diets (high grain levels for milk production or rapid growth), urea should not make up more than 13 percent of the total crude protein in the diet. Excessive amounts of urea may result in the formation and absorption of toxic levels of ammonia. Rumen microorganisms require an adaptation period (2 to 3 weeks) to diets containing urea. Thus, abrupt changes in using this material should be avoided.

Mineral Requirements

Milk production requires large quantities of calcium and phosphorus. A 600-kg cow producing 30 kg of milk daily requires 20 g of calcium and 10 g of phosphorus for maintenance and another 72 g of calcium and 48 g of phosphorus for milk production. In general, calcium to phosphorus ratios on the order of 1:1 to 2:1 appear to be satisfactory. Fat additions to feeds for high-producing cows increase the calcium requirement to offset soap formation in the digestive tract.

Salt and trace minerals are normally provided as a part of mixed concentrates, supplements, or in free-choice form.

Vitamin Requirements

Because of the use of large amounts of high-quality roughages in dairy diets, the fat-soluble vitamins (A, D, E, and K) are usually present in sufficient amounts. The exception is when hay is stored for long periods (a year or more) and the carotene content is reduced. Consumption of hay lacking carotene may result in a vitamin A deficiency and the related negative effects on performance and in tissue damage. When animals are not exposed to sunlight or do not have access to sun-cured hay, vitamin D requirements may not be met. Finally, calves under 6 weeks of age, because of their nonfunctioning rumina, may require supplemental water-soluble vitamins. Except in these young calves, water-soluble vitamins are not routinely added to feeds for dairy cattle.

Water Requirements

Restricted water consumption by lactating cows is quickly reflected in reduced milk yield. Water requirements for lactating cows are greater relative to body weight than for any other farm animal. A water shortage is compounded by lower feed intake. In hot, dry climates, if water is limited, feed consumption drops to near zero after 4 days (NRC, 1989a). As with all animals, water intake is related to dry matter intake and increases with rising environmental temperatures. Studies on the effect of cooling water for cows do not show a positive impact on productivity. The same is true for warming water in winter, as long as freezing is prevented. An approximation of water demand in environments ranging from -17 to $27°C$ is 3.5 to 5.5 kg of water per kilogram of dry matter diet.

Metabolic Disorders Commonly Encountered in Dairy Enterprises

Certain metabolic disorders are common in dairy cattle and appear to be related to the demands associated with pregnancy, parturition, and lactation. Such disorders are described in NRC (1989a) and summarized below:

Udder Edema

Udder edema may develop before calving. The problem is thought to involve an imbalance between increased blood flow to the area without a compensating increase in lymph removal from the affected tissue area. Some conclude that udder edema is related to excessive salt consumption, and a restricted level may be helpful.

Milk Fever

Milk fever usually occurs within 48 hours postpartum. It is associated with high calcium intake during the dry period and may be reduced by limiting calcium intake before calving and increasing it at calving time. The fundamental problem involved is the inability to mobilize adequate amounts of calcium quickly in the postpartum period. The problem may result from feeding too much calcium prepartum.

Ketosis (Acetonemia)

Ketosis is a disorder that usually occurs during the first 6 weeks postpartum, with a peak incidence of around 3 weeks. The cause is apparently an inadequate supply of glucose precursors to meet the demands associated with lactation. Control involves liberal concentrate feeding after calving in a balanced ration with an adequate roughage level. The use of propylene glycol as a feed additive during the susceptible period has proven useful (Schultz, 1971). Supplemental niacin has also been shown to be beneficial (Fronk and Schultz, 1979).

Fat Cow Syndrome

Fat cow syndrome results from excessive fatness in cows and occurs after calving. It is characterized by the development of fatty livers, and affected cows may exhibit downer syndrome. Proper control of the condition of cows and adequate levels of dietary roughage are necessary for recovery.

Fatty Liver

The fat level in the liver usually increases at the onset of lactation; however, cows that are fed excessive amounts of energy during the last half to third of the dry period are more susceptible. Ensuring proper levels of energy throughout the dry period is the correct management approach to minimize this problem.

DESIGNING THE DIETARY ENVIRONMENT FOR HORSES

Energy Requirements

The horse, a nonruminant herbivore, depends primarily on carbohydrates as a source of energy. The animals have a prececal digestive tract like monogastrics that empties into a cecum of large capacity. The cecum maintains a microbial population that performs digestive functions much like those of the rumen in cattle and sheep. Thus, half or more of readily digestible carbohydrates such as starches and compound sugars are digested and the end product sugars absorbed by the small intestine before reaching the cecum. The undigested portion of these carbohydrates and the complex carbohydrates such as the cellulose and hemicellulose components of roughages (e.g., hay, pasture, and straw) pass into the cecum for further digestion by fermentation processes. The end products, volatile fatty acids, are absorbed and utilized for energy. The cecal microbial population serves the host animal through synthesis of amino acids and many vitamins in addition to performing the digestive fermentation required to allow horses to effectively utilize roughages.

The NRC (1989b) suggests that energy requirements for horses be expressed as DE or total digestible nutrients (TDN), since few ME and NE values have been determined for the feedstuffs of horses.

In addition to the several factors influencing energy requirements of all animals (e.g., size, temperature, and body or growth composition), the requirement for horses must often include intensity and duration of work, weight of rider, condition of the running surface, and degree of fatigue. Since data are not available for all of these influences and the interactions among them, energy requirements are considered as general guidelines (NRC, 1989b).

Energy Requirements for Maintenance

Scientists developing nutrient requirement data for horses, unlike for the other species, utilize the actual weight for calculations rather than metabolic size. Thus, the following equations for estimating energy requirements are based on this relationship for horses confined to stalls (NRC, 1989b; Pagan and Hintz, 1986a):

$$DE \text{ (Mcal/day)} = 0.975 + 0.021W$$

$$W = \text{weight of the horse (kg)}$$

For nonworking horses moving about as normal activity and weighing less than 600 kg, the equation is as follows:

$$DE \text{ (Mcal/day)} = 1.4 + 0.03W$$

For nonworking horses moving about as normal activity and weighing more than 600 kg, the equation is as follows:

$$DE \text{ (Mcal/day)} = 1.82 + 0.0383W - 0.000015W^2$$

Energy Requirements for Mares

Breeding Mares   NRC (1989b), based on the work of Henneke et al. (1984), emphasizes the importance of condition in reproductive performance. Thin mares must be fed at a level to be gaining weight at the time of breeding, or reproduction will fail. Mares in good to high levels of condition conceive readily if they are on maintenance energy levels or even if loss of weight occurs during breeding.

Henneke et al. (1983) proposed the use of condition descriptions and scores in relation to conditioning mares for breeding. On a scale of 1 (poor) to 9 (extremely fat), these workers described a midpoint score of 5 (moderate) as a baseline level for adjusting dietary energy levels. The moderate level is described as having the following characteristics: the back is flat (no crease or ridge), ribs are not visually distinguishable, fat is easily felt around the tail head and is beginning to feel spongy, withers appear rounded over spinous processes, and the shoulders and neck blend smoothly with the body. Mares exhibiting such features are considered to be in strong breeding condition. For mares below this level, energy allowances should be 10 to 15 percent above the reported requirements in order to produce some weight gain. Mares scoring above this level should have an energy allowance 10 to 15 percent below the reported requirement to adjust the condition downward.

Pregnant Mares   Energy requirements for pregnancy are not greatly above preconception levels; however, during the last 3 months (ninth, tenth, and eleventh), maintenance energy levels should be increased, respectively, by 1.11, 1.13, and 1.20 times (NRC, 1989b).

Lactating Mares   NRC (1989b) assumes a relationship between mare weight and level of milk production and provides energy requirements for weight categories of females for use in establishing feeding levels to meet lactation requirements. Daily energy levels are recommended for mares during the first 3 months of lactation and then from 3 months postpartum to weaning.

These recommendations are determined by the following calculations (NRC, 1989b):
First 3 months postpartum:

$$\text{Daily DE for females 200 to 299 kg of weight (Mcal/day)} = DE_m + (0.04W \times 0.792)$$

$$\text{Daily DE for females 300--900 kg of weight (Mcal/day)} = DE_m + (0.03W \times 0.792)$$

From 3 months postpartum to weaning:

$$\text{Daily DE for females 200 to 299 kg of weight (Mcal/day)} = DE_m + (0.03W \times 0.792)$$

$$\text{Daily DE for females 300 to 900 kg of weight (Mcal/day)} = DE_m + (0.02W \times 0.792)$$

Feeding programs for lactating mares may include only good-quality pasture or roughage of excellent quality. If less-than-abundant good-quality roughage is available, supplemental concentrate feeding is required, especially during the first 3 months of lactation. Because horses have a limited capacity for roughage and yet a need for a bulky diet, NRC (1989b) suggests the following guidelines: The animal should receive a minimum of 1 kg of roughage dry matter per 100 kg body weight daily, along with concentrates at the level necessary to meet the daily energy requirement. Horses are able to consume about 2 to 2.5 percent of body weight equivalent in dry matter daily in the form of good-quality roughage or when grazing. As the quality of roughage declines, less is consumed. This guide is particularly important in planning the nutritional environment for lactating females.

Energy Requirements for Growth
The determination of energy allowances for growth in young horses is by the following formula (NRC, 1989b):

$$DE \text{ (Mcal/day)} = DE_m \text{ (Mcal/day)} + (4.81 + 1.17 \times -0.023X^2)(ADG)$$

$$ADG = \text{average daily gain (kg)}$$

$$X = \text{age of animal in months}$$

The provided determination recognizes an increasing energy requirement per unit of gain with advancing age (higher level of fat in tissue gain). The specific relationships between rate of development and potential performance or soundness characteristics are not established. Table 7.9 suggests levels of gain for growing horses based on mature weight categories (NRC, 1989b).

Energy Requirements for Work or Performance
Anderson et al. (1983) proposed the following formula as a means of estimating the energy required for performance in horses.

TABLE 7.9   Energy Requirements for Different Levels of Growth for Young Horses Having Different Potential for Mature Size

| | Potential Mature Size | | | |
| | 400 kg | | 600 kg | |
Age	Daily Gain (kg)	Mcal DE[a]/day	Daily Gain (kg)	Mcal DE/day
Weaning, 6 mo				
Moderate growth	0.55	12.9	0.75	17.0
Rapid growth	0.70	14.5	0.95	19.2
Yearling				
Moderate growth	0.40	15.6	0.65	22.7
Rapid growth	0.50	17.1	0.80	25.1
2-year-old				
Not in training	0.15	15.3	0.30	23.5
In training	0.15	21.5	0.30	32.3

[a]DE, digestible energy
Source: NRC, *Nutrient Requirements of Horses* (Washington, D.C.: National Academy Press, 1989), p. 43–44, by permission

$$DE\ (Mcal/day) = 5.97 + 0.021W + 5.036X - 0.48X^2$$

$$W = \text{body weight of the horse (kg) and } X = Z \times km \times 10^{-3}$$

$$Z = \text{weight of horse, rider, and tack (kg)}$$

The preceding formula is suggested for intense effort in a burst of speed and not for endurance. Other formulas have been developed for horses expending energy for hard work over extended periods. One of these formulas for calculating the DE requirement above maintenance (kcal/kg of weight of horse, rider, and tac) follows (NRC, 1989b; Pagan and Hintz, 1986b):

$$DE\ (kcal/kg/hr) = \frac{[e^{(3.02 + 0.0065Y)} - 13.92] \times 0.06}{0.57}$$

$$Y = \text{speed in m/min}$$

NRC (1989b) concludes that general guidelines for the energy requirement for work (performance) are estimated as follows:

$$DE \text{ for light work} = DE_m \times 1.25$$
$$DE \text{ for medium work} = DE_m \times 1.50$$
$$DE \text{ for intense work} = DE_m \times 2.00$$

In the case of draft horses, the suggestion is that an energy allowance of 10 percent above maintenance per hour of work be provided.

In the case of the athletic horse, two questions arise as to dietary preparation for speed and endurance. In general, the questions take the same form as in relation to human athletes. The first issue is the effect of an elevated caloric content of the diet on speed and endurance. Considerable research has been done in increasing energy levels of the feed by fortification with fats or oils. The results suggest that horses will consume added fat or oil (if of good quality) at quite high levels (5-20 percent of the diet). In general, fat additions in horse diets do not increase total energy intake significantly, and, as a result, there is not good evidence that this is an effective practice to support enhanced performance. The second issue that arises in dietary preparation for racing is the one of carbohydrate loading or raising the muscle glycogen level. The objective in this case is to increase muscle glycogen stores above that accomplished by the routine training-exercise regimen. Increasing the routine daily level of readily digestible carbohydrates in the diet above normal for racing animals does not necessarily result in carbohydrate loading. Feeding high levels of carbohydrate after an exhaustive exercise activity may increase muscle glycogen levels, however, the net effect of efforts to increase speed or endurance through management of dietary carbohydrate is not well established. Therefore, there appears to be some uncertainty as to effective methods for accomplishing carbohydrate loading for athletic horses. Thus, the issue of the effectiveness in performance horses is subject to disagreement among scientists.

Energy Requirements for Stallions

Young stallions should be developed at moderate rates of growth to prevent excessive condition. Because intact males grow faster than females or castrated animals, an added energy allowance may be desirable. Mature stallions should be kept at near maintenance levels of energy during the nonbreeding season, with an additional allowance provided during an active breeding season. This requirement represents an increase in DE of about 20 percent above DE_m.

Protein Requirements

Protein quality expressed as kinds and amounts of amino acids is not a major consideration for horses except in the case of young foals. Several studies suggest that lysine is the first limiting amino acid for these young animals (NRC, 1989b).

Although horses tolerate and make some use of nonprotein-containing material such as urea, there appears to be no reason to include this material in their diet. Doing so may reduce the growth rate of young animals and lower milk production in mares. These effects suggest clearly that only natural protein sources are preferred forms of supplemental protein for horses.

Mineral Requirements

The calcium to phosphorus ratio (preferred ratio is 1:1) is important in horse nutrition. The level of calcium may exceed that of phosphorus; however, the phosphorus level should not exceed that of calcium.

Sodium and chlorine are provided by salt added to the concentrates at a rate of 0.5 to 1 percent or through free choice as iodized or trace mineralized salt.

Trace minerals are usually provided by fortifying the concentrate mix with a premix or offering them as a part of a mineral mix or trace mineralized salt on a free-choice basis.

Vitamin Requirements

Because horses normally receive substantial quantities of good-quality forage as pasture or high-quality hay, supplemental fat-soluble vitamins are not required. If hays are stored for a year or more, supplemental vitamins A and E may be necessary.

Due to the presence of a microbial population in the cecum and its ability to synthesize water-soluble vitamins, such dietary additives are not required in most cases.

Water Requirements

As with other species, a relationship exists among the size of the animal, dry matter intake, environment, lactation, exercise, and water requirements. Horses require 2 to 3 L of water per kilogram of dry matter intake.

Horses perspire profusely as a means of body temperature regulation. Thus, both water and electrolyte balance are issues in providing a proper nutritional environment for horses. Intense work may increase water loss up to 300 percent (NRC, 1989b); associated with this water loss are losses of NaCl. Thus, adequate water, salt, and rest are the normal corrective actions. Electrolyte additions to water may be helpful for horses working under hot, humid conditions. Some trainers may administer electrolytes before races as a routine practice, but its benefits are not well supported scientifically.

DESIGNING THE DIETARY ENVIRONMENT FOR POULTRY

Energy Requirements

The most common quantitative expression of dietary energy for poultry is kilocalories of metabolizable energy per kilogram of diet.

Commercial poultry production systems are largely based on *ad libitum* feeding in which the birds have access to feed at all times. The birds then consume feed to satisfy their energy requirements. Because voluntary consumption determines the level of total feed consumed, feeds are formulated to provide quantities of all other nutrients at levels relative to the energy content of the diet and expressed as a percentage of the diet, weight per kilogram of the diet, or units per kilogram of the diet. For this reason, nutrient requirements are reported for diets with specific levels of energy as well as stage of growth or stage of production for the birds involved in the system. For example, in diets for growing Leghorn-type chickens, a typical level of energy is 2,850 to 2,900 kcal ME/kg. Four-week-old birds are expected to consume about 260 g of feed weekly. The same bird at 16 weeks consumes about 430 g of a diet containing 2,900 kcal ME/kg. Since the older bird is consuming more feed, the percentage protein requirements as a portion of the total diet are approximately 18 and 15, respectively. The lysine requirement drops from 0.85 to 0.45 percent, and the nonphytate phosphorus requirement is reduced from 0.40 to 0.30 percent of the diet as age progresses. Careful management is essential to ensure that daily feed consumption is at a level to meet nutrient needs. Required concentrations of all nutrients, however, are not reduced. The concentration of many vitamins expressed as milligrams or international units may be held constant at both ages for these birds.

The previously mentioned relationships establish the fact that the energy concentration of the poultry diet is a key factor in designing the total dietary environment available. As a result, NRC (1994) recommendations reflect a desirable level of energy in the feed and feed consumption requirements on a weekly basis for growing and laying birds for different ages and types as well as various stages or levels of production. Daily required energy (kcal ME/hen) levels are reported (NRC, 1994) for laying hens at different weights and different rates of egg production. Such data provide a useful illustration of the relative energy demand to support production and reflect the critical nature of dietary adequacy in environmental design for poultry. For example, chickens weighing 1.5 kg with egg production rates of 60 and 80 percent have respective daily energy requirements of 251 and 276 kcal ME/hen/day, or a difference of about 10 percent.

Poultry, like other animals, require quantities of energy to meet requirements for both maintenance and production. The daily energy requirements for laying chickens were determined by NRC (1981a, 1994) utilizing the following formula, which accounts for these factors and the influence of environmental temperature and change in body weight:

$$ME/hen \ daily = W^{0.75}(173 - 1.95T) + 5.5\Delta W + 2.07EE$$

W = body weight (kg)
T = ambient temperature (°C)
ΔW = change in body weight (g/day)
EE = daily egg mass (g)

The energy levels recommended (table 7.10) for various types of poultry and different stages of development and production reflect concentration in the diet and thus are not daily requirements (NRC, 1994). This approach establishes the proper ratios for the various nutrients and ensures that such relative amounts of the various nutrients are consumed at different levels of daily consumption of a given diet.

The thermal environment in which poultry exist has important influences on how the diet should be altered to compensate. A reduction in ambient temperature increases the maintenance requirement and likely results in increased feed intake. High environmental temperatures are associated with reductions in feed intake at the rate of about 1.5 percent for each degree increase above the TNZ (NRC, 1994). Since nutrient content is established based on the weight of the feed (e.g., by percentage or mg/kg), significant changes in total consumption necessitate altering the concentration of the various nutrients accordingly. For growing chickens and layers, *ad libitum* daily consumption of high-energy feeds is used to support efficient production. In the case of breeders (chickens or turkeys), holding diets may be used to control the condition of females after mature body weight is reached. Such diets are lower in energy and protein than those for development.

Protein and Amino Acid Requirements

The requirement for protein is usually expressed as a percentage of the diet. However, nutritional adequacy is assured only by formulating feeds to meet the requirement for individual amino acids. Some amino acids, referred to as nonessential amino acids, are produced within the body if the proper nitrogen forms are available. Amino acids that must be present in the diet because they are not synthesized in adequate amounts in the body are referred to as essential amino acids. Protein requirements, then, have two components, expressed by NRC (1994) as follows: (1) the essential amino acids needed by the bird because it cannot synthesize them or synthesize them rapidly enough, and (2) sufficient protein to supply either the nonessential amino acids themselves or the amino nitrogen for their synthesis.

Required dietary protein and amino acid levels are established by the level of feed consumption, which, in a moderate thermal environment (16 to 24°C), is largely determined by the level of energy in the feed. Major influences on the pro-

TABLE 7.10 Recommended Levels of Energy in Poultry Diets

		kcal ME/kg Diet
Growing Leghorn-type chickens		
0–12 wk		2,850[a]
12 wk to first lay (18–22 wk)		2,900[a]
Laying hens		2,900
Breeders		2,900
Broilers		
0–8 wk		3,200
Turkeys		
Female	*Male*	
0–4 wk	0–4 wk	2,800
4–8 wk	4–8 wk	2,900
8–11 wk	8–12 wk	3,000
11–14 wk	12–16 wk	3,100
14–17 wk	16–20 wk	3,200
17–20 wk	20–24 wk	3,300
Turkey breeding hens (holding and breeding)		2,900
Geese		
0–4 wk		2,900
After 4 wk		3,000
Ducks		
0–2 wk		2,900
2–7 wk		3,000
Breeders		2,900
Pheasant		
Starting (0–8 wk)		2,800
Growing (9–17 wk)		2,700
Breeding		2,800
Quail		
Bobwhite		2,800
Japanese		2,900

[a]For brown egg layers the level is approximately 50 kcal ME/kg diet lower
Source: NRC, *Nutrient Requirements of Poultry* (Washington, D.C.: National Academy Press, 1994), p. 20, 27, 36, 40, 42, 44, 45, by permission

tein and amino acid requirements are rate of growth and egg production. Because the potential for these traits is established by genetic makeup, nutrient requirements vary among breeds and strains.

Feed formulation technology recognizes some specific and important amino acid relationships in designing the diet, including (NRC, 1994):

Methionine-cystine: The requirement for cystine can be met by cystine or methionine.

Phenylalamine-tyrosine: The requirement for tyrosine can be met by tyrosine or phenylalamine.

Glycine-serine: Glycine and serine can be utilized interchangeably.

Feed formulation procedures also recognize the existence of certain antagonistic relationships among some amino acids if they are fed in improper ratios. These antagonisms occur among amino acids that are structurally related. Because of these relationships, relative amounts of amino acids in poultry feeds are considered in establishing recommended levels of each.

Some amino acids may contribute to dietary vitamin adequacy. Methionine may partially compensate for deficiencies of choline or vitamin B_{12} by providing methyl groups for vitamin synthesis. Tryptophan may be converted to niacin. In designing diets, however, these relationships should not be relied on to meet the bird's needs.

Amino acid availability is an issue of importance because ingredient and feed processing techniques may alter availability favorably or unfavorably. Heat treatment of soybean meal inactivates the antitrypsin factor present in raw soybeans, which enhances the digestibility of soybean protein. Excessive heating of blood meal or meat scrap reduces protein digestibility. These problems reflect the importance of quality control and assessment in selecting dietary ingredients.

Changes in protein and amino acid requirements for turkeys provide an illustration (table 7.11) of the influence of stage of growth on requirements for these nutrients (NRC, 1994).

Mineral Requirements

Both major and trace minerals are essential in the diet of poultry. The metabolic functions of the various minerals are reviewed briefly in the section treating dietary design for swine.

Among the food animal species, poultry have relatively high rates of growth and bone formation. Layers have high mineral requirements associated with eggshell formation. Rapid growth rates require special attention to calcium and phosphorus levels in the diet, and egg production has a dramatic influence on the demand for calcium. The demand for dietary calcium to support egg production is shown by an increase in the calcium requirement from 0.60 percent of the diet for maturing pullets at 14 to 20 weeks of age to 3.40 percent with the onset of laying. This requirement is illustrated more dramatically by daily calcium intake. Maturing pullets at 20 weeks of age consume about 460 g of feed containing 0.60 percent calcium per week, or a total of 2.76 g calcium. At 30 weeks, around the peak in egg production, feed consumption is about 770 g per week. At 3.40 percent calcium, the weekly intake of this element is 26.18 g, nearly a tenfold increase in demand for the nutrient.

The availability of dietary phosphorus is especially important in poultry nutrition. The phosphorus consumed by poultry as a component of plant materials in the diet is in organic form. The inorganic component in typical poultry feeds is provided largely from phosphorus supplements such as dicalcium phosphate. The phosphorus present in plant materials is typically 60 to 70 percent in the form of phytin phosphorus, which is largely unavailable as a source of phosphorus for poultry. The nonphytin phosphorus in the diet (30 to 40 percent) is considered to be usable. Thus, the phosphorus content of ingredients of

TABLE 7.11   Protein and Lysine Requirements of Turkeys

Age of Turkey (wk)	Protein Requirement (Percentage)	Lysine Requirement (Percentage)
Male, 0–4	28	1.6
Female, 0–4		
Male, 12–16	19	1.0
Female, 11–14		
Male, 20–24	14	0.65
Female, 17–20		

Source: NRC, *Nutrient Requirements of Poultry* (Washington, D.C.: National Academy Press, 1994), p. 36, by permission

plant origin must be discounted appropriately in the formulation of the bird's diet. The level of phosphorus available in the phytin form is generally assumed to be about 50 percent.

Minerals normally considered in formulating poultry feeding programs are calcium, phosphorus, sodium, chlorine, potassium, magnesium (normally present in sufficient amounts in feedstuffs), manganese, zinc, iron, iodine, and selenium. In addition, trace minerals may be considered. Trace minerals should be added based on special situations and in geographical regions known to produce feeds deficient in one or more of these elements.

Vitamin Requirements

Poultry require most fat- and water-soluble vitamins. Vitamins A, D, E, and K of the fat-soluble group and thiamin, riboflavin, niacin, biotin, pantothenic acid, pyridoxine (usually present in adequate amounts in most practical diets), folacin, vitamin B_{12}, and choline of the water-soluble group are necessary in formulating poultry diets.

Water Requirements

Water is usually provided *ad libitum* to ensure an adequate supply. The quantity needed depends on the temperature, humidity, level of production involved, and body conservation. NRC (1994) generally assumes that birds consume an amount of water approximately double the weight of feed consumed. Actual consumption is related to feed intake but is also influenced by environmental temperature, humidity, and level of salt in the diet.

Water consumption by poultry in typical floor litter housing requires consideration beyond that related to meeting the needs of the birds. Excess water consumption and excretion contribute to a level of moisture in litter that in turn creates sanitation problems and increases ammonia production in the facility. The primary causes of high water consumption in addition to environmental temperature and humidity are concentrations of minerals in feed or water. The added water is required for elimination of the excess minerals consumed. Water wastage caused by birds or faulty equipment is an added management issue of importance in controlling the moisture content of litter. Water containing high levels of dissolved solids is expected to cause watery droppings and excessively wet litter and poor performance.

MANAGING THE FEEDING PROGRAM IN RELATION TO COMPOSITION AND SUPPLY OF FEEDSTUFFS

The component of the environment represented by feed and water is obviously critical in terms of meeting the needs of both the animals and the economic considerations of the enterprise involved. Some of the same elements of climate that influence an animal's environment have an impact on the plant population of a region. These effects may alter both the composition and the quality of feed resources. Seasons of the year may have unique impacts on animals; because of their influence on stage of growth, they affect the nutrient composition of plant materials that are ultimately used for feed. The composition of plant material either harvested by the animal or harvested and stored for later use influences the kinds and amounts of other feed materials to be used as supplements in order to provide an appropriate dietary environment in total.

Limited feed supplies create special situations that require consideration of options in feed and nutrient sources to meet the animal's requirements. Economic and biological considerations are necessary also in evaluating cost-benefit factors in relation to performance penalties that may be associated with some choices in providing the diet. This issue brings to the forefront the fact that animals are flexible in terms of the level of some nutrients in the diet and may remain in a satisfactory state of health and well-being even though maximum performance is sacrificed.

The water supply and location may impose profound management restrictions in the use of feed resources in livestock systems that involve grazing. Water intake level, as discussed earlier, is strongly related to level of dry matter feed intake. Therefore, water restrictions may also impose limits on feed intake and on performance.

Proper pasture or range management can have dramatic influences on available feed, quality of the resource, animal performance and well-being, and conservation of the natural resources involved. Improper management can result in loss of soil through increased rates of erosion and in population changes of plants making up the grazing resource.

The preceding examples illustrate the complexity of interactions among the resource base for animal feeds, the products resulting from those resources, and nutrition as a part of the animal's total environment and related management.

Factors Influencing the Composition or Nutritive Value of Feedstuffs

Type of Animal
The nutritive value of feedstuffs can vary significantly with the type of animal. Ruminants, because of their digestive tract capacity and the microbial population in the rumen, can utilize feeds high in cellulose, whereas swine, for example, do

not have this capability. Including large amounts of fiber in feeds for growing and finishing pigs results in reduced performance and higher than necessary costs because the production goal is generally to produce maximum growth. The young pig does not have the capacity to consume sufficient energy in the form of high-fiber diets to support the desired level of performance. Mature sows, on the other hand, because of the need to limit energy intake and control weight, are commonly fed higher-fiber feeds, and the total allowance of feed may be limited as well. Horses, with their large ceca containing microorganisms, make good utilization of high-fiber feeds such as pasture products and hay. Because the total capacity of the digestive tract is limited in horses, higher-quality roughages are normally supplied even to mature animals of this species, although the goal in dietary design is often basically one of maintaining body weight.

Stage of Plant Growth and Seasonal Influences

The most highly nutritious plant materials are those characterized by rapid growth and seed maturation. Increasing maturity of the nongrain portion of plants is associated with increasing levels of cellulose and lignin. For example, when grasses reach maturity, the nutrient value declines rapidly. Loss of leaves and weathering cause further reductions in nutritive value. Ruminants and horses may be able to subsist on such mature, weathered materials; however, supplemental protein, energy, and minerals are normally essential. There are some exceptions to the general rule that a protein supplement is always required for such animals grazing on dry grasses. Some species of grasses, including short grasses such as the gramas, have greater retention of crude protein after maturity. Tall grasses, such as the andropogons, lose protein rapidly after maturity but serve as the primary source of energy throughout the Midwest and Plains states for much of the nation's beef cow herd. Supplemental protein and mineral management is a major factor in wintering these millions of animals each year. The most common sources of such supplemental protein are cottonseed meal and soybean meal, with lesser amounts of other oilseed meals.

Climate-Soil-Plant Interaction

Soil deficiencies or excesses of minerals may influence the composition of feedstuffs. Deficiencies of phosphorus and excesses of selenium in soils are common examples. Climate may affect the nutritive value of plants as well as the total yield of the crop. For example, grains produced under drought conditions normally have higher than normal protein content resulting from a higher ratio of reproductive components of the grain to the starch content. Very-high-yielding pastures commonly produce a greater total weight gain in a group of animals expressed as gain per area of land, but individual performance may be reduced because of differences in plant nutrient composition.

Storage Conditions

Influences of storage conditions include losses of nutrients through chemical degradation, spoilage due to microbial damage, and development of toxic effects that may be associated with molds. Animals often reject feeds that have been stored for extended periods or those maintained in unsatisfactory conditions such as excessive levels of moisture that foster mold growth. Such reductions in intake obviously have negative effects on animal performance.

Processing Method
Prime examples of the effects of processing are grinding, rolling, pelleting, or flaking of grains to enhance utilization. Consumption of coarse hays may be improved by chopping to reduce particle size. In some cases it may be advantageous to mix grains and roughages in complete feeds to ensure more uniform consumption of nutrients from available feed components; grinding or chopping the roughage is essential for properly mixing the components.

Plant Population in Pasture and Harvested Forages
In continuous grazing systems animals graze selectively, which influences the balance of nutrient intake. When animals are managed under intensive, rotational grazing systems, they have less opportunity to choose plant species in pastures because the animals are forced to consume virtually all of the available plant material in an area. When the crop or range is harvested as hay or silage, the animals consuming the products have less opportunity to sort plant components because the plants are mixed in the process of harvesting and the animals often consume food more rapidly in competitive feeding environments.

Nonnutritive Materials
Nonnutritive feed additives may influence the consumption and digestibility of feedstuffs. Growth promotants may increase nutritive demand and the level of feed consumption. Other additions may have a primary influence on digestibility. Such additives may actually reduce intake since each unit of feedstuff consumed has a higher digestible nutrient value and is thus utilized to a higher degree.

Factors Influencing the Availability of Feed and the Animal's Access to Feed Resources

Pasture and Range Management
A large body of research literature and numerous texts relating to pasture and range management are available. The intent of this discussion is to outline some major factors for consideration in providing an adequate nutritional environment for animals maintained in systems heavily based on improved pastures or native pastures or on range lands.

Major considerations are the type and composition of plants making up the grazing resource, quantity of plant products available, and supplemental nutrients required. Because of variations in climate, plant populations, and animal needs, range management is a very dynamic enterprise. To casual observers it appears to be otherwise because a group of animals grazing on a good pasture is incorrectly assumed to be the epitome of low-technology agriculture. However, the range manager must make decisions about interactions among animal performance, improvement in both plant populations and quantity of the feed resource, sustainability of the soil and plant resources, and often multiple use by domestic animals and wildlife. These same factors are important in providing an adequate nutritional environment for grazing animals.

Improvement Practices   Improvement practices normally involve enhancement of both quality and quantity of the feed resource and considerations about the longer-term protection of the soil and plant resource base. In the more humid regions and in irrigated areas such improvements involve soil cultivation, fertilization, and reseeding to make a major change in the plant population. In more arid regions practices such as overseeding and careful animal management are the more common approaches.

Maintenance of the Grazing Resource   The management of pasture and range is directed toward conserving the soil resource and desirable plant populations and controlling the stage of growth to enhance nutrient value. Maintaining a desirable plant population and enhancing the nutrient value are largely matters of timing and the level of defoliation allowed. Protecting the soil requires attention to ground cover and plant population diversity for erosion control, compaction by animals, trail patterns, and maintenance of riparian zones.

Utilization Technologies and Animal Management Strategies   Extensive pasture-range systems and more intensive pasture systems in humid regions have historically involved continuous grazing of a given area for extended periods. In these systems herds of animals are provided rather large grazing areas to seek and consume plants at will. Because animal density is such that animals do not normally experience intense competition, selective grazing occurs in both selecting plants from the variety that may exist in the population and selecting parts of individual plants. In general, under this grazing environment, animals are expected to select the safest, most nutritious diet as long as the supply of plant materials is adequate. In this system the primary concern in providing a good nutritional environment is to ensure adequate plant growth and supplemental nutrients as influenced by the total feed supply available and seasonal or stage-of-growth influences.

More recently cattle and sheep producers have been utilizing systems involving intensive rotational grazing systems. In many cases such systems increase the carrying capacity of a given land area, resulting in greater production or output per unit of land. Individual animal performance may be lower, however. Intensive rotational systems typically involve a number of small pastures, and herds of animals are introduced to such fields in high density and for relatively short periods. In the process, the forage present in a field is consumed rapidly, competitively, and in many cases more completely than in extensive systems. In considering the management of such intensive systems to provide an adequate nutritional environment for animals, certain factors must be considered: short-term climatic effects such as these influence growth rate of the plants involved, the consequences of the animal's difficulty in effectively selecting different plants or different plant parts, and the proper time for introduction and removal of animals from a field. Thus, while the intensification of grazing may hold potential for significant increases in output from a land area, the management demands to ensure an adequate nutritional environment for the animal are increased dramatically.

Given the increased interest in intensive rotational grazing systems, it is appropriate to illustrate some of the decision-making processes involved to ensure an adequate nutritional environment for the animals. The first and clearly important assessment is stocking rate (the number of animals to be involved in the rotational grazing program for a given area of pasture). A contingency feed resource must be available if the stocking rate is overestimated. The next decision-making criteria are the stage of growth of the forage material at which grazing of the particular field will be initiated and the length of time required for that field to regrow to an appropriate level following grazing. For example, assume a herd of fifty beef cows will be grazing 80 acres of improved grass pasture during a growing period from May 10 to October 15. Eight 10-acre pastures will be provided. If the system utilizes a 4-day grazing rotation period, a 28-day rest or regrowth period (calculated: [4 (8 − 1)]) will be allowed for each pasture as the rotation progresses. Stage of the growing season will influence the regrowth time required, as will climate. Thus, the length of grazing period may be altered by season. Drought conditions or excessive rain and muddy conditions will disturb the pattern and require alternate grazing areas or supplemental feeding. Variations of this system could be illustrated, but the provided information is sufficient to demonstrate that highly intensive grazing systems require much more frequent and detailed management decisions. If the management function does not follow the sophistication of the system, then the animal's environment may be inadequate and well-being compromised. Table 7.12 summarizes management issues related to an intensive grazing plan for an example herd of beef cows with calves (Russell, 1997). This example illustrates clearly that intensive grazing requires refined management and planning to ensure adequate dietary resources.

Physical Constraints

Facilities   Adequate facilities are obviously important in intensive livestock systems. In extensive systems such as ranching, cross-fencing to control grazing and access to watering facilities, as examples, may be important in devising the most adequate environment from a given set of resources.

Distances   Animals that must move excessive distances for feed, water, or shelter are being maintained in an inadequate environment.

Stress-Related Barriers   Management to control or at least mitigate factors such as aggression, dominance, and other social influences is important. Climatic influences such as freezing of the water supply and severe thermal conditions that prevent proper feeding are examples.

TABLE 7.12 **Management Plan for Intensive Grazing for a Herd of Eighty Spring-Calving Beef Cows Nursing Calves in The Upper Midwestern United States**

Season	Best Management	Risks	Risk Management
Mid-spring	Fertilize grass pastures with nitrogen[a]; initiate grazing when forage is 4 inches tall; rotate animals to new paddocks every 1 to 2 days to synchronize with rapidly growing forage	Poor forage growth caused by cool temperatures	Graze an additional pasture of stockpiled perennial forage or a cold-tolerant small grain species; feed stored feeds
		Poor forage growth and soil compaction caused by grazing under muddy conditions	Graze an additional sacrifice pasture that may be renovated easily
Late spring	Rotate animals to new paddocks every 1 to 2 days until grasses begin to show seed heads	Poor forage quality caused by excess forage growth	Remove excess forage as hay from one-third to one-half of the total pasture acres; graze extra animals such as fall calves or replacement heifers with cows to remove excess forage
Summer	Rotate animals to new paddocks when half of the forage is removed to synchronize rotation with slow-growing forage and reduce grazing selectivity by cows; there should be an adequate number of paddocks to assure a minimum rest period of 30 days between grazing episodes[b]	Poor forage growth of cool-season species in midsummer heat	Incorporate legume forage species into pasture with cool-season grass species; seed warm-season grass species into several paddocks for midsummer grazing
		Poor summer forage growth caused by drought	Incorporate drought-tolerant legume forage species into pastures with cool-season grasses; remove extra animals from the pasture; creep feed calves; early wean calves; feed stored feed to cows in one paddock of the pasture

TABLE 7.12   (Continued)

Season	Best Management	Risks	Risk Management
Early fall	Rotate animals to a new paddock every 1 to 2 days to synchronize rotation with rapid growth of cool-season species	Reduced persistence of legume forage species	Move cows to paddocks containing only grass and allow legume pastures to rest for at least 30 days before a killing frost
		Poor forage growth caused by overstocking	Wean calves; graze forage stockpiled on hay fields; graze corn crop residues; feed stored feeds

[a]Application of nitrogen fertilizer to mixed grass-legume pastures will reduce the persistence of desirable legume forage species
[b]Increasing the number of paddocks to eight significantly increases the length of the pasture rest interval; increasing the number of paddocks beyond eight has little effect on length of pasture rest period but increases the efficiency of forage utilization
Source: Courtesy of J. R. Russell, Iowa State University

Contingency Management of Forage Supply

Major contingencies in caring for animals involve shortage of feed resources, toxic ingredients in feed or water, power failures, and disease. These potential problems must be considered in any plan to ensure proper management and animal well-being. Factors such as reserve feed supplies, feed quality and safety control, biological security, and alternative power sources are essential in developing strategies for assuring a proper feed supply.

CHAPTER 8

Designing the Social Environment of Animals

The most controversial area in considering basic factors for planning animal environments is likely that relating to design of the social environment. The fact that animals exhibit social characteristics is not debatable. Anyone having extended contact with either domestic or wild animals is very much aware of such social orientation. Most farm animals are referred to as contact or social species. In general, these species are docile, easily trained, and socialize with humans, conspecifics, and, in some cases, members of other non-human species. These animals obviously demonstrate characteristics that rendered them adaptable to domestication. Species that for the most part remain wild reflect spatial requirements, territorial protection, and other characteristics that do not favor domestication. Regardless of these differences, both domesticated and wild animals have a great number of social characteristics that have been researched to varying degrees.

In reviewing some basic factors that should be recognized in the design of the social environment, one is challenged first as to how a discussion of these factors should be organized and second to ensure objectivity, thus avoiding anthropomorphism. One's objectivity may also be influenced by biases related to the acceptance of practices of long-standing use in the industry and those that are primarily economic considerations. Potential problems associated with objectivity extend to the public policy arena, which should prompt a serious effort to encourage a continuing dialogue among all parties interested in animal well-being to more clearly understand its basis in science. In addition, such concerns about objectivity should encourage both the public and the scientific community to place higher priority on research to develop a more extensive knowledge base for well-informed public policy decisions. This improved information base will also assist animal managers in making better decisions about animal care, including those relative to social needs.

Several aberrant behaviors are associated with inadequacies in animals' social environment. These same behaviors have characteristics that humans readily empathize with and, in many cases deplore. As a result, the level of concern on the part of the public and the scientific community in matters relating to animal care is increasing. Unfortunately, neither group is assigning sufficient priority to the issue in terms of funding and scientific effort, given the importance of the interdependence between animals and humans. Insufficient objective knowledge relative to social issues in animal well-being poses a threat that policies may have negative influences on this important interdependence. In this regard, it is useful to remind ourselves of both historical and current symbiotic relationships. Imagine a world without pets or zoos. Imagine a world without animals to convert materials that are not digestible or otherwise useful to humans into foods of the highest biological food value. Imagine a world of farming without animals to utilize rotational forage crops as an integral part of soil-conservation practices; a truly sustainable, renewable agriculture is dependent on soil-plant-animal relationships. Imagine a world without people and animals working together as teams in resource management and use, sport, and entertainment. The list of such relationships could be greatly expanded; however, these examples are enough to remind us of critical relationships in great variety. Recognition of this interdependence makes it imperative that we have the best possible scientific information for future decisions on which to base the continuation of this critical symbiotic relationship. Thus, the practices used in animal agriculture must not only meet the animals' needs, they must comply with the social ethic relative to animal care. Giving full attention to animals' requirements will ensure that the human-animal partnership continues.

In terms of organization of this discussion of the social environment, a reasonable approach is to consider relevant topics in two classes: The first involves factors that might be related to a change in the animal's condition as these relate to social relationships. The second is the effect of depriving the animal of certain activities that may directly or indirectly impact its social interactions.

ALTERED CONDITION THAT MAY INFLUENCE SOCIAL RELATIONSHIPS OF ANIMALS

Genetic Selection

Livestock producers historically have indicated that certain animals or lines within a herd are more docile and as a result are more easily managed. Faure and Mills (1998) reviewed research with Japanese quail selected for either low fear or high social reinstatement. These authors show that selection for or against fear and sociality is possible using simple measures and that animals from each divergent line were significantly different in their responses in less than twenty generations. Selection against fear responses and certain social characteristics will have obvious implications to animal welfare. However, as Faure and Mills suggest, selection to remove any fear response is highly unlikely as this is such a critical motivation in species survival. Dickson et al. (1970) reported heritability estimates ranging from 47 to 53 percent for handling ease of cows during milking. The fact remains that many breeders of livestock are convinced that selection for disposition is important. If they are correct, then the genetics of individuals can influence social stability within a group of animals and is a consideration in environmental design.

Removal of Horns

Elimination of horns (either by physical removal or by selection for the polled gene) has certain positive effects, such as reducing damage by bruising and the possibility of more serious injury resulting from goring of other animals. Questions regarding the impact on social characteristics revolve around whether this procedure alters an animal's dominance characteristics. There is little doubt that animals with horns have an advantage in terms of dominance. Animals experiencing loss of dominant position are negatively affected, at least temporarily. In some cases the effect may be one of voluntary separation from the group because of harsh treatment imposed by other members. Loss of hierarchical position because of dehorning is not likely to occur if all animals in a group are dehorned. Thus, in terms of members of such groups, no negative social impacts are likely to result from the management practice. Such animals substitute other approaches to establishing a social order. Evidence suggests that dominant animals express this characteristic in more than one trait. For example, animals that dominate in feeding behavior also dominate in sexual behavior (Fowler and Jenkins, 1976). In certain conditions, producers feel that the presence of horns reduces herd losses to predators. In most livestock production systems, however, the benefits derived from dehorning outweigh possible negative effects.

Castration

Castration clearly influences the physical characteristics of animals, largely eliminates their ability to copulate, and prevents them from reproducing. Castrated (spayed) heifers do not exhibit the high level of reproductive activity in the feedlot associated with estrus. (Spaying is not, however, a common practice in beef production currently.) Steers are less aggressive than bulls, and therefore groupings are more stable in terms of level of aggression. However, bulls gain more rapidly and more efficiently than steers and produce leaner beef. If bulls are marketed before attaining 14 to 18 months of age, the eating quality of beef is very comparable to that of steers. If pigs are left as intact males beyond approximately 120 days of age or 220 lb, the pork is likely to have objectionable odor. This negative effect eliminates some of the flexibility in management decisions for the pork producer as opposed to the beef producer.

Questions arise as to possible negative effects of castration on animal well-being. From a physical standpoint, castration, if properly accomplished, has largely positive effects in that an improved group environment results from reduced aggressiveness and related competition, fighting, and excessive physical activity. Both animals and the enterprise appear to benefit. The question then is related to the impact of castration on the social well-being of the animal. The animal is clearly deprived of reproductive capability and activity to a large extent, although some attempts at mating behavior may still occur. There is no evidence that castration has longer-term negative effects on animals since it results in no particular aberrant behaviors. The lack of such behaviors suggests that significant stress of a psychological character probably does not result from the practice. Altered sexual behavior is, of course, characteristic. Interestingly, psychological castration of individuals may occur as a result of dominance by another male (Houpt, 1991). However, this condition is reversed shortly upon removal of the dominant animal. This phenomenon suggests that dominance may dictate the design of the environment in terms of breeding management if it is desirable to balance the number of offspring by sires used in cow herds with multiple bulls.

Hormonal Treatment

The use of various preparations with endocrine-like influences is common in the livestock industry. These preparations may be used to stimulate growth, alter product composition, enhance milk production, enhance feed utilization, synchronize estrus, or control parturition. In many cases, similar products are used in human medicine for basically the same purposes. A large body of data covers the biological and performance effects of such materials. Such treatments do not appear to be associated with any negative psychological effects. Outside of the altered reproductive activity associated with specific compounds, some implants have been shown to increase agonistic behavior in groups of feedlot steers. Effects include a higher level of aggression and a higher rate of success by dominant animals in such encounters. However, such effects are variable based on the type of implant used (Stricklen et al., 1979).

Tail Clipping or Docking

Cannibalism can be appropriately considered a social behavior problem. It is generally thought to be associated with more intensive, densely populated, crowded housing conditions. Thus, the routine practice of docking the tails of pigs is considered a positive influence on behavioral problems because it largely eliminates injury and possible death loss associated with the vice. Tails of lambs and some dogs are routinely docked. Negative effects on the animals so treated are not apparent unless infection occurs as a result of the procedure. The risk of infection underscores the importance of proper procedures in the process.

Tail docking, common in some dairy operations, is done to eliminate problems associated with collection of matted dirt and manure in the tail switch. Views as to merits and effects in terms of animal well-being are mixed. If the procedure is properly accomplished without subsequent infection, the practice is not likely to interfere with the animal's general well-being.

Shearing

Removing the wool from sheep or hair from other animals clearly alters the animal's condition temporarily. Positive and negative biological effects are related primarily to adaptability to the thermal environment. There is no evidence of social impact on adults (Lynch et al., 1992). Shearing ewes with lambs may temporarily cause problems in lambs identifying their dam, however. Removal of the wool from the ventral regions of the preparturient ewe may be an important factor in preventing wool pulling by young lambs, which appear to be attracted to soiled wool in these areas. The practice may also prevent matting of wool with manure, which can result in parasite infestation in the area and possible infection.

Cosmetic Surgery

Cosmetic surgery such as ear cropping in dogs, tail cropping in horses, and similar practices to alter appearance do not appear to have social influences on the animals involved based on common knowledge. There is little research in this area on which to base objective conclusions, however.

Beak Trimming

Removal of a portion of the beak of laying and broiler chickens, turkeys, and ducks is a common practice to minimize damage to birds from aggressive attacks occurring in normal confined production systems. Evidence suggests that the procedure is painful and that the beak may remain tender for an extended period (Breward, 1984; Duncan et al., 1989; Lee and Craig, 1991). As a result, efforts have been made to determine whether genetic lines of birds can be developed that are less aggressive and therefore less likely to damage cage mates. Such an undertaking is possible and has been demonstrated by Craig and Muir (1993, 1996) and Muir (1996). Improved equipment utilizing laser technology is available and presently used widely in the industry. Good environmental management, including light level control, minimizes the problem of aggressive pecking. Irritating environmental deficiencies such as excessive heat, excessive light, high dust level, and crowding should be considered as possible factors contributing to this behavior. Improved conditions minimize the need for beak trimming.

Toe Trimming

Toe trimming or toe removal is performed in male broiler breeders and turkeys to minimize injury in birds engaging in agonistic behavior and to lessen tissue damage to females during mating in natural service. Toe removal is used widely in commercial turkey production to reduce injury associated with aggressive behavior. The practice is generally considered to be sufficiently important in minimizing injury to justify the procedure. Such damage is likely to be reflected in reduced carcass grade in market birds, although variable results have been reported (Owings et al., 1972; Proudfoot et al., 1979; Moran, 1985). The recommended practice is to remove the first joint of one or more toes at 1 day of age. Modern technology involves an instrument utilizing microwave energy for simultaneous removal and cauterization of the area.

DEPRIVATION OF ACTIVITIES THAT MAY INFLUENCE SOCIAL RELATIONSHIPS OF ANIMALS

Many environments deprive animals of some particular element. Though an environmental component may be desirable, it does not necessarily follow that its absence results in a stressful condition. In some cases animals adapt readily to elements of the environment that initially may be considered stressors. In other cases the length of time an animal can resist stressors may be limited and the ability to withstand the effects thus exhausted. This condition will eventually be reflected in one or more stress effects impairing an animal's health and well-being. Socially oriented stressors are often in the form of some denial or lack of the existence of one or more elements in the environment that may be important to expression of normal social characteristics. As a result, psychological effects that may result in characteristic behavioral responses occur. Even so, such responses may not have a negative influence on other animals, caretakers, or economic factors involved in maintaining animals. The purpose of this discussion is to assess impacts that may be involved in common situations in which the animal's environment limits socialization or the expression of certain social characteristics. It is often assumed that factors such as isolation, separation, limited space, denial of reproductive behavior, and restricted movement are psychological stressors. Such assumptions are generalizations that may or may not be true in all situations. There is, however, some objective information relating to each of the mentioned environmental influences.

Isolation and Physical Separation

Animals maintained in isolation are much more likely to develop forms of abnormal behavior referred to as stereotypies. Some of these behaviors are injurious to the animal, and others are more likely to damage equipment. In such cases management decisions usually come down on the side of providing an environment to minimize both animal suffering and economic loss. In other cases, such as route pacing, tongue rolling, and bar chewing, there may be little physical impact on animals or facilities. Such behaviors may be annoying and suggest that different management strategies might be desirable, but they are not necessarily detrimental to animal performance. Rearing animals in isolation may also result in abnormal behaviors or failure of certain functions when such animals encounter other animals later. Failure of males in terms of reproductive competence, for example, can have serious economic effects on the success of animal enterprises.

Sows maintained in gestation stalls, though not isolated from visual contact, are separated from conspecifics and restricted in movement by space and stall configuration. Research with sows maintained in this way suggests that access to straw in the stall tends to reduce stereotyped activity. This reduction in stereotyped behavior is apparently due to the sow manipulating the straw. However, stereotypies related to oral/nasal/facial behavior do occur when sows are maintained on soil or on pasture (Dailey and McGlone, 1997). Studies have been conducted to assess the comparative reproductive performance of sows maintained in stalls and those maintained in groups. This body of research is inconclusive to date (Barnett and Hemsworth, 1991). Comparisons of sows maintained with neck or girth tethers with those maintained in groups indicate negative effects of tethering (McGlone et al., 1994; Janssens et al., 1994). Although stalled sows tend to show more stereotyped activity than group-housed sows, more fighting and sow injury occur in the group environment. Good group management may reduce the incidence of fighting by incorporating practices such as avoiding the introduction of strange animals and providing a good feeding environment to minimize competition. Earlier discussion suggested that limited feeding of sows during gestation was essential to maintaining good physical condition of the sow and enterprise economy. This practice results in a very strong competitive environment at feeding time and an unacceptable level of fighting and injury in many cases when sows are maintained in groups.

Cattle maintained in isolation or in groups in close confinement frequently chew stalls and equipment. Veal calves are normally maintained in individual stalls for half to all of the typical 16-week feeding period. The occurrence of stereotyped activity is common among such calves, even though pens are usually designed to provide visual and physical contact. Some of these behaviors develop as replacement behaviors for nursing, since the calves are not normally allowed to nurse their dams for significant periods after birth and feeding may be by drinking rather than by artificial nipple systems. Even in nipple feeding the time involved in consumption does not appear to satisfy the nursing time desired by calves. To investigate the stress imposed on veal calves by various housing systems, Friend et al. (1985) studied veal calves housed in either yards, hutches, pens or stalls. They found that increasing degrees of confinement increased triiodothyronine, thyroxine, and adrenal responsiveness of veal calves. Thus, the degree of confinement does appear to affect the physiology of veal calves indicating that they are experiencing chronic stress in these systems.

Veal calves may be tethered in individual stalls. Producers feel that individual penning of dairy calves, because of the very early weaning, provides an improved environment for starting calves because of better health and nutritional management. Although veal calves are commonly tethered and reared in stalls, replacement female calves are now typically reared in hutches inside small pens. Some veal production systems involve starting the calves in individual stalls the first half of the total production period and then penning in groups from about 8 to 16 weeks. Any advantage of this system is not well established by research evaluating calf performance. Objective data reflecting influences of grouping for the last half of the period on stereotypic activity is unfortunately limited. Thus, the controversy about the well-being of veal calves produced in these systems is likely to continue until such systems are evaluated more thoroughly by objective measures of stress and psychological well-being.

Isolation during the development of young males has been shown to influence adequacy of subsequent sexual performance. Such failure may be related to the absence of early sexual-related experience. Very young animals engage in mounting activity that is presumed to influence, at least in some individuals, future sexual competence. Such activity occurs far in advance of puberty in many species. Pigs exhibit this activity before 2 months of age (Fraser and Broom, 1990); Andrew (1966) reported that 20 to 30 percent of test chicks demonstrated mounting behavior by 16 days of age. Rearing bulls with others of the species, either male or female, appears to correct the problem of sexual incompetence related to rearing in isolation. The ram's sexual capability, on the other hand, is enhanced by rearing with females. Such capability may be impaired if, during adolescent rearing, contact is with males only (Orgeur and Signoret, 1984; Price, 1984). Subsequent sexual competence of young boars may be enhanced by rearing with conspecifics (Hemsworth et al., 1977).

It is not clear whether isolation of young males or rearing in a unisex environment influences the occurrence of homosexual tendencies. However, Price and Smith (1984) and Zenchak and Anderson (1980) reported that there may be a tendency for young males to establish early relationships with other males in a unisex environment that may have, at least temporarily, a negative impact on their association with females later.

Isolation eliminates many learning experiences that result from interactions with other animals. Development of sexual abilities and interests, discussed earlier, is but one example. Other deficiencies may result from effects on exploration and transfer of information among animals that results from play, foraging, establishment of social order, and so forth. Some evidence suggests that lambs reared in isolation are less exploratory and less likely to interact normally with other animals when later contact is made (Zito et al., 1977). Pigs typically show a very high rate of exploration that, if limited, may result in abnormal behavior.

Separation of animals prevents mutual grooming behavior and may result in some replacement activity. Such activities as hair pulling and excessive licking may be associated effects.

Horses demonstrate many undesirable behaviors related to isolation and inactivity. Some of these activities, such as rubbing, stall kicking, and stall chewing, may result in injury and damage. The only consistent recommendation in the literature related to minimizing these activities some combination of allowing more space, exercise, and social interaction. Thus, environmental design should accommodate strategies for this approach.

Space Limitations and Crowding

The concept of personal space is important to animals. In crowded conditions animals attempt to preserve some distance. If such separation is not possible, an effort is made to maintain space around the head. Fraser and Broom (1990) concluded that this effect may not be seen during short periods of crowding such as during driving or working activity, but over the longer term crowded animals seek a head space of about 1 m in radius. They also concluded that animals prefer square corners and square pens to those of round configuration. This preference is presumed to be related to the perception of greater space in configurations with corners. However, since aggression may be more damaging in corners, facility design characteristics should include configurations to prevent animals from being trapped in sharp corners by dominant pen mates. Such observations point to the importance of space considerations in environmental design that recognize both quantitative and qualitative aspects of the space occupied.

The physical effects of crowding animals in a group environment may be associated with limited access to feed, water, adequate air movement, and resting space. Social effects may relate to aggressive tendencies and severe impacts resulting from competitive relationships. These effects can be of a magnitude to have negative influences on animal performance, well-being, and economics. Eubank and Bryant (1972) point out that animals in confinement may not be able bring closure to agonistic encounters. Koolhaas et al. (1997) suggest that because of homeostatic motivation for aggressive behavior directed to control of resources the negative effects of this behavior may be greatest in confined systems. This may result in dominant individuals expending more energy in maintaining dominance than in systems where the submissive animal can practice avoidance more effectively. Sows in gestation stalls, as an example of secondary effects, may sit down rather than lie down in normal fashion. This positioning may be due primarily to the space allowance limiting the ability to negotiate the normal up-and-down movements required for the process. Negative health impacts such as urinary tract infections may be associated with the behavior.

Space limitations may hinder normal excretory behavior. For example, in many swine production environments, it is important to train animals to dung in certain areas of the pen for sanitation and ease of facility cleaning. Crowding may preclude the ability of pigs to perform this otherwise rather normal and routine practice.

Stereotyped behaviors associated with housing of birds in small cages and housing of zoo animals serve as examples of those that are likely related to both restrictions in the animal's movement and limited interactions with conspecifics species. In practical conditions, confounding of effects of restricted space and limited interaction of a social nature is common.

Appleby (1997) suggests that space limitations may have negative influences on the effectiveness of specific stimuli such as display behaviors as well as behaviors of a more general nature.

Finally, in any discussion of restricted movement, confined housing, penning, and stalling, one is confronted with questions about the importance of physical exercise to an animal's well-being. Unfortunately, no clear-cut answer exists. In a good nutritional environment, adult domestic animals normally spend 60 to 90 percent of their time resting and engage in very little vigorous exercise voluntarily. In few cases could voluntary exercise observed in animals be termed of an aerobic nature. Conditioning horses and dogs for racing and demanding hunting activities requires a forced exercise regimen. Animals normally voluntarily accomplish short-term exercise, stretching, and other muscular activity. Those maintained in small pens usually run, jump, or show some signs of animated movement upon release. In most cases this level of activity would be described as very limited. It cannot be concluded, however, that this kind of activity is necessarily of significance to the animal's well-being. On the other hand, there appears to be very little objective data that would establish just how important this level of activity is to the well-being of farm animals or pets. Numerous stereotypic activities characteristic of confined animals are prevented or reduced by allowing animals a reasonable level of activity outside of small stalls or crates. Thus it is reasonable to conclude that the level of well-being of at least some animals is enhanced by such activity or the opportunity to interact with a more enriched environment.

Restrictions in Grouping Such as Herding or Flocking

Species and individuals differ in terms of their responses to factors involved in grouping. Movement of strange animals into a group results in disruption of the established social order and associated agonistic encounters to reestablish such an order. Species have characteristic requirements to maintain some distance from others. Some species have strong flocking instincts, and individual animals may be extremely uncomfortable in being separated from the group. In some cases this instinct is so strong that animals will virtually risk life and limb to return to the group. Other animals are comfortable being distant from their group but have a fairly definite tolerance in the magnitude of that distance. Such animals will make every effort to maintain this acceptable distance. Most animals seek isolation during parturition, even though they have strong herding tendencies at other times. Another important characteristic related to grouping is group size and the related ability of animals to maintain identity of their group mates. Beyond this size, more internal aggressive conflict is exhibited within the group. This increase results from the fact that members of the group cannot identify all individuals in the group and thus confront strangers more frequently apparently because the group size has exceeded the ability of the animals to make such identity. The usual efforts to establish dominance relationships follow these encounters. Thus, grouping and group size are important characteristics of animals and must be recognized in developing an appropriate environment.

Restricted or Absence of Reproductive Behavior

Denial of social activities during rearing, including play that mimics sexual activity, may influence the future sexual competence of males. Price et al. (1991) reported the importance of the young ram's association with ewes in the development of mating behavior. The importance of rearing males with conspecifics is discussed in the section on isolation effects. The significance of early association with females during rearing is especially important in rams but less so with other species of farm animals. The importance of rearing females with those of the opposite sex is not as important as that for males.

In adult sheep, association with males before the breeding season is a critical management practice to hasten the onset of estrus and it also tends to synchronize estrus in a group of ewes. The presence of a boar or the odor of a boar just before breeding is an important influence in the breeding activity of sows. Thus, denial of such influence may have negative economic impacts through lowered reproductive efficiency. The mentioned effects demonstrate the great importance of considering an appropriate social environment to foster normal reproductive activity in livestock production.

Denial of sexual behavior through castration or separation of the sexes does not have negative influences on animals maintained for high levels of productivity. Castrated males, however, typically gain at slower rates than intact male conspecifics. The differential performance related to sex condition is clearly associated with the influence of testosterone in the intact males.

The question as to the importance of actual reproductive experience, producing young, and mothering in terms of well-being arises frequently, especially in the case of pet management. Evidence that such denial is an important issue in the psychological well-being of farm animals is limited.

General Considerations in Designing the Social Environment

Ample evidence suggests that socially oriented needs of animals exist and should be prominent considerations in planning the total environment provided for animals.

An enormous body of information provides an understanding of many aspects of animal behavior. Behaviors are carefully classified and described in a generally consistent manner. That body of work often describes some behavior as normal and some as abnormal. One could take the view that all behaviors are normal because they are typical results of stimuli or lack of stimuli, and behavioral scientists have accomplished much in associating behaviors with rather specific environmental stimuli including those that influence socially oriented behaviors. Animals vary greatly in their response to the presence or absence of such stimuli. Thus, all animals in a group may not show so-called abnormal behavior. This variation in response prompts efforts toward genetic selection for environmental compatibility. Genetic selection is an important tool in this regard, however, timely considerations in designing environments for current use must rely on existing information and sensitivity on the part of managers to ensure animal well-being.

In considering the social environment for animals, one can become entangled in a variety of points of view: What are needs? What are real or basic needs? What minimum needs must be met? At what level must they be met? Questions also arise as to the ethics of many dimensions of animal ownership, management, and control. Pet owners, livestock producers, zoo managers, and others are influenced by these views and consider them as they plan for the animals under their care. In the final analysis, however, they must decide how they will provide for the animals. They have to consider how to meet physical needs as facilities and equipment are altered, replaced, or built. Objective information on which to base decisions about the physical environment is readily available. The same is true in terms of providing nutritional resources. Such information, being constantly expanded and refined, provides the basis for improving environments.

In considering the social environment, one is challenged to find the best approach, but we can establish a reasonable goal for such a consideration and systematically approach that goal. The goal should be to enhance the quality of the animal's social environment by considering avenues to minimize abnormal behavior. The level of understanding of many such behaviors and probable cause and effect issues provided currently by behavioral science is both impressive and utilitarian. In many cases answers about prevention and treatment are known. Thus, a review of these issues and a thorough understanding of them will undoubtedly increase the sensitivity of animal owners and managers to these issues and in turn lead to a greater effort to provide an enhanced social environment. Here, again, as facilities and systems are planned, such issues will take higher priority. As a result, our responsibility in ensuring the proper care and well-being of animals will be discharged more effectively.

A number of aberrant behaviors related directly or indirectly to the animal's social environment are discussed in detail in part 3. The section dealing with cause and effect relationships in aberrant behavior provides a description of such behaviors and possible corrective environmental changes. In considering the desirability of and necessity for such management changes, one must again remember the philosophy that the occurrence of aberrant behavior suggests that the social or physical environment is inadequate.

GLOSSARY

Aberrant behavior   Deviation from the usual.

Acclimation   Becoming adjusted to changes in environmental characteristics.

Acclimatization   Physiological adjustment by an organism to environmental change. Usually used to refer to adjustments to changes in climate. Changes may occur in the subcutaneous vascular anatomy of an animal, which influences the effectiveness of insulation provided by peripheral tissue.

Acetylcholine   A neurotransmitter in motor neurons and those of the parasympathetic nervous system. It also serves as the neurotransmitter in preganglionic nerve cells, which connect with postganglionic nerve cells innervating the sympathetic nervous system through the release of norepinephrine by the postganglionic fiber.

Acquired characteristics   Characteristics developed as a result of environmental influences. One of two influences on an individual's characteristics, the other being genetic. Phenotypic expression is a reflection of inherited and environmental (acquired) influences combined.

ACTH   *See* adrenocorticotropic hormone.

Action pattern   A behavioral sequence by an animal that may be related genetically to the species. Other action patterns may be due to environmental factors, leading to a repetitive characteristic of a behavioral scenario. Fixed-action patterns are common to a species and may be characteristic when first performed by the individual.

Acute stress   Short-term stress that results from a sudden environmental change of short duration. The same stressor, if continued, may result in chronic or longer-term stress.

Adaptation   Adjustment to environmental conditions by modification of an organism. Species may show adaptation through natural genetic selection. Artificial selection can also aid in adapting a population to a given environment. Individual animals may accommodate short-term environmental influences, which may also aid in coping and potentially in their survival. Traits that aid in an animal's fitness are described as having adaptive benefits.

Ad libitum   A term used to describe free and unrestricted access of animals to some resource, often feed and water.

Adrenal cortex   The outer layer of the adrenal gland that is the source of glucocorticoids and mineralocorticoids. This tissue surrounds the adrenal medulla or the core of the gland.

Adrenal gland   An endocrine gland heavily involved in stress responses. The adrenal cortex produces adrenocorticoids; the adrenal medulla produces epinephrine and norepinephrine. Anatomical location is near the kidneys.

Adrenaline   A hormone of the adrenal medulla, also called epinephrine.

Adrenal medulla   The core tissue of the adrenal gland giving rise to epinephrine.

Adrenergic   A term used to refer to the adrenal medulla's effect in the function of the sympathetic nervous system. Epinephrine, for example, has adrenergic effects.

Adrenergic system   The stress pathway involving the hypothalamus, the sympathetic nervous system, and the adrenal medulla.

Adrenocortical   A term used to refer to the action and/or products of the adrenal cortex.

Adrenocorticotropic hormone (ACTH)   The hormone released by the anterior pituitary in response to corticotropic-releasing hormone (CRH) and arginine vasopressin (AVP), which stimulates the adrenal cortex to release corticosteroids.

Adrenomedullary   A term used to refer to actions and/or products of the adrenal medulla.

Aerophagia   Abnormal activity characterized by exaggerated sucking in, swallowing, and discharging of air. An undesirable behavior most common in horses.

Afferent neuron   A nerve fiber conducting impulses to the central nervous system. Afferent nerve fibers transmit impulses from the periphery to the spinal cord, which then conducts the impulses along afferent spinal tracts to the brain. Compare with efferent neuron.

Aggregation (behavioral)   Grouping of animals that assemble because of some common goal or resource access. Examples are watering and feeding.

Aggregation (neural)   The collection of neural information by components of the brain. The hypothalamus is an example of a body in the central nervous system that performs aggregation of information and coordination of responses.

Aggression   Behavior addressed to another animal or to humans that is intended to harm or restrict the subject in some way. Controlling resources, defending a territory, restricting access to mates, and dams protecting young are examples.

Agonist   A factor that enhances the effect of another factor. Its opposite is an antagonist, or a factor that limits or reduces the effects of another factor.

Agonistic behavior   Aggressive and submissive behaviors demonstrated toward another individual. Such action occurs normally in the establishment of dominance in which both aggressive and submissive behaviors serve the development and maintenance of the social order.

Allogrooming   Grooming activity directed toward another animal. The mutual grooming observed in horses is an example.

Allomimetic behavior   Following or mimicking are forms. Other forms include cooperation between individuals. Mimetic refers to relationship.

Alpha (male or female)   The dominant animal in a group; the animal that is first in the social order.

Altricial   Relative immaturity at birth. Such animals require careful and extended care by parents or caretakers. Human young are described as altricial because much nurturing is required early in the infant's life. Dogs and cats are classed as altricial, whereas farm animals such as horses, cattle, and sheep are classified as precocial because of their more advanced development at birth.

Amino acid   An organic acid containing an amino group. Compounds that are synthesized into combinations to form proteins or parts of proteins (e.g., peptides).

Amygdala   A neural center of the limbic system within the brain that is involved in control of at least some behaviors of a social nature. Some control of aggressive behavior is associated with this structure.

Amygdala control   Refers to neural control of certain types of behavior by amygdaloid neuclei.

Anomalous behavior   Exaggerated levels of response to stimuli. Tonic immobility and lethargy are examples. Behaviors that may appear to be similar to a normal activity but are performed excessively may be referred to as anomalous.

Anorexia   Behavior reflecting lack of or severe restriction in food intake.

Anosmia   Condition of absent or ineffective olfactory sense.

Anthropomorphism   Ascribing human traits to animals. Attempts to interpret animal behavior in the context of the human experience. Examples: The ram acts proudly. The pig acts happy. (These are interpretations in the human context and do not necessarily reflect the animal's viewpoint.)

Auditory   Related to the sense of hearing. Describes signals or stimuli transmitted by sound.

Autonomic nervous system   The nervous system component consisting of sympathetic (thoracolumbar) and parasympathetic (craniosacral) divisions. The sympathetic component is typically involved with stimulation of activity, whereas the parasympathetic component is typically more involved with conservation and restorative processes within the body, although the effects can be opposite to this depending on receptors at the effector site. Compare with somatic nervous system.

Aversion   Characteristic of animals attempting to avoid contact with other animals, humans, and, in some cases, physical or chemical factors.

Avoidance system   A developed behavioral system involving an aversion response.

Axon   The nerve fiber of the neuron (nerve cell) that transmits the neural impulse to its synapse with the next neuron in a series or with receptors in effector tissues.

Bar biting   An aberrant behavior observed in sows characterized by mouthing and biting bars or other structural members of gestation stalls, for example. Generally associated with restraint, lack of freedom of movement, and barren environments (lack of complexity) such as stall confinement. Bar biting may also occur in other farm animal species. Cribbing in horses may result from similar environmental conditions, but the form of biting is usually with the front teeth without taking the bar or other object into the mouth proper.

Baroreceptor   Neural receptors sensing changes in pressure within body tissues.

Biological efficiency   An expression of the relative use and conservation of resources in biological systems. A higher reproduction rate, for example, could enhance biological efficiency by a greater output per breeding animal maintained.

Biostimulation   Stimulation of animals by associated individuals.

Bonding   Formation of a close and extended relationship between individuals. Early establishment of parent-offspring relationships is critical to survival of young.

Bovine   Cattle or closely related animal.

Bovine somatotropin (BST)   Bovine growth hormone.

Brainstem   The portion of the brain linking the spinal cord to the cerebrum and cerebellum. It is made up of the medulla, pons, and midbrain along with numerous nuclei and neural tracts.

Bronchi   Large air passages of the lungs. Sometimes referred to as the windpipe. Smaller bronchus passages connecting to various parts of the lungs have specific nomenclature (e.g., primary and segmental).

Broodiness   Maternal behavior in hens related to hatching eggs and caring for offspring.

Cardiac insufficiency   Below-normal performance of the heart due to reduced rate or pumping pressure and volume.

Causal factor   Any factor leading to development, alteration, or execution of a behavior or reaction.

Central nervous system   The combination of brain, its various components, and the spinal cord.

Cerebellum   Functionally defined as a coordination center for activity of the central nervous system as it is involved with motor functions throughout the body. A component of the hindbrain.

Cerebral cortex   The outer tissue layer of the cerebrum (two hemispheres) or forebrain. It serves in integration and analysis of information and decisions as to actions.

Cerebrum   The portion of the brain made up of the two cerebral hemispheres connected by the corpus callosum. The cerebral cortex is the outer part of this component of the brain. Higher order of thought processes are believed to occur here.

Chemical messenger   Substance, produced by a cell, that influences the activity of another cell. Various hormones and neurotransmitters are examples.

Cholinergic   Descriptive of effect by action of acetylcholine as a neurotransmitter.

Chronic stress   Stress resulting from long-term, continued influence of a stressor.

Circadian   Cyclic character of a system with events recurring within a period of a day (24 hours). Compare with diurnal (within the light part of a day) and nocturnal (within a night).

Circannual   Cyclic character of a system with events recurring within a period of 1 year.

Cognition   Characterized by awareness, perception, understanding, and reasoning.

Colostrum   Milk produced the first few days after parturition. The milk contains important antibodies that assist the newborn in developing immunity.

Comfort shift   Change in posture.

Conditioning   The term refers to the development of association responses to stimuli. An example is the feeding activity response animals develop in response to sounds associated with the process of preparing and delivering feed.

Condition of the animal   Commonly used to refer to general health and in some cases degree of fatness of the animal. The term may also be used to refer to acquired characteristics of an animal as opposed to those of genetic origin.

Conspecifics   Animals of the same species, generally at a similar stage of the life cycle.

Constitution of the animal   Sometimes used to refer to the genetic component of an animal as it relates to fitness. The term is no longer in common use but appears in some literature related to genetic selection.

Controller   An animal in a group that influences group activity and what the group is likely to pursue.

Coping   Efforts of an animal and its biological systems to meet demands of the environment to maintain normal functioning and stability of the whole individual.

Coprophagy   Eating of feces.

Cortex   The outer layer of a multiple-layered organ (e.g., cerebral cortex and adrenal cortex).

Corticotropic-releasing hormone (CRH)   A neuropeptide arising in the hypothalamus that stimulates the anterior pituitary to release ACTH.

Crowding   Descriptive of spatial relationships. May have negative impacts on animal behavior and functions. Excessive population density is referred to as overcrowding.

Depression   A mental state that may be reflected as unresponsiveness or lack of interest in surrounding activities or things.

Dietary density   Concentration of nutrients in relation to total feed. Animals consuming greater amounts of feed may require lower density in terms of protein, for example, or a lower-percentage protein in the diet.

Displacement activity   An activity by an animal that appears to be out of context. These may be incomplete behaviors that result from frustration associated with not being able to complete a goal-oriented function.

Display   A prominent, noticeable action that serves as a stimulus to attract attention. Common in the reproductive activity of birds and some animals. A form of communication.

Diurnal variation   Variation occurring within a day. The fact that the body temperature of an animal may be higher in the afternoon than in the morning is an example. Generally refers to the daytime period, whereas nocturnal refers to nighttime. Compare with circadian.

Dominance   A social and behavioral characteristic that results in achieving priority over other animals. Access to resources is normally regulated in this manner and may present unique problems in some animal management systems. Social stability in a group of animals is dependant on the establishment of a dominance order.

Dopamine A neurotransmitter used by specific neurons in the central nervous system. It is synthesized from tyrosine and is an intermediate in the synthesis of norepinephrine and epinephrine. Classed as a catecholamine.

Dopaminergic Related to nerve cells that respond to dopamine as the neurotransmitter or to the effect of dopamine.

Drive A common term used to relate an individual's dedication to achieving a goal. Motivation.

Dysfunction Disturbance, failure, or abnormality in a system or organ.

Dystocia Refers to a difficult birth process.

Ecology The science related to interactions between organisms and their environment.

Efferent neuron A nerve fiber that conducts impulses away from the central nervous system. Neural impulses moving from brain centers, through the spinal tracts, to more remote target tissues such as glands and muscles. Compare with afferent neuron.

Endocrinology Science of the endocrine glands, their products (hormones), and their physiological effects.

Endogenous Refers to any factor from within the body that has biological effects, as opposed to a factor that enters the body from the outside or exogenous.

Endorphin Opioid peptides that bind to receptors located throughout the body and central nervous system. Forms such as α and β have different physiological functions.

Enkephalins Opoid peptide found in the brain and thought to serve as endogenous neurotransmitters and analgesics. Different forms (metenkephalin and leuenkephalin) have different physiological effects.

Environment The collection of all factors with which an organism has contact.

Epimiletic Describes caregiving behaviors.

Epinephrine A hormone produced by the adrenal medulla and transmitted through the bloodstream to effector cells throughout the body. Important in the biological response to stress because it prepares the various systems for coping. Also called adrenaline.

Epizootic Generally used to reflect widespread occurrence or above-normal frequency of a disease or abnormality.

Equine Horses and asses.

Esophagus The tube or passage between the stomach and the pharynx through which food passes.

Estrus The component of the female reproductive cycle described as a period of sexual receptivity. May also be referred to as heat.

Ethogram A record of behavioral characteristics of a species or individual animal.

Ethology The scientific study of behavior from a biological perspective concentrating on observation and description.

Evolution Modification of successive generations based on survival of genes related to fitness in animals.

Exogenous Factors that enter or contact the body from external sources. Biological effects may result. Compare to endogenous.

Exploration A normal behavior that provides the animal with information for current and future use. Exploration is used to formulate a memory map of resources and environmental limits, boundaries, and hazards.

Extrapyramidal system Descending nerve innervation pathways other than the corticospinal (pyramidal) system.

Feedback system   An internal message system by which the output component of the system controls or regulates the level of continued output from the source tissue or gland, for example. Neural or humoral feedback often functions to control continuation or initiation of biological events. Behaviorally, feedback from an activity may limit continuation.

Feed-forward system   A term used to suggest that decisions related to available information result in taking actions to meet anticipated influences.

Fight-flight syndrome   The response syndrome described by W. B. Cannon. It represents the response of the sympathetic adrenomedullary (SA) system characterized by release of adrenaline and its effects as a response to emergencies.

Flehmen   A physical activity predominantly of males (but may occur in females) in many species related to mating behavior. It is characterized by extension of the head and neck, raising the upper lip, and short, rapid respiration. The behavior is related to sampling of odor from the body or urine for function of the vomeronasal organ in the process of evaluating the receptivity of the female. May be referred to as gaping in swine.

Flight distance   The distance at which an animal will initiate an escape as its space is intruded upon by a person or another animal. This is an important consideration in animal handling and is usually depicted as a radius dimension.

Foraging   Seeking out and consuming food.

Frustration   A state that results from an inability to achieve a goal and may lead to aberrant behavior. A psychological stressor.

Ganglion   A tissue mass that contains a concentration of nerve cell bodies and synapses. A given ganglion is a structure of nerve cells serving given target areas, tissues, or glands. Examples are caudal mesenteric ganglia serving the bladder and celiac ganglia serving the digestive organs.

Gastric   Descriptive of factors related to or arising from the stomach (e.g., gastric juice and gastric digestion).

General adaptation syndrome (GAS)   The syndrome described by Hans Selye as the hypothalamus-pituitary-adrenocortical (HPA) system responding to stress. It involves a sequence of physiological changes in animals that results from exposure to a stressful situation. In general, the sequence includes alarm, resistance, and exhaustion.

Genotype   Genetic makeup or characteristics of animals. Compare with phenotype, which reflects the combination of genetic and environmental influences.

Glucocorticoids   Products and secretions of the adrenal cortex. These compounds alter metabolic functions related to stress responses. Examples are cortisol, aldosterone, and corticosterone.

Gluconeogenesis   The biochemical process whereby noncarbohydrate molecules such as amino acids are converted to glucose and then utilized through glucose pathways as an energy source.

Glucose   The simple sugar (hexose) that is the end product of starch digestion and makes up a significant part of blood sugar. Glucose and fructose molecules combined make up the disaccharide sucrose. The disaccharide maltose is made up of two molecules of glucose.

Glycogen   The carbohydrate referred to as animal starch made up of glucose molecules. This compound is the principal carbohydrate reserve in the body.

Glycogenesis   The synthesis of glycogen from glucose.

Glycolysis   The biochemical process whereby glycogen is broken down and may be ultimately converted to lactic acid in muscle metabolism.

Gonad   Gamete-producing tissues (testes and ovaries).

Grooming   A behavior that manipulates and cleans the animal's surface areas.

Group distance   The distance a group of animals will attempt to maintain as a separation from other groups. Compare with individual distance and flight distance.

Group effect   Facilitation among animals by common participation.

Growth hormone   The hormone produced and released by the anterior pituitary gland that stimulates IGF-1, nitrogen retention, and related growth.

Habituation   A form of learning involving a decline in response to a stimulus or becoming accustomed to a factor that formerly might have been a stressor.

Head pressing   An aberrant behavior in horses or cattle in which the forehead is held in contact with a surface for extended periods of time.

Head resting   A component of male reproductive behavior in swine, cattle, horses, and sheep. The male presses his head on the back of the female as a means of testing whether she is receptive to mating.

Head shaking   A behavior observed in poultry characterized by rapid head turning movements. It can be expressed as an aberrant behavior.

Head tossing   A behavior observed in horses that involves a sequence of vertical head movements. The action can be rather violent. It can be expressed as an aberrant behavior.

Hierarchy   A social priority system among animals. May be established by fighting or other means of domination.

Hippocampal control   Neural control related to functions of the hippocampus, a component of the limbic system.

Hippocampus   A component of the limbic system.

Homeostasis   A tendency to uniformity or stability in body states. Examples are thermal homeostasis and glucose homeostasis. Homeostatic motivation is involved in some behaviors.

Homeothermy   Maintenance of normal, stable body temperature.

Home range   The area used regularly by an animal or a group. May be, but is not usually, a territory for defense. Nonterritorial animals or groups of animals may make common use of utilities such as feed and water resources in home ranges that overlap.

Hormone   Substance (chemical messenger), produced by a cell, that is delivered by body fluids and influences the activity of another cell. Such a system is described as humoral as opposed to neural, which refers to the transmission of nerve impulses.

Hypertrophy   Increase in size of a tissue or organ caused by enlarging of the constituent cells.

Hypophysis   Also called the pituitary gland; located below the hypothalamus.

Hypothalamus   The part of the forebrain that is vitally involved in homeostatic systems. Thus, it serves critically in stress responses.

Imprinting   A rapid type of learning occurring very early in life (critical in the first few hours of the life of the newborn and its parent in some species). Recognition behavior and following are examples of the results of imprinting.

Ingestive behavior   Animal behavior related to food consumption.

Initiator   The animal in a group that starts or signals the start of an action that may result in group action or action on the part of another individual.

Innervation   The neural action stimulus and its avenues involved in relaying nerve impulses to a target tissue or organ.

Instinct   Behavior that is inherited. Such behaviors typically have fixed-action patterns characteristic of a given species. Nesting behavior in sows and nursing behavior in all young animals are examples. Innate is used synonymously.

Intersucking   Behavior directed to other animals by sucking appendages. Most common among early weaned animals. Milk ingestion is an important stimulus.

Intromission   The insertion of a part (e.g., penile penetration).

Involution   Shriveling of an organ or tissue (e.g., uterus after parturition).

Lacrimal gland   The gland that produces tears.

Lactose   The primary milk sugar (disaccharide). Consists of molecules of glucose and galactose.

Leader   An animal that leads a group. A leader may not always be the initiator of group activities, however.

Libido   Level of sexual desire or demonstration of sexual activity.

Lignophagy   Animal behavior characterized by chewing wood.

Lipolysis   Breakdown of fat.

Lipolytic   Descriptive of an agent that causes the breakdown of fat.

Lipoprotein   A combination of lipid and protein in a molecule with properties common to proteins. Most lipids in the blood are in this type of molecule as either α or β forms.

Malfunction   Abnormal function of an animal or part.

Mismatch   The difference between what an animal expects to happen and what actually happens that may result in stress arising from frustration.

Muscle fiber (red)   Fibers characteristically demonstrating aerobic metabolic activity.

Muscle fiber (white)   Fibers characteristically demonstrating anaerobic metabolic activity. More prominent in muscle showing signs of the porcine stress syndrome (PSS).

Myoglobulinuria   Presence of myoglobin in the urine.

Need   A requirement that may result in behavior to acquire a particular resource. Contrast to want.

Neonatal   Pertains to the first few weeks of life. Animals in this stage are referred to as neonates.

Neural memory   Memory functions of the central nervous system.

Neurobiology   The field of science involving study of the nervous system.

Neurochemical   Describes chemical compounds that stimulate neural tissue (e.g., the transfer of an electrical nervous activity from one nerve cell to the next at the synapse by the neurotransmitter acetylcholine or norepinephrine).

Neuroendocrine   System in which a neural stimulation prompts action by an endocrine tissue receptor (e.g., sympathetic nervous system impulses stimulating the adrenal medulla to release epinephrine).

Neuroethology   The study of the neural control of behavior.

Neurohormone   A hormone (neuromodulator) produced by nerve cells. These are normally considered to have humoral influence on neural tissue. Dopamine is an example. Some compounds previously classified as neurotransmitters are now considered neurohormones.

Neurology   The field of science dealing with the nervous system and its function and disorders.

Neuron   A cell that conducts electrical impulses to another cell or to the ultimate target (receptor) tissue. A neuron normally consists of dendrite(s), the cell body, and axon(s).

Neuropsychobiology   Study of relationships between the nervous system and neural stimuli and psychological phenomena.

Neurotransmitter   A chemical messenger that transmits neural activity (by generating impulses) from one neuron to the next. The transmitter is derived from the cell body of the neuron and moves along the axon for release at the terminal end of that cell. This release allows the neural message to proceed through the synapse to innervate the connecting fibers of other cells or target effector cells. Examples of neurotransmitters are acetylcholine, norepinephrine, some amino acids (glycine, glutamic acid, and gamma-aminobutyric acid [GABA]), and some neuropeptides.

Noradrenergic system   The response system associated with norepinephrine (noradrenaline). Involved in the sympathetic nervous system response.

Norepinephrine   A neurotransmitter derived from norepinephrine-containing cells in the brainstem that connect with the cerebral cortex, cerebellum, and spinal cord and from nerve cells of the peripheral nervous system. It is the postganglionic neurotransmitter serving the sympathetic nervous system and is therefore a primary component in the stress response. Norepinephrine is an intermediate in the formation of epinephrine and is found in association with the adrenal medulla.

Observational learning   Learning resulting from one animal watching another.

Olfaction   Sense of smell. Function of the olfactory nerve.

Ontogeny   The process of complete development of an organism from a single cell to a specified stage of life.

Pair bonding   A developed relationship between two individuals.

Pairing   Tendency to form specific partnerships by two individuals.

Pale, soft exudative pork (PSE)   An abnormal condition of pork muscle associated with the porcine stress syndrome (PSS). PSS is inherited as a simple recessive characteristic.

Pancreas   The gland giving rise to insulin, which regulates carbohydrate metabolism, and to pancreatic secretions, which flow into the duodenum for digestive functions.

Pandiculation   A stretching and yawning behavior.

Paralytic myoglobinuria   The presence of myoglobin in the urine resulting from muscle degeneration. It occurs following rest periods in extremely active horses whose muscles have accumulated high levels of glycogen.

Parasympathetic nervous system   A component of the autonomic nervous system that is important in balancing the effects of the sympathetic nervous system and in a number of regulatory measures related to conservation and restorative processes.

Parathyroid   A gland located beside the thyroid gland. The parathyroid gland secretes parathyroid hormone (PTH), which is concerned primarily with calcium and phosphorus metabolism.

Parturition   The birth process.

Pathological conditions   Diseased or abnormal conditions.

Peck order   A dominance order. Often involved in establishing priority access to feed, water, shelter, and resting space. Also referred to as social order.

Peptidergic   Descriptive of nervous transmissions that use peptides as neurotransmitters.

Phenotype   Characteristics of an animal reflecting both genetic and environmental influences.

Pheromone   Chemical messenger released by an animal into its environment. It is an olfactory stimulus to another animal, usually in the same species.

Photoperiod   Biological rhythms associated with light such as day length. An important influence in the reproductive process of some species. Some species respond to increasing day length; others to decreasing day length.

Pica   Craving of inappropriate items for ingestion. Consumption of things that are not normally considered a part of the animal's diet.

Pineal gland   An endocrine gland located ventrally to the cerebrum. This gland produces melatonin, which influences gonadal function in some species. Melatonin is important in animals in which reproductive cycles are influenced by photoperiod. Levels of production and release are inversely related to length of day or light. Managing day length may be a significant tool in regulating reproduction in some species. The gland also contains serotonin, a neurotransmitter, which is an intermediary product in the synthesis of melatonin.

Pituitary gland   *See* hypophysis.

Polydipsia   Consumption of excessive quantities of water.

Porcine   Swine.

Porcine somatotropin (PST)   Porcine growth hormone.

Porcine stress syndrome (PSS)   An abnormality associated with a recessive characteristic of genetic origin that reduces an animal's ability to cope with stress. Pork produced by animals with this genetic makeup may be pale in color, soft, and exudatitive (PSE). This undesirable characteristic is controlled through genetic selection.

Postnatal   The period or events occurring after birth. (syn., postpartum.)

Postpartum   The period after parturition. (syn., postnatal.)

Precocial   Animals born with advanced abilities at birth. Farm animals, for example, are born with eyes open, ability to walk, and a general ability to accomplish activity to meet their needs.

Predisposition   A condition described as increased susceptibility to disease, for example.

Preening   Grooming of feathers.

Prenatal   The period of pregnancy and fetal development.

Prolactin   The hormone released from the anterior pituitary that stimulates lactation. It is also involved in maternal behavior in birds.

Psychoimmunobiology   The field of study dealing with relationships between the mind and the immune system.

Psychological stressors   Factors influencing mental faculties that stimulate stress responses.

Pylorus   The aperture through which material passes from the stomach to the duodenum; it is controlled by the pyloric sphincter.

Rate of passage   A term used to describe the rate of movement of digestive tract contents through the gastrointestinal tract.

Redirected behavior   Activity directed at an individual or object other than the one taking the action that generated the response. For example, an aggressive attack toward an animal other than the attacker or object that follows receipt of an attack by a dominant counterpart represents redirected aggression. A direct return attack on the dominant animal is likely to hold great risk of further damage.

Reinforcement   An influence that increases the probability that an animal will respond in a desired way. Examples are various rewards associated with animal training and performance. Reinforcement may be positive, such as a reward for a desired behavior, or negative, such as application of an aversive stimulus until the desired response is completed.

Ritual   A behavior that has acquired a routine sequence. May have a communication function. Examples are the various behaviors involved in mate attraction during reproductive scenarios.

Rumen parakeratosis   An abnormal histology of the rumen lining that typically involves matting and dysfunction of rumen papillae.

Salivary glands   Glands giving rise to saliva. The individual glands are parotid, sublingual, and submaxillary.

Scent marking   Applying scent-bearing secretions or excretions of the animal to objects as a communication medium. The system may be used to mark territories of animals dominating an area or serve as assistance in locating mates.

Sensitization In a behavioral sense, an increasing response to a continuing stimulus.

Sensory A function of the various sense organs (e.g., smell, taste, and sight).

Serotonin A neurotransmitter-neurohormone produced by tissues of the central nervous system. An intermediate in the synthesis of melatonin, which is involved in photoperiod influences on reproduction in some species.

Set point A biological term used to describe a characteristic point that a system will attempt to achieve. Examples are weight of an animal or a given level of some specific component of an animal.

Social dominance A level achieved in the hierarchy of a group of animals that establishes a priority for meeting the dominant animal's needs or wants. Dominance achieved and maintained over another individual.

Social facilitation The influence of one animal on another that enhances an activity or the achievement of a goal by others or the group.

Social organization Characteristics of a group in terms of size, composition, and individual relationships involved.

Somnolence A state characterized by lack of response to stimuli.

Spacing behavior Characteristics of animals relating to distances maintained from others or in utilization of physical space.

Spatial intrusion Movement into the personal space of an animal.

Spatial requirements A term used to reflect the space or distance from other animals that a given animal appears to need for a comfortable environment. Spatial intrusion by another animal may be a stressor.

Spinal cord The central neural communication system running through the vertebral column (spinal column). A component of the central nervous system containing nerve cells and tracts.

Spinal fluid More appropriately called the cerebrospinal fluid, which circulates within the elastic meningeal sac; serves as a protective covering for nerve cells and as a medium of exchange of metabolic materials.

Stereotypies Behaviors involving repeated movements performed in a constant manner. The behaviors may appear to serve no known purpose in terms of goal achievement. Several aberrant behaviors are classified as stereotypies.

Stress A state resulting from the influence of a stressor that causes a response system to engage as a means of coping. Extended stress may deplete response systems and as a result progressively reduce the animal's ability to cope.

Stressor Any influence that results in stress.

Stress response Biological or psychological response to a stressor. Examples are the HPA and SA systems in metabolic responses. Behavioral responses of both voluntary and involuntary character occur.

Symbiosis A term used to describe dependent relationships in which each party benefits from the other.

Sympathetic nervous system A component of the autonomic nervous system vitally important in the stress response. Actions relate to stimulating the animal's biology to prepare for and respond to stressors.

Teleology Refers to the capability of an individual to see the end result of behavior. Example: The cat licked the wound to prevent infection. (Most likely the cat licked the wound as an innate behavior, which also may have prevented infection.)

Territoriality Defense of a given area. May be characteristic of an individual or species. Farm animal species are not typically territorial.

Thalamus A part of the diencephalon. The primary neural relay to the cerebral cortex.

Thigmotaxis Movement in response to physical contact with another animal or object.

Thyroid   The endocrine gland that produces and releases thyroxine (T_4) and triiodothyronine (T_3), which regulate metabolic rate.

Thyroid-stimulating hormone (TSH)   A hormone, released by the anterior pituitary, that activates the thyroid gland to release thyroid hormones.

Tonic immobility   A behavior, characterized by no movement, that is normally the result of a stimulus such as fear of a predator. Most commonly observed in poultry.

Vasoconstriction   Reduction in the size of blood vessels in response to stimuli, which leads to heat conservation and possible reduction in blood loss from peripheral injury. Regulation of vessel diameter by smooth muscle in the vascular system is critical in the regulation of blood pressure.

Vasodilation   Expansion in the size of blood vessels to enhance blood flow for surface heat exchange and to meet needs for more rapid movement of metabolites. Vasodilation is also a critical factor in the regulation of blood pressure.

Vasopressin   A hormone stimulating vasoconstriction. It is produced in the hypothalamus and stored in the posterior pituitary. Antidiuretic hormone.

Visual stimulation   View of an object or activity that serves as a stimulus. Important in group activity. Ingestive and reproductive behaviors are examples typically influenced by animals observing one another.

Well-being   State that encompasses factors such as good health, a sense of safety and comfort, protection from damaging environments including both physical and psychological influences, and capability to engage in normal behaviors. The concept of a harmonic relationship between the animal and its environment in quality of living is an attractive one in evaluating welfare. Often used synonymously with welfare.

Whetting   Rubbing of horns, nails, and claws or repeated and habitual movement of any body part against a fixed surface.

REFERENCES

Adams, K. L., D. H. Baker, and A. H. Jensen. 1980. Effect of supplemental heat for nursing pigs. *J. Anim. Sci.* 50:779.

Ahlquist, R. P. 1948. A study of adrenotropin receptors. *Am. J. Physiol.* 153:586.

Alcock, J. 1993. *Animal Behavior.* 5th ed. Sinauer Associates, Sunderland, Mi.

Anderson, P. A., G. D. Potter, J. L. Krider, and C. C. Courtney. 1983. Digestible energy requirement of horses. *J. Anim. Sci.* 56:91.

Andrew, R. J. 1966. Precocious adult behavior in the young chick. *Anim. Behav.* 12:64.

Appleby, M. C. 1997. Life in a variable world: Behavior, welfare and environmental design. *Appl. Anim. Behav. Sci.* 54:1.

Ballieux, K. E., and C. J. Heijnen. 1987. Stress and the immune system. In *Biology of Stress in Farm Animals: An Integrative Approach.* Edited by P. R. Wiepkema and P. W. M. van Adrichem. Dordrecht, The Netherlands: Martinus Nijhoff, 29.

Banks, P. A., and T. B. Koen. 1989. Intermittent lighting regimens for laying hens. *Poult. Sci.* 68:739.

Barker, J., S. Curtis, O. Hogsett, and F. Humenik. 1986. Safety in swine production systems. *Pork Industry Handbook* AS-572. Ames, Ia.: Iowa State University Cooperative Extension Service.

Barnett, J. L., P. H. Hemsworth, G. M. Cronin, E. A. Newman, and T. H. McCallum. 1991. Effects of design of individual cage-stalls on the behavioural and physiological responses related to the welfare of pregnant pigs. *Appl. Anim. Behav. Sci.* 32:23.

Bernard, C. 1878. *Lecons sur les phenomenes de la vie commons. Aux Animaux et aux vegetaux.* Librarie Paris: J. B. Bailliere et Fils.

Blackshaw, J. K., F. J. Thomas, and J. A. Lee. 1997. The effect of a fixed or free toy on growth rate and aggressive behavior. *Appl. Anim. Behav. Sci.* 53:203.

Blaxter, K. L. 1977. Environmental factors and their influence on the nutrition of farm livestock. In *Nutrition and the Climatic Environment.* Edited by W. Haresign, H. Swan, and D. Lewis. London: Butterworths, 1.

Borg, R., and D. Halverson. 1985. *Turkey Management Guide.* St. Paul, Minn. Minnesota Turkey Grower's Association.

Brambell, Commission. 1965. *The report of the technical committee to inquire into the welfare of animals kept under intensive livestock husbandry systems.* Her Majesty's Stationary Office Command Report No. 2836, London.

Breward, J. 1984. Cutaneous nociceptor in the chicken beak. *J. Physiol.* (Lond.) 346:56.

Broom, D. M., and M. J. Potter. 1984. Factors affecting the occurrence of stereotypies in stall housed dry sows. In *Proceedings of the International Congress on Applied Ethology for Farm Animals.* Edited by J. Unshelm, G. van Patten, and K. Zeeb. Darmstadt, Germany: K.T.B.L., 229.

Broom, D. M. 1986. Indicators of poor welfare. *Br. Vet. J.* 142:524–526.

Brumm, M. C., and D. P. Shelton. 1993. Interaction of reduced nocturnal nursery temperatures and dietary lysine on weaned pig performance to slaughter. *J. Anim.* (Supp. 1 Ab. 640) 71:250.

Butcher, J. E. 1970. Is snow adequate as a water source for sheep? *Nat. Wool Grower* 60:28.

Cannon, W. B. 1932. *The Wisdom of the Body.* New York: W. W. Norton.

Carr, L., and T. A. Carter. 1985. Housing and management of poultry in hot and cold climates. In *Stress Physiology in Livestock,* Vol. 3, *Poultry.* Edited by M. K. Yousef. Boca Raton, Fla.: CRC Press, 73.

Carter, T. A. 1981. Hot weather management of poultry. *Poultry Science and Technology Guide* No. 30. Agricultural Extension Service, North Carolina State University, Raleigh.

Casaday, R. B., R. M. Myers, and J. E. LeGates. 1953. The effect of exposure to high ambient temperature on spermatogenesis in the dairy bull. *J. Dairy Sci.* 36:14.

Chase, C. C. Jr., R. E. Larsen, R. D. Randel, A. C. Hammond, and E. L. Adams. 1995. Plasma cortisol and white blood cell responses in different breeds of bulls: A comparison of two methods of castration. *J. Anim. Sci.* 73:975.

Clutton-Brock, J. 1981. *Domesticated Animals from Early Times.* London: British Museum/W. Heinemann.

CDGCAA. 1988. *Guide for the Care and Use of Agricultural Animals in Agricultural Research and Teaching.* 1st ed. Consortium for Developing a Guide for the Care and Use of Agricultural Animals in Agricultural Research and Teaching. Champaign, Ill.: FASFAS.

Craig, J. V. 1981. *Domestic Animal Behavior.* Englewood Cliffs, N.J.: Prentice-Hall.

Craig, J. V., and J. A. Craig. 1985. Corticosteroid levels in White Leghorn hens as affected by handling, laying house environment, and genetic stock. *Poult. Sci.* 64:809.

Craig, J. V., and W. M. Muir. 1993. Selection for reduction of beak-inflicted injuries among caged hens. *Poult. Sci.* 72:411.

Craig, J. V., J. A. Craig, and J. Vargas. 1986a. Corticosteroids and other indicators of hens' well-being in four laying-house environments. *Poult. Sci.* 65:856.

Craig, J. V., J. Vargas, and G. A. Milliken. 1986b. Fearful and associated responses of White Leghorn hens: Effects of cage environments and genetic stocks. *Poult. Sci.* 65:2199.

Craig, J. V., and W. M. Muir. 1996. Group selection for adaptation to multiple-hen cages: Beak related mortality, feathering, and body weight responses. *Poult. Sci.* 75:294.

Crawford, M. A. 1974. *The Case for New Domestic Animals.* Oryx 12:351.

Curtis, S. E. 1983. *Environmental Management in Animal Agriculture.* Ames, Ia.: Iowa State University Press.

Curtis, S. E. 1985a. Physiological responses and adaptation in swine. In *Stress Physiology in Livestock.* Edited by M. K. Yousef. Boca Raton, Fla.: CRC Press.

Curtis, S. E. 1985b. What constitutes animal well-being? In *Animal Stress.* Edited by G. P. Moberg. Bethesda, Md.: American Physiology Society, 1.

Curtis, S. E., and G. L. Morris. 1982. Opperant supplemental heat in swine nurseries. In *Livestock Environment: Proceedings of the 2nd International Livestock Environment Symposium.* St. Joseph, Mo.: American Society of Agricultural Engineering, 295.

Dailey, J. W., and J. J. McGlone. 1997. Oral/nasal/facial and other behaviors of sows kept individually outdoors on pasture, soil or indoors in gestation crates. *Appl. Anim. Behav. Sci.* 52:25.

Dammrich, K. 1987. Organ change and damage during stress: Morphological diagnosis. In *Biology of Stress in Farm Animals: An Integration Approach.* Edited by P. R. Wiepkema and P. W. M. Van Adrichem. The Netherlands: Martimus Nijhoff Publishers.

Dantzer, R., and G. Mittleman. 1993. Functional consequences of behavioral stereotypy. In *Stereotypic Animal Behaviour: Fundamentals and Application to Welfare.* Edited by A. B. Lawrence and J. Rushen. Wallingford, Oxon, UK: CAB International.

Dawkins, M. S. 1980. *Animal Suffering: The Sciences of Animal Welfare.* London: Chapman & Hall.

Dechambre, E. 1949. La theorie de la foetalisation et la formation des races de chiens et de porcs. *Mammalia* 13:129.

de Passile, A. M., J. Rushen, and M. Janzen. 1997. Some aspects of milk that elicit non-nutritive sucking in the calf. *Appl. Anim. Behav. Sci.* 53:167.

DeShazer, J. A., and D. G. Overhults. 1982. Energy demand in livestock production. In *Livestock Environment: Proceedings of the 2nd International Livestock Environment Symposium*. St. Joseph, Mo.: American Society of Agricultural Engineering.

DeSmet, S. M., H. Pauwels, S. DeBie, D. I. Demeyer, J. Callewier, and W. Eeckhout. 1996. Effect of halothane genotype, breed, feed withdrawal, and lairage on pork quality of Belgian slaughter pigs. *J. Anim. Sci.* 74:1854.

Dickson, D. P., G. R. Barr, L. P. Johnson, and D. A. Wieckart. 1970. Social dominance and temperament in Holstein cows. *J. Dairy Sci.* 53:904.

Duncan, I. J. H., 1996. Animal welfare defined in terms of feelings. J. Anim. Sci. Suppl. 27:29.

Duncan, I. J. H., and T. B. Poole. 1990. Promoting the welfare of farm and captive animals. In *Managing the Behavior of Animals*. Edited by P. Monaghan and D. Wood-Gush. London: Chapman & Hall, 193–232.

Duncan, I. J. H., G. Slee, and P. Kettlewell. 1986. Comparison of the stressfulness of harvesting broiler chickens. *Br. Poult. Sci.* 27:109.

Duncan, I. J. H. 1981. Animal Rights—Animal Welfare: A scientist's assessment. Poult. Sci. 60:489.

Duncan, I. J. H., G. S. Slee, E. Seawright, and J. Breward. 1989. Behavioral consequences of partial beak amputation (beak trimming) in poultry. *Br. Poult. Sci.* 30:479.

Ewbank, R., and M. J. Bryant. 1972. Aggressive behavior amongst groups of domesticated pigs kept at various stocking rates. *Anim. Behav.* 20:21.

Fagen, R. 1981. *Animal Play Behavior.* New York: Oxford Press.

Farrell, D. J., and S. Swain. 1977. Effects of temperature treatments on the energy and nitrogen metabolism of fed chickens. *Br. Poult. Sci.* 18:735.

Faure, J. M., and A. D. Mills. 1998. Improving the adaptability of animals by selection. In *Genetics and the Behavior of Domestic Animals*. Edited by T. Grandin. San Diego: Academic Press.

Fowler, D. G., and L. D. Jenkins. 1976. The effects of dominance and infertility of rams on reproductive performance. *Appl. Anim. Ethol.* 2:327.

Fox, M. W. 1986. *The Case for Animal Experimentation.* Berkeley and Los Angeles: University of California Press.

Fraser, A. F., and D. M. Broom. 1990. *Farm Animal Behavior and Welfare.* 3d ed. London: Bailliere Tindall.

French, N. P., R. Wall, and K. L. Morgan. 1994. Lamb tail docking: A controlled field study of the effects of tail amputation on health and productivity. *Vet. Rec.* 134:463.

Frey, R. G. 1980. *Interests and Rights: The Case against Animals.* New York: Clarendon Press.

Friend, T. H., G. R. Dellmeier, and E. E. Gbur. 1985. Comparison of four methods of calf confinement. I. Physiology. *J. Anim. Sci.* 60:1095.

Fritschen, R. D., and A. J. Muehling. 1989. Flooring for swine. *Pork Industry Handbook* AS-486. Ames, Ia.: Iowa State University, Cooperative Extension Service.

Fronk, J. J., and L. H. Schultz. 1979. Oral nicotinic acid as treatment for ketosis. *J. Dairy Sci.* 62:1804.

Gilbert, S. F. 1991. *Developmental Biology.* 3d ed. Sunderland, Mass.: Sinauer Associates.

Goldman, L., G. D. Coover, and S. Levine. 1973. Biodirectional effects of reinforcement shifts on pituitary adrenal activity. *Physiol. Behav.* 10:209.

Gonyou, H. W., R. P. Chapple, and G. R. Frank. 1992. Productivity, time budgets, and social aspects of eating in pigs penned in groups of five or individually. *Appl. Anim. Behav. Sci.* 34:291.

Gonzalez-Jiminez, E., and K. L. Blaxter. 1962. The metabolism and thermoregulation of calves in the first month of life. *Br. J. Nutr.* 16:199.

Grandin, T. 1988. *Livestock Handling Guide.* Bowling Green, Ky.: Livestock Conservation Institute.

Grandin, T. 1992a. *Proper Handling Techniques for Non-Ambulatory Animals.* Bowling Green, Ky.: Livestock Conservation Institute.

Grandin, T. 1992b. *Livestock Trucking Guide.* Bowling Green, Ky.: Livestock Conservation Institute.

Grandin, T. 1995. Behavioural principles of cattle handling under extreme conditions. In *Livestock Handling and Transport.* Edited by T. Grandin. Wallingford, UK: CAB International.

Gregory, N. G., L. J. Wilkins, S. D. Austin, C. G. Belyavin, D. M. Alvey, and S. A. Tucker. 1992. Effect of catching method on the prevalence of broken bones in end-of-lay hens. *Avian. Pathol.* 21:717.

Hafez, E. S. E., and M. F. Bouissou. 1975. The behavior of cattle. In *The Behavior of Domestic Animals.* 3d ed. Edited by E. S. E. Hafez. London: Bailliere Tindall.

Hafez, E. S. E., and J. P. Signoret. 1969. The behavior of swine. In *The Behavior of Domestic Animals.* 2d ed. Edited by E. S. E. Hafez. Baltimore, Md.: Williams & Wilkins.

Hahn, G. L. 1985. Management and housing of farm animals in hot environments. In *Stress Physiology in Livestock.* Vol. 2. Edited by M. K. Yousef. Boca Raton, Fla.: CRC Press, 151.

Hahn, G. L., and D. D. Osburn. 1969. Feasibility of summer environmental control for dairy cattle based on expected production losses. *Trans. Am. Soc. Agric. Eng.* 12:448.

Hale, E. B. 1969. Domestication and the evolution of behavior. In *The Behavior of Domestic Animals.* 2d ed. Edited by E. S. E. Hafez. London: Bailliere, Tindall, and Cassell, 22.

Hansen, R. S. 1976. Nervousness and hysteria of mature female chickens. *Poult. Sci.* 55:531.

Harlow, H. F., and M. K. Harlow. 1962. Social deprivation in monkeys. *Sci. Am.* 207:137.

Harmon, J., and H. Xin. 1995. *Environmental Guidelines for Confinement Swine Housing.* Ext. PM-1586a. Ames, Ia.: Iowa State University.

Harrison, R. 1964. *Animal Machines.* London: Vincent Stuart.

Hart, B. L. 1985. *The Behavior of Domestic Animals.* New York: W. H. Freeman.

Heffner, H. E., and R. S. Heffner. 1992. Auditory perception. In *Farm Animals and the Environment.* Edited by C. Phillips and D. Piggins. Wallingford, UK: CAB International, 159–184.

Hemsworth, P. H., and D. B. Galloway. 1979. The effect of sexual stimulation on sperm output of the domestic boar. *Anim. Reprod. Sci.* 2:387.

Hemsworth, P. H., R. G. Beilharz, and D. B. Galloway. 1977. Influence of social conditions during rearing on the sexual behaviour of the domestic boar. *Anim. Prod.* 24:245.

Henneke, D. R., G. D. Potter, and J. L. Krider. 1984. Body condition during pregnancy and lactation and reproductive efficiency of mares. *Theriogenology* 21:897.

Henneke, D. R., G. D. Potter, J. L. Krider, and B. F. Yentes. 1983. Relationship between condition score, physical measurement, and body fat percentage in mares. *Eq. Vet. J.* 15:371.

Herd, D. B., and L. R. Sprott. 1986. *Body condition, nutrition, and reproduction of beef cows.* Texas A&M University Extension Bulletin 1526.

Hoff, E. H., R. Guillemin, and L. Guillemin. 1974. Claude Bernard. *Lectures on the Phenomena of Life Common to Animals and Plants.* Translation. Springfield, Ill.: Charles C. Thomas.

Holden, P., R. Ewan, M. Jurgens, T. Stahly, and D. Zimmerman. 1996. *Life Cycle Swine Nutrition.* Ext. PM-489. Ames, Ia.: Iowa State University.

Holmes, C. W., and W. H. Close. 1977. The influence of climatic variables on energy metabolism and associated aspects of productivity in the pig. In *Nutrition and the Climatic Environment.* Edited by W. Haresign, H. Swan, and D. Lewis. London: Butterworths.

Hou, S. M., M. A. Boone, and J. T. Long. 1973. An electrophysiological study on the hearing and vocalization in *Gallus Domesticus*. *Poult. Sci.* 52:159.

Houpt, K. A., D. M. Zahorik, and J. A. Swartzman-Andert. 1990. Taste aversion learning in horses. *J. Anim. Sci.* 68:2340.

Houpt, K. A. 1991. *Domestic Animal Behavior*. Ames: Iowa State University Press.

Houpt, K. A., and S. Lieb. 1993. Horse handling and transport. In *Livestock Handling and Transport*. Edited by T. Grandin. Wallingford, UK: CAB International.

Hughes, B. O., and I. J. H. Duncan. 1988. The notion of ethological need models of motivation and animal welfare. *Anim. Behav.* 36:1696.

Hurnik, J. F. 1992. Animal responses to the environment: Behavior. In *Farm Animals and the Environment*. Edited by C. Phillips and D. Piggins. Wallingford, UK: CAB International.

Hurnik, J. F., A. B. Webster, and P. B. Siegel. 1995. *Dictionary of Farm Animal Behavior*. 2d ed. Ames: Iowa State University Press.

Hy-Line. 1996. *Hy-Line Variety W-36 Management Guide: Chick, Pullet, Layer*. West Des Moines, Iowa: Hy-Line International.

Jacobs, G. H. 1993. The distribution and nature of colour vision among mammals. *Biol. Rev.* 68:413.

Janssens, C. J. J. G., F. A. Helmond, and V. M. Wiegant. 1994. Increased cortisole response to exogenous adrenocorticotropic hormone in chronically stressed pigs: Influence of housing conditions. *J. Anim. Sci.* 72:1771.

Jewell, P. 1969. Wild mammals and their potential for new domestication. In *The Domestication and Exploitation of Plants and Animals*. Edited by P. Ucko and D. Dimbleby. London: Duckworth, 101.

Johnson, H. D. 1985. Physiological responses and productivity of cattle. In *Stress Physiology in Livestock (Vol. 2)*. Edited by M. K. Yousef. Boca Raton, Fla.: CRC Press.

Kannan, G., and J. A. Mench. 1996. Influence of different handling methods and crating periods on plasma corticosterone concentrations in broilers. *Br. Poult. Sci.* 37:21.

Kendrick, K. M. 1992. Perception of the environment by farm animals: Cognition. In *Farm Animals and the Environment*. Edited by C. Phillips and D. Piggins. Wallingford, UK: CAB International.

Kilgour, R. 1987. Learning and the training of farm animals. In *The Veterinary Clinics of North America*, Vol. 3, No. 2, *Farm Animal Behavior*. Edited by E. O. Price. Philadelphia: W. B. Saunders.

Klemm, W. R., C. J. Sherry, L. M. Schake, and R. F. Sis. 1983. Homosexual behavior in feedlot stress: An aggression hypothesis. *Appl. Anim. Ethol.* 11:187.

Klemm, W. R., C. J. Sherry, R. F. Sis, L. M. Schake, and A. B. Waxman. 1984. Evidence of a role for the vomeronasal organ in social hierarchy in feedlot cattle. *Appl. Anim. Behav. Sci.* 12:53.

Koolhaas, J. M., S. F. deBou, and B. Bohas. 1997. Motivational systems or motivational states: Behavioral and physiological evidence. *Appl. Anim. Behav. Sci.* 53:131.

Krzack, W. E., H. W. Gonyou, and L. M. Lawrence. 1991. Wood chewing by stabled horses: Diurnal pattern and effects of exercise. *J. Anim. Sci.* 69:1053.

Ladewig, J., and E. von Borell. 1988. Ethological methods alone are not sufficient to measure the impact of environment on animal health and well being. In *Proceedings of the International Congress on Applied Ethology in Farm Animals, Skara 1988*. Edited by J. Unshelm, G. Van Putten, K. Zeeb, and I. Ekesbo. Darmstadt, Germany: Kuratorium fur Technik und Bauwesen in der Landwirtschaft.

Ladewig, J., A. M. de Passille, J. Rushen, W. Schouten, E. M. C. Terlouw, and E. von Borell. 1993. Stress and physiological correlates of stereotypic behavior. In *Stereotypic Animal Behavior*. Edited by A. B. Lawrence and J. Rushen. Wallingford, UK: CAB International.

Lawrence, A. B., and E. M. C. Terlouw. 1993. A review of behavioral factors involved in the development and continued performance of stereotypic behaviors in pigs. *J. Anim. Sci.* 7:2815.

Lay, D. C. Jr., T. H. Friend, C. L. Bowers, K. K. Grissom, and O. C. Jenkins. 1992a. Behavioral and physiological effects of freeze or hot-iron branding on crossbred cattle. *J. Anim. Sci.* 70:330.

Lay, D. C. Jr., T. H. Friend, C. L. Bowers, K. K. Grissom, and O. C. Jenkins. 1992b. A comparative physiological and behavioral study of freeze and hot-iron branding using dairy cows. *J. Anim. Sci.* 70:1121.

Lay, D. C. Jr., T. H. Friend, K. K. Grissom, C. L. Bowers, and M. E. Mal. 1992c. Effects of freeze or hot-iron branding of Angus calves on some physiological and behavioral indicators of stress. *Appl. Anim. Behav. Sci.* 33:137.

Leadon, D., C. Frank, and W. Backhouse. 1989. A preliminary report on studies on equine transit stress. *J. Eq. Vet. Sci.* 9(4):200.

LeDividich, J., and J. Noblet. 1981. Colostrum intake and thermoregulation in the neonatal pig in relation to environmental temperature. *Biol. Neonate* 40:167.

LeDividich, J., P. Herpin, P.A. Geraert, and M. Vermorel. 1992. Cold stress. In *Farm Animals and the Environment*. Edited by C. Phillips and D. Piggins. Wallingford, UK: CAB International, 3–25.

Lee, H.-Y., and J. V. Craig. 1991. Beak trimming effects on the behavior patterns, fearfulness, feathering, and mortality among three stocks of White Leghorn pullets in cages or floor pens. *Poult. Sci.* 70:211.

Lofgreen, G. P., and W. N. Garrett. 1968. A system for expressing net energy requirements and feed values for growing and finishing beef cattle. *J. Anim. Sci.* 27:793.

Lynch, J. J., G. N. Hinch, and D. B. Adams. 1992. *The Behavior of Sheep*. Wallingford, UK: CAB International.

Mabry, J. W., M. T. Coffey, and R. W. Seely. 1983. A comparison of an 8- versus 16-hour photoperiod during lactation on suckling frequency of the baby pig and maternal performance. *J. Anim. Sci.* 57:292.

Mader, D. R., and E. O. Price. 1984. The effects of sexual stimulation in the sexual performance of Hereford bulls. *J. Anim. Sci.* 59:294.

Mader, T. L., J. M. Dahlquist, and J. B. Gaughan. 1997. Wind protection effects and airflow patterns in outside feedlots. *J. Anim. Sci.* 75:26.

Mal, M. E., T. H. Friend, D. C. Lay, S. G. Vogelsang, and O. C. Jenkins. 1991. Physiological responses of mares to short term confinement and social isolation. *J. Eq. Vet. Sci.* 11:96.

Mason, G. J. 1993. Forms of stereotypic behavior. In *Stereotypic Animal Behavior*. Edited by A. B. Lawrence and J. Rushen. Wallingford, UK: CAB International, 7–40.

Mason, J., and P. Singer. 1980. *Animal Factories*. New York: Crown Publishers.

Matthews, L. R., and J. Ladewig. 1994. Environmental requirements of pigs measured by behavioural demand functions. *Anim. Behav.* 47:713.

Maton, A., J. Daelemaus, and J. Lambrecht. 1985. *Housing of Animals*. Amsterdam: Elsevier.

McCracken, K. J., and B. J. Caldwell. 1980. Studies on diurnal variations of heat production and the effective lower critical temperature of early-weaned pigs under commercial conditions of feeding and management. *Br. J. Nutr.* 43:321.

McGartney, M. G. 1971. Reproduction of turkeys as affected by age at lighting and light intensity. *Poult. Sci.* 50:661.

McGlone, J. 1993. What is animal welfare? J. Agric. Environ. Ethics 6(Special Suppl. 2):26.

McGlone, J. J., and S. E. Curtis. 1985. Behavior and performance of weanling pigs in pens equipped with hide areas. *J. Anim. Sci.* 60:20.

McGlone, J. J., J. L. Salak-Johnson, R. I. Nicholson, and T. Hicks. 1994. Evaluation of crates and girth tethers for sows: Reproduction performance, immunity, behavior, and ergonomic measures. *Appl. Anim. Behav. Sci.* 39:297.

Mellor, D. J., and L. Murray. 1989. Effects of tail docking and castration on behaviour and plasma cortisol concentrations in young lambs. *Res. Vet. Sci.* 46:387.

Mench, J. A. 1993. Animal well-being and research priorities: Fair '95 and some personal observations. *Food Animal Well-Being Conference Proceedings.* Washington, D.C.: U.S. Department of Agriculture, and West Lafayette, Ind.: Purdue University.

Meyer, V. M., L. B. Driggers, K. Ernest, and D. Ernest. 1991. Swine growing and finishing units. *Pork Industry Handbook* AS-435 (Rev). Ames, Ia.: Iowa State University Cooperative Extension Service.

Milligan, J. D., and G. I. Christison. 1974. Effect of month of the year on performance of feeder cattle in Canada. *Can. J. Anim. Sci.* 54:605.

Minick, J. A., D. C. Lay Jr., S. P. Ford, L. M. Hohenshell, N. J. Bienson, and M. E. Wilson. 1997. Differences in maternal behavior between Meishan and Yorkshire gilts. *J. Anim. Sci.* 75(Suppl. 1):38.

Minton, J. E., T. R. Coppinger, P. G. Reddy, W. C. Davis, and F. Blecha. 1992. Repeated restraint and isolation stress alter adrenal and lymphocyte functions and some leukocyte differentiation antigens in lambs. *J. Anim. Sci.* 70:1132.

Moberg, G. P. 1985a. Biological response to stress: Key to assessment of well-being. In *Animal Stress.* Edited by G. P. Moberg. Bethesda, Md.: American Physiological Society.

Moberg, G. P. 1985b. Influence of stress on reproduction: Measure of well-being. In *Animal Stress.* Edited by G. P. Moberg. Bethesda, Md.: American Physiological Society.

Moberg, G. P. 1993. Using risk assessment to define domestic animal welfare. *J. Agric. Environ. Ethics* 6(Special Suppl. 2):1.

Moberg, G. P. 1996. Suffering from stress: An approach for evaluating the welfare of an animal. *Acta. Agric. Scand.* 27:46.

Molony, V., and J. E. Kent. 1997. Assessment of acute pain in farm animals using behavioral and physiological measurements. *J. Anim. Sci.* 75:266.

Moran, E. T., Jr. 1985. Effect of toe clipping and pen population density on performance and carcass quality of large turkeys reared sexes separately. *Poult. Sci.* 64:226.

Morrow-Tesch, J., and B. Jones. 1997. Effect of castration on behavior and physiological measures of stress in young calves. *Proceedings of the 31st International Congress of the International Society for Applied Ethology.* Lennoxville, Quebec. International Society for Applied Ethology.

Moule, G. R., and G. M. H. Waites. 1963. Seminal degeneration in the ram and its relation to the temperature of the scrotum. *J. Reprod. Fert.* 5:433.

Mount, L. E. 1959. The metabolic rate of the newborn pig in relation to environmental temperature and to age. *J. Physiol.* 147:333.

Mount, L. E. 1960. Influence of huddling and body size on the metabolic rate of young pigs. *J. Agric. Sci.* 55:101.

Mount, L. E. 1963. The thermal insulation of the new-born pig. *J. Physiol.* 168:698.

Mount, L. E. 1964a. Radiant and convective heat loss from the new-born pig. *J. Physiol.* 173:96.

Mount, L. E. 1964b. The tissue and air components of thermal insulation in the newborn pig. *J. Physiol.* 170:286.

Mount, L. E. 1966. The effect of wind-speed on heat production in the new-born pig. *Q. J. Exp. Physiol.* 51:18.

Muir, W. M. 1996. Group selection for adaptation to multiple hen cages: Direct responses. *Poult. Sci.* 75:447.

Murphy, J. P., D. D. Jones, and L. L. Christian. 1991. *Pork Industry Handbook (PIH-60).* Ext. AS-496. Ames, Ia.: Iowa State University Cooperative Extension Service.

MWPS. 1987. *Structures and Environment Handbook.* 11th ed. Ames, Ia.: Midwest Plan Service, Iowa State University.

MWPS-3. 1994. *Sheep Housing and Equipment Handbook.* 4th ed. Ames, Ia.: Midwest Plan Service, Iowa State University.

MWPS-6. 1987. *Beef Housing and Equipment Handbook.* 4th ed. Ames, Ia.: Midwest Plan Service, Iowa State University.

MWPS-7. 1995. *Dairy Freestall Housing and Equipment Handbook.* 5th ed. Ames, Ia.: Midwest Plan Service, Iowa State University.

MWPS-8. 1983. *Swine Housing and Equipment Handbook.* 4th ed. Ames, Ia.: Midwest Plan Service, Iowa State University.

Nielsen, L. H., L. Mogensen, C. Krohn, J. Hindhede, and J. T. Sorensen. 1997. Resting and social behavior of dairy heifers housed in slatted floor pens with different sized bedding areas. *Appl. Anim. Behav. Sci.* 54:307.

Nicol, C. J., and C. Saville-Weeks. 1993. In *Poultry Handling and Transport.* Edited by T. Grandin. Wallingford, UK: CAB International.

North, M. O., and D. D. Bell. 1990. *Commercial Chicken Production Manual.* New York: Van Nostrand Reinhold.

NPPC. 1992. *Swine Care Handbook.* Des Moines, Iowa: National Pork Producers Council.

NRC. 1981a. *Effect of Environment on Nutrient Requirements of Domestic Animals.* Washington, D.C.: National Academy Press.

NRC. 1981b. *Nutritional Energetics of Domestic Animals and Glossary of Energy Terms.* Washington, D.C.: National Academy Press.

NRC. 1984. *Nutrient Requirements of Beef Cattle.* Washington, D.C.: National Academy Press.

NRC. 1985. *Nutrient Requirements of Sheep.* Washington, D.C.: National Academy Press.

NRC. 1988. *Nutrient Requirements of Swine.* Washington, D.C.: National Academy Press.

NRC. 1989a. *Nutrient Requirements of Dairy Cattle.* Washington, D.C.: National Academy Press.

NRC. 1989b. *Nutrient Requirements of Horses.* Washington, D.C.: National Academy Press.

NRC. 1994. *Nutrient Requirements of Poultry.* Washington, D.C.: National Academy Press.

NRC. 1996. *Nutrient Requirements of Beef Cattle.* Washington, D.C.: National Academy Press.

Occupational Safety and Health Administration. 1989. *OSHA Safety and Health Standards.* Washington, D.C.: OSHA, U.S. Department of Labor.

Orgeur, P., and J. P. Signoret. 1984. Sexual play and its functional significance in domestic sheep. *Physiol. Behav.* 33:111.

Owings, W. J., S. L. Balloun, W. W. Marion, and G. M. Thomson. 1972. The effect of toe clipping turkey poults on market grade, final weight and percent condemnation. *Poult. Sci.* 51:638.

Pagan, J. D., and H. F. Hintz. 1986a. Equine energetics I. Relationship between body weight and energy requirements in horses. *J. Anim. Sci.* 63:815.

Pagan, J. D., and H. F. Hintz. 1986b. Equine energetics II: Energy expenditure in horses during submaximal exercise. *J. Anim. Sci.* 63:822.

Payne, C. G. 1966. Practical aspects of environmental temperatures for laying hens. World *Poult. Sci.* 22:126.

Pettigrew, J. E. 1981. Supplemental dietary fat for peripartal sows: A review. *J. Anim. Sci.* 53:107.

Pickett, B. W., J. L. Voss, and E. L. Squires. 1977. Impotence and abnormal sexual behavior in the stallion. *Theriogenology* 8:329.

Piggins, D. 1992. Visual perception. In *Farm Animals and the Environment.* Edited by C. Phillips and D. Piggins. Wallingford, UK: CAB International, 131.

Pinel, J. P. J. 1993. *Biopsychology.* 2d ed. Needham Heights, Ma.: Allyn and Bacon.

Plomin, R., and C. S. Bergman. 1991. The nature of nurture: Genetic influences on environmental measures. *Behav. and Brain Sci.* 14:373.

Price, E. O. 1978. Genotype versus experience effects on aggression in wild and domestic Norway rats. *Behaviour* 64:340.

Price, E. O. 1984. Behavioral aspects of animal domestication. *Quart. Rev. Biol.* 59:1.

Price, E. O. 1992. Personal communication.

Price, E. O. 1998. Behavioral genetics and the process of animal domestication. In *Genetics and the Behavior of Domestic Animals.* Edited by T. Grandin. San Diego: Academic Press, 31.

Price, E. O., G. C. Dunn, J. A. Talbot, and M. R. Dally. 1984. Fostering lambs by odor transfer: The substitution experiment. *J. Anim. Sci.* 59:301.

Price, E. O., and V. M. Smith. 1984. The relationship of male-male mounting to mate choice and sexual performance in male dairy goats. *Appl. Anim. Behav. Sci.* 13:71.

Price, E. O., S. J. R. Wallach, and M. R. Dally. 1991. Effects of sexual stimulation on the sexual performance of rams. *Appl. Anim. Behav. Sci.* 30:333.

Proudfoot, F. G., H. W. Hulan, and W. F. DeWitt. 1979. Response of turkey broilers to different stocking densities, light treatments, toe clipping and intermingling the sexes. *Poult. Sci.* 58:28.

Provenza, F. D. 1996. Acquired aversions as the basis for varied diets of ruminants foraging on range lands. *J. Anim. Sci.* 74:2010.

Regan, T. 1983. *The Case for Animal Rights.* Berkeley and Los Angeles: University of California Press.

Richardson, B. S. 1992. The effect of behavioral state on fetal metabolism and blood flow. *Semin. Perinatol.* 16:227.

Rivier, C. 1993. Effect of peripheral and central cytokines on the hypothalamic-pituitary-adrenal axis of the rat. In *Corticotropin-Releasing Factor and Cytokines: Role in the Stress Response.* Edited by Y. Tache and C. Rivier. New York: New York Academy of Sciences.

Robertson, I. S., J. E. Kent, and V. Molony. 1994. Effect of different methods of castration on behaviour and plasma cortisol in calves of three ages. *Res. Vet. Sci.* 56:8.

Robison, S. R., C. H. Wong, S. S. Robertson, P. W. Nathanielsz, and W. P. Smotherman. 1995. Behavioral responses of the chronically instrumented sheep fetus to chemosensory stimuli presented in utero. *Behav. Neurosci.* 109:551.

Rollin, B. E. 1981. *Animal Rights and Human Morality.* New York: Prometheus Books.

Rollin, B. E. 1995. *Farm Animal Welfare: Social, Bioethical, and Research Issues.* Ames, Ia.: Iowa State University Press.

Rollin, B. E. 1989. *The Unheeded Cry: Animal Consciousness.* Oxford, UK: Oxford University Press.

Ross, M. G., and M. J. Nijland. 1997. Fetal swallowing: Relation to amniotic fluid regulation. *Clin. Obst. Gynecol.* 40:352.

Rothenbuhler, N. 1964. Behavioral genetics of nest cleaning in honey bees. IV. Responses of F1 and backcross generation to disease-killed brood. *Amer. Zoologist* 4:111.

Rowan, A. N. 1993. Animal well-being: Some key philosophical, ethical, political, and public issues affecting food animal agriculture. *Food Animal Well-Being Conference Proceedings.* Washington, D.C.: U.S. Department of Agriculture, and West Lafayette, Ind.: Purdue University.

Ruckebusch, Y. 1972. The relevance of drowsiness in farm animals. *Anim. Behav.* 20:637.

Rushen, J. 1984. Stereotyped behavior, adjunctive drinking and feeding periods of tethered sows. *Anim. Behav.* 32: 1059.

Rushen, J. 1991. Problems associated with the interpretation of physiological data in the assessment of animal welfare. *Appl. Anim. Behav. Sci.* 32:349.

Rushen, J., A. B. Lawrence, and E. M. C. Terlouw. 1993. The motivational basis for stereotypies. In *Stereotypic Animal Behaviour: Fundamentals and Application to Welfare.* Edited by A. B. Lawrence and J. Rushen. Wallingford, Oxon, UK: CAB International.

Russell, J. R. 1997. Personal communication.

Savory, C. J., E. Seawright, and A. Watson. 1992. Stereotyped behaviors in broiler breeders in relation to husbandry and opioid receptor blockade. *Appl. Anim. Behav. Sci.* 32:349.

Schultz, L. H. 1971. Management and nutritional aspects of ketosis. J. Dairy Sci. 54:962.

Schwartzkopf-Genswein, K. S. 1996. *Behavioural and physiological effects of hot-iron and freeze branding on beef cattle.* M.S. Thesis. University of Saskatchewan–Saskatoon.

Scott, G. B., and P. Moran. 1993. Fear levels in laying hens carried by hand and by mechanical conveyors. *Appl. Anim. Behav. Sci.* 36:337.

Selye, H. 1976. *The Stress of Life.* New York: McGraw-Hill.

Shutt, D. A., L. R. Fell, R. Connell, and A. K. Bell. 1988. Stress responses in lambs docked and castrated surgically or by the application of rubber rings. *Aust. Vet. J.* 65:5.

Singer, P. 1975. *Animal Liberation.* New York Review of Books. New York: Random House.

Stafford, C., and R. Oliver. 1991. *Horse Care and Management.* London: J. A. Allen, 53.

Stahly, T. S., G. L. Cromwell, and M. P. Aviotti. 1979a. The effect of environmental temperature and dietary fat supplementation of the performance and carcass characteristics on growing and finishing swine. *J. Anim. Sci.* 49:1478.

Stahly, T. S., G. L. Cromwell, and M. P. Aviotti. 1979b. The effect of environmental temperature and dietary lysine source and level on the performance and carcass characteristics of growing swine. *J. Anim. Sci.* 49:1242.

Stevenson, J. S., D. S. Pollmann, D. L. Davis, and J. P. Murphy. 1983. Influence of supplemental light on sow performance during and after lactation. *J. Anim. Sci.* 56:1282.

Stolba, A., N. Baker, and D. G. M. Wood-Gush. 1983. The characterisation of stereotyped behaviour in stalled sows by informational redundancy. *Behaviour* 87:157.

Stricklin, W. R., C. E. Heisler, and L. L. Wilson. 1980. Heritability of temperament in beef cattle. *J. Anim. Sci.* 51, Suppl. 1:109.

Stricklin, W. R., L. L. Wilson, H. B. Graves, and E. H. Cash. 1979. Effects of concentrate level, protein source and growth promotant: Behavior and behavior-performance relationships. *J. Anim. Sci.* 49:832.

Stull, C. L. 1992. Welfare parameters in commercial veal facilities. *J. Anim. Sci.* (Suppl 1):163.

Stull, C. L., and D. A. McMartin. 1992. *Welfare parameters in veal calf production facilities.* University of California–Davis: Cooperative Extension, School of Veterinary Medicine.

SVCAWS. 1996. *Report of the Scientific Veterinary Committee, Animal Welfare Section, on Welfare of Laying Hens.* Brussels: Commission of the European Communities, Directorate-General for Agriculture VI/8660/96.

SVCAWS. 1995. *Report of the Scientific Veterinary Committee, Animal Welfare Section, on the Welfare of Calves.* Brussels: Commission of the European Communities, Directorate-General for Agriculture. VI/5891/95 PVH/kod.

Swanson, J. C. 1995. Farm animal well-being and intensive production systems. *J. Anim. Sci.* 73:2744.

Swenson, M. J. 1993. Ch 3. Physiological properties and cellular and chemical constituents of blood. In *Dukes' Physiology of Domestic Animals.* 11th ed. Edited by M. J. Swenson and W. O. Reece. Ithaca, N.Y.: Cornell University Press.

Thwaites, C. J. 1985. Physiological responses and productivity of sheep. In *Stress Physiology in Livestock,* Vol. 2. Edited by M. K. Yousef. Boca Raton, Fla.: CRC Press.

Toates, F. 1987. The relevance of models of motivation and learning to animal welfare: In *Biology of Stress in Farm Animals: An Integrative Approach.* Edited by P. R. Wiepkema and P. W. M. Van Adrichem. Dordrecht, The Netherlands: 153.

Trelouw, E. M. C., A. B. Lawrence, and A. W. Illius. 1991. Influences of feeding level and physical restriction on development of stereotypies in sows. *Anim. Behav.* 42:981.

Verstegen, M. W. A., G. Mateman, H. A. Brandsma, and P. I. Haartsen. 1979. Rate of gain and carcass quality in fattening pigs at low ambient temperatures. *Livest. Prod. Sci.* 6:51.

Verstegen, M. W. A., H. A. Brandsma, and G. Mateman. 1982. Feed requirements of growing pigs at low environmental temperatures. *J. Anim. Sci.* 55:88.

Verstegen, M. W. A., J. M. F. Verhagen, and L. A. den Hartog. 1987. Energy requirements of pigs during pregnancy. *Livest. Prod. Sci.* 16:75.

Warburton, D. M. 1987. The neuropsychobiology of stress response control. In *Biology of Stress in Farm Animals: An Integrative Approach.* Edited by P. R. Wiepkema and P. W. M. van Adrichem. Dordrecht, The Netherlands: Martinus Nijhoff, 87.

Webster, A. J. F. 1970. Direct effects of cold weather on the energetic efficiency of beef production in different regions of Canada. *Canada J. Anim. Sci.* 50:563.

Webster, A. J. F. 1974. Heat loss from cattle with particular emphasis on effects of cold. In *Heat Loss from Animals and Man.* Edited by J. L. Monteith and L. E. Mount. London: Butterworths, 205.

Webster, A. J. F. 1976. The influence of climatic environment on metabolism of cattle. In *Principles of Cattle Production.* Edited by H. Swan and W. H. Broster. London: Butterworths, 103.

Webster, J., C. Saville, and D. Welchman. 1986. *Improved husbandry systems for veal calves.* London: Farm Animal Care Trust, 26.

Wemelsfelder, F. 1993. The concept of animal boredom and its relationship to stereotyped behavior. In *Stereotypic Animal Behaviour: Fundamentals and Application to Welfare.* Edited by A. B. Lawrence and J. Rushen. Wallingford, Oxon, UK: CAB International.

Wiepkema, P. 1993. Forward. In *Stereotypic Animal Behavior.* Edited by A. B. Lawrence and J. Rushen. Wallingford, UK: CAB International, ix.

Willham, R. L., D. F. Cox, and G. G. Karas. 1963. Genetic variation in a measure of avoidance learning in swine. *J. Comp. Physiol. Psychol.* 56:294.

Willham, R. L., D. F. Cox, and G. G. Karas. 1964. Partial acquisition and extinction of an avoidance response in two breeds of swine. *J. Comp. Physiol. Psychol.* 57:117.

Wood-Gush, D. G. M. 1972. Strain differences in response to sub-optimal stimuli in the fowl. *Anim. Behav.* 20:72.

Yousef, M. K. 1984. *Stress Physiology in Livestock,* Vol. 3. Boca Raton, Fla.: CRC Press.

Zenchak, J. J., and G. C. Anderson. 1980. Sexual performance levels of rams as affected by social experiences during rearing. *J. Anim. Sci.* 50:167.

Zeuner, F. E. 1963. *A History of Domesticated Animals.* New York: Harper & Row.

Zimmerman, D. R., M. Wise, A. P. K. Jones, R. D. Aldrich, and R. K. Johnson. 1980. Testicular growth in swine as influenced by photoperiod and ovulation rate selection in females. *J. Anim. Sci.* 57:292.

Zito, C. A., L. L. Wilson, and H. B. Graves. 1977. Some effects of social deprivation on behavioral development of lambs. *Appl. Anim. Ethol.* 3:367.

SUGGESTED READING

On Animal Behavior

Albright, J., and C. Arave. 1997. *The Behaviour of Cattle.* Wallingford, UK: CAB International.

Alcock, J. 1993. *Animal Behavior: An Evolutionary Approach.* Sunderland, Mass.: Sinauer Associates.

Appleby, M. C., B. O. Hughes, and H. E. Elson. 1992. *Poultry Production Systems: Behaviour, Management, and Welfare.* Wallingford, UK: CAB International.

Arnold, G. W., and M. L. Dudzinski. 1978. *Ethology of Free-Ranging Domestic Animals.* New York: Elsevier Scientific Publishing.

Broom, D. M. 1981. *Biology of Behaviour.* Cambridge, Mass.: Cambridge University Press.

Craig, J. V. 1981. *Domestic Animal Behavior.* Englewood Cliffs, N.J.: Prentice-Hall.

Cregier, S. E. 1989. *Farm Animal Ethology: A Source Book.* North York, Ontario: Captus Press.

Consortium. 1998. *Guide for the Care and Use of Agricultural Animals in Agricultural Research and Teaching.* 2nd. Ed. Savoy, Ill.: FASFAS, Association Headquarters, (1111 N. Dunlap Ave. 61874).

Dawkins, M. S. 1986. *Unraveling Animal Behaviour.* Essex, England: Longman Group.

Fox, M. W., ed. 1968. *Abnormal Behavior in Animals.* Philadelphia: W. B. Saunders.

Fox, M. W. 1984. *Farm Animals: Husbandry, Behavior, and Veterinary Practice.* Baltimore, Md.: University Park Press.

Fraser, A. F. 1968. *Reproductive Behaviour in Ungulates.* New York: Academic Press.

Fraser, A. F. 1992. *The Behaviour of the Horse.* Wallingford, UK: CAB International.

Fraser, A. F., and D. M. Broom. 1990. *Farm Animal Behaviour and Welfare.* London: Balliere Tindall.

Grandin, T., ed. 1993. *Livestock Handling and Transport.* Wallingford, UK: CAB International.

Grier, J. W., and T. Burk. 1992. *Biology of Animal Behavior.* 2d ed. St. Louis, Mo.: Mosby–Year Book.

Hafez, E. S. E., ed. 1975. *The Behavior of Domestic Animals,* 3d ed. London: Balliere Tindall.

Hart, B. L. 1985. *The Behavior of Domestic Animals.* New York: W. H. Freeman.

Hinde, R. A. 1970. *Animal Behaviour: A Synthesis of Ethology and Comparative Psychology.* New York: McGraw-Hill.

Houpt, K. A. 1991. *Domestic Animal Behavior for Veterinarians and Animal Scientists,* 2d ed. Ames, Ia.: Iowa State University Press.

Hurnik, J. F., A. B. Webster, and P. B. Siegel. 1995. *Dictionary of Farm Animal Behaviour.* Ames, Ia.: Iowa State University Press.

Immelmann, K., and C. Beer. 1989. *A Dictionary of Ethology.* Cambridge, Mass.: Harvard University Press.

Kiley-Worthington, M. 1977. *Behavioural Problems of Farm Animals.* London: Oriel Press.

Kilgour, R., and D. C. Dalton. 1984. *Livestock Behaviour: A Practical Guide.* Auckland: Methven.

Lawrence, A. B., and J. Rushen, eds. 1993. *Stereotypic Animal Behaviour: Fundamentals and Application to Welfare.* Wallingford, UK: CAB International.

Lehner, P. N. 1979. *Handbook of Ethological Methods.* New York: Garland STPM Press.

Lorenz, K. Z. 1981. *The Foundations of Ethology.* New York: Springer-Verlag New York.

Lynch, J. J., G. N. Hinch, and D. B. Adams. 1992. *The Behaviour of Sheep.* Wallingford, UK: CAB International.

Martin, P. 1986. *Measuring Behaviour.* Cambridge, Mass.: Cambridge University Press.

McFarland, D. 1985. *Animal Behaviour.* Harlow: Longman Scientific and Technical.

McFarland, D. 1989. *Problems of Animal Behaviour.* New York: Wiley.

Monaghan, P., and D. Wood-Gush, eds. 1990. *Managing the Behavior of Animals.* London: Chapman & Hall.

Paterson, D., and M. Palmer. 1989. *The Status of Animals: Ethics, Education, and Welfare.* Wallingford, UK: CAB International.

Phillips, C. J. C. 1993. *Cattle Behaviour.* Ipswich, UK: Ipswich Farming Press.

Phillips, C., and D. Piggins, eds. 1992. *Farm Animals and the Environment.* Wallingford, UK: CAB International.

Price, E. O., ed. 1987. Farm animal behavior. In *The Veterinary Clinics of North America,* Vol. 3, No. 2. London: W. B. Saunders.

Rollin, B. E. 1995. *Farm Animal Welfare: Social, Bioethical, and Research Issues.* Ames, Ia.: Iowa State University Press.

Sainsbury, D. 1986. *Farm Animal Welfare.* London: Collins.

Syme, G. J., and L. A. Syme. 1979. *Social Structure in Farm Animals: Developments in Animal and Veterinary Sciences,* Vol. 4. Amsterdam: Elsevier.

Toates, F. 1986. *Motivational Systems.* Cambridge, Mass.: Cambridge University Press.

Waring, G. H. 1983. *Horse Behavior.* Park Ridge, N.Y.: Noyes Publications.

Wood-Gush, D. G. M. 1983. *Elements of Ethology: A Textbook for Agricultural and Veterinary Students.* London: Chapman & Hall.

Zayan, R., and I. J. H. Duncan, eds. 1987. *Cognitive Aspects of Social Behaviour in the Domestic Fowl.* Amsterdam: Elsevier.

On Stress Physiology

Appleby, M. C., and B.O. Hughes, eds. 1997. *Animal Welfare.* Wallingford, UK: CAB International.

Burchfield, S. R., ed. 1985. *Stress: Psychological and Physiological Interactions.* Washington, D.C.: Hemisphere.

Carlson, N. R. 1991. *Physiology of Behavior.* 4th ed. Boston: Allyn & Bacon.

Consortium. 1998. *Guide for the Care and Use of Agricultural Animals in Agricultural Research and Teaching.* 2nd. Ed. Savoy, Ill.: FASFAS, Association Headquarters, (1111 N. Dunlap Ave. 61874).

Crousos, G. P., D. L. Loriaux, and P. W. Gold, eds. 1988. *Mechanisms of Physical and Emotional Stress: Advances in Experimental Medicine and Biology,* Vol. 245. New York: Plenum Press.

Curtis, S. E. 1983. *Environmental Management in Animal Agriculture.* Ames, Ia.: Iowa State University Press.

Davenport, J. 1985. *Environmental Stress and Behavioural Adaptation.* London: Croom Helm.

Dawkins, M. S. 1980. *Animal Suffering: The Science of Animal Welfare.* London: Chapman & Hall.

DeKloet, E. R., V. M. Wiegant, and D. deWied. 1987. *Neuropeptides and Brain Function: Progress in Brain Research,* Vol. 72. Amsterdam: Elsevier.

Folk, G. E. 1974. *Textbook of Environmental Physiology.* 2d ed. Philadelphia: Lea & Febiger.

Grandin, T., ed. 1993. *Livestock Handling and Transport.* Wallingford, UK: CAB International.

Hafez, E. S. E. 1968. *Adaptation of Domestic Animals.* Philadelphia: Lea & Febiger.

Hemsworth, P. H., and G. J. Coleman. 1998. *Human-Livestock Interactions: The Stockperson and the Productivity and Welfare of Intensively Farmed Animals.* Wallingford, UK: CAB International.

Henry, J. P., and P. M. Stephens. 1977. *Stress, Health, and the Social Environment: A Sociobiologic Approach to Medicine. Topics in Environmental Physiology and Medicine.* New York: Springer-Verlag.

Leshner, A. I. 1978. *An Introduction to Behavioral Endocrinology.* New York: Oxford University Press.

Levine, S., ed. 1972. *Hormones and Behaviour.* New York: Academic Press.

Moberg, G. P., ed. 1985. *Animal Stress.* Bethesda, Md.: American Physiological Society.

National Research Council. 1996. *Guide for the Care and Use of Laboratory Animals.* Washington, D.C.: National Academy Press.

Phillips, C., and D. Piggins, eds. 1992. *Farm Animals and the Environment.* Wallingford, UK: CAB International.

Rollin, B. E. 1995. *Farm Animal Welfare: Social, Bioethical, and Research Issues.* Ames, Ia.: Iowa State University Press.

Wiepkema, P. R., and P. W. M. van Adrichem, eds. 1987. *Biology of Stress in Farm Animals: An Integrative Approach.* Dordrecht, The Netherlands: Martinus Nijhoff.

Yousef. M. K., ed. 1985. *Stress Physiology in Livestock,* Vol. 1. Boca Raton, Fla.: CRC Press.
Yousef, M. K., ed. 1985. *Stress Physiology in Livestock,* Vol. 2. Boca Raton, Fla.: CRC Press.
Yousef, M. K., ed. 1985. *Stress Physiology in Livestock,* Vol. 3. Boca Raton, Fla.: CRC Press.

INDEX